# New Physics

**by**

**Trevor G. Underwood**

By the same author:

> "*Quantum Electrodynamics – annotated sources. Volumes I and II*." (April 2023);
>
> "*Special Relativity.*" (June 2023);
>
> "*General Relativity.*" (November 2023);
>
> "*Gravity.*" (March 2024);
>
> "*Electricity & Magnetism.*" (May 2024);
>
> "*Quantum Entanglement.*" (June 2024);
>
> "*The Standard Model.*" (September 2024)

all distributed by Lulu.com at Lulu Bookstore.

Published by Trevor G. Underwood
18 SE 10th Ave.
Fort Lauderdale, FL 33301

ISBN: 979-8-218-52975-8 (hardcover)
Library of Congress Control Number: 2024921561

Printed and distributed by Lulu Press, Inc.

627 Davis Dr.
Ste. 300
Morrisville, NC 27560
http://www.lulu.com/shop

# CONTENTS.

Page no.

65 **List of elementary particles in the Standard Model.**

65 **52 *elementary particles* and their *anti-particles* of which only the *electron* and the *photon* are stable.**

66 **(2) The speed of light in vacuum is constant for all observers, regardless of the motion of light source or observer.**

66 **Einstein's theory of Special Relativity.**

66 **Underwood, T. G. (2023). *Special Relativity*,** Conclusion, p. 381:

67 **Underwood, T. G. (2023). *General Relativity*,** Conclusion, p.474.

68 **Einstein, A. (September, 1905). Zur Elektrodynamik bewegter Körper. (On the electrodynamics of moving bodies.)** *Ann. Phys.*, 322, 10, 891-921; http://www.physik.uni-augsburg.de/annalen/history/einstein-papers/1905_17_891-921.pdf.; *Einstein's theory of Special Relativity*.

91 **Einstein, A. (May, 1912). Lichtgeschwindigkeit und Statik des Gravitionsfeldes. (The Speed of Light and the Statics of the Gravitational Field.)** *Ann. Phys.*, 38, 355-69; translation in A. Beck (translator), D. Howard (consultant). (1996). *The Collected Papers of Albert Einstein,* Volume 4: The Swiss Years: Writings, 1912-1914. (English translation), Princeton University Press, Princeton, Vol. 4, Doc. 3, 95-106; https://einsteinpapers.press.princeton. edu/vol4-doc/151; Einstein (1911) showed that the validity of one of the fundamental laws of his *theory of Special Relativity*, namely, the law of the *constancy of the speed of light*, could claim to be *valid only for space-time domains of constant gravitational potential*. Einstein noted that despite the fact that this result *excluded the general applicability of the Lorentz transformation*, it should not deter us from pursuing the consequences of that path. Here he took that further by demonstrating that the Lorentz transformation could not be established for infinitely-small space-time regions either *as soon as one abandons the universal constancy of c*.

104 **(3) All elementary particles, including electromagnetic waves (photons), are quantized.**

104 **Relativistic Quantum Electrodynamics, Quantum Field Theory, Renormalization.**

104 **Underwood, T. G. (2023).** ***Quantum Electrodynamics*****,** Volume II, Preface, pp. 33-5.

106 **Schwinger, J. (November, 1948). Quantum Electrodynamics. I. A Covariant Formulation.** *Phys. Rev.*, 74, 10, 1439-61; https://doi.org/10.1103/PHYSREV.74.1439111; lack of convergence in current formulations of quantum electrodynamics indicates that revision of electrodynamic concepts at ultra-relativistic energies is necessary. The question is whether all *divergencies* can be isolated in unobservable *renormalization* factors. This paper is occupied with the formulation of a completely covariant electrodynamics.

## 135 (4) All elementary particles have mass, apart from the photon and gluons.

135 **Masses of elementary particles in the Standard Model.**

136 **Higgs, P. W. (October, 1964). Broken Symmetries and the Masses of Gauge Bosons.** *Phys. Rev. Lett*., 13, 16, 508–9; https://journals.aps.org/prl/pdf/10.1103/PhysRevLett. 13.508; Brout and Englert showed that *gauge vector fields*, abelian and non-abelian, could acquire *mass* if empty space were endowed with a particular type of structure that one encountered in material systems. Other physicists, Peter Higgs and Gerald Guralnik, C. R. Hagen and Tom Kibble had reached similar conclusions at about the same time. The *Brout–Englert–Higgs* (BEH) *mechanism* is believed to give rise to the *masses* of all the elementary particles in the *Standard Model*. This includes the *masses* of the W and Z *bosons*, and the *masses* of the *fermions*, i.e. the *quarks* and *leptons*. In a previous paper, Higgs had shown that the *Goldstone theorem*, that Lorentz-covariant field theories in which spontaneous breakdown of symmetry under an internal Lie group occurs contain zero-mass particles, *failed if and only if the conserved currents associated with the internal group were coupled to gauge fields*. The purpose of the present note was to report that, *as a consequence of this coupling, the spin-one quanta of some of the gauge fields acquired mass*; the longitudinal degrees of freedom of these particles (which would be absent if their *mass* were zero) go over into the *Goldstone bosons when the coupling tends to zero. The model was discussed mainly in classical terms*; nothing was proved about the quantized theory. Higgs noted that it should be understood, therefore, that *the conclusions which were presented concerning the masses of particles were conjectures based on the quantization of linearized classical field equations.*

**Relativity.)** *Koniglich Preußische Akademie der Wissenschaften, Sitzungsberichte* (Berlin), 142–52; translation by W. Perrett & G. B. Jeffery in A. Engel (translator), E. Schuckling (consultant). (1997). *The Collected Papers of Albert Einstein*, Volume 6: The Berlin Years: Writings, 1914-1917, Princeton University Press, Princeton, Doc. 43, 421-32; https://einsteinpapers.press.princeton.edu/vol6-trans/433; describes Einstein's struggles with supplementing the *relativistic differential equations* by *limiting conditions* at *spatial infinity* in order to regard the universe as being of infinite spatial extent.

168 **Weyl, H. (April, 1929). Gravitation and the electron.** *PNAS*, 15, 4, 323-34, https://doi.org/10.1073/pnas.15.4.323; also in Weyl, H. (May, 1929). Elektron und Gravitation. (Electron and gravitation.) *Zeit. Phys.*, 56, 330-52; https://doi.org/10.1007/BF01339504; attempt to incorporate Dirac's theory into the scheme of *general relativity*, introduces *gauge invariance* of *theory of coupled electromagnetic potentials* and Dirac *matter waves*, explains why "anti-symmetric" Pauli-Fermi statistics for electrons lead to "symmetric" Bose-Einstein statistics for photons, barrier which hems progress of quantum theory is quantization of *field equations*.

181 **Einstein, A. (March, 1921). Eine naheliegende Ergänzung des Fundaments der allgemeinen Relativitätstheorie. (On a natural addition to the foundation of the general theory of relativity.)** *Sitzungsberichte*, 261-4; translation in A. Engel (translator), E. Schuckling (consultant). (2002). *The Collected Papers of Albert Einstein*, Volume 7: The Berlin Years: Writings, 1918-1921, Princeton University Press, Princeton, Doc. 54, 224-8; https://einsteinpapers. press.princeton.edu/vol7-trans/240; Einstein's comments on Hermann Weyl's attempt to supplement the *general theory of relativity* by adding a further condition of invariance.

186 **(8) Elementary and composite particles with the same electric charge attract each other, and elementary and composite particles with opposite electric charge are repulsed, through the electromagnetic interaction or electromagnetic force, according to Coulomb's law.**

186 **Electromagnetism.**

186 **Coulomb's Law.**

186 **Ampère's Force Law for Magnetism.**

206 **Preliminary list of elementary particles in New Physics.**

206 **Reduction of number of elementary particles (and antiparticles) from 52 in the Standard Model to 14 in New Physics.**

207 **Paul Dirac – 1933 Nobel Lecture, December 12, 1933. *Theory of electrons and positrons.*** In his Nobel Lecture, Dirac described the current state of his theory of *electrons* and *positrons*. Dirac recognized that "*there exists no relativistic quantum mechanics (that is, one valid for large velocities) which can be applied to particles with arbitrary properties*"

212 **Fermi, E. (March, 1934) Versuch einer Theorie der β-Strahlen. I. (Attempt at a theory of β rays. I.)** *Zeit. Phys.,* 88, 161–77; https://doi.org/10.1007/BF01351864; Fermi proposed the first theory of the *weak interaction*, known as *Fermi's interaction*, in which four fermions directly interact with one another. A quantitative theory of *β decay* is proposed, in which the existence of the *neutrino* is assumed, and the emission of *electrons* and *neutrinos* from a nucleus in the β case is treated with a method similar to that of the emission of a quantum of light from an excited atom in radiation theory. Formulas for the lifetime and for the shape of the emitted continuous β radiation spectrum are derived and compared with experience. Tree Feynman diagrams describe the interaction remarkably well. Unfortunately, loop diagrams cannot be calculated reliably because Fermi's interaction is *not renormalizable*.

221 **(2) The speed of light in a vacuum is constant relative to the emitter.**

221 **Avoidance of length contraction and time dilation for a moving observer by replacing Einstein's theory of Special Relativity with Walter Ritz's theory that the speed of light in a vacuum is constant relative to the emitter.**

221 **Underwood, T. G. (2023). *Special Relativity*,** Conclusion, p. 381, 1st paragraph.

221 **Walter Ritz's theory that the speed of light in a vacuum is constant relative to the emitter.**

221 **Walther Heinrich Wilhelm Ritz (February 22, 1878 – July 7, 1909).**

226 **Ritz, W. (August, 1908). Recherches critiques sur les theories electrodynamiques de Cl. Maxwell et de H. A. Lorentz. (Critical research on the electrodynamic theories of Cl. Maxwell and H. A. Lorentz.)** *Archives des Sciences phys. et nat.*, 4, 26, 209-36; reprinted in (1911). Société Suisse de Physique, ed., *Gesammelte Werke Walther Ritz Œuvres*. Gauthier-Villars, Paris,

pp. 427-46; https://ia904708.us.archive.org/3/items/gesammeltewerkeo00ritzuoft/gesammeltewerkeo00ritzuoft.pdf; translation by T. G. Underwood; in this paper Ritz presents a summary of his criticisms of the Maxwell and Lorentz's theories and the resulting experimental uncertainties.

229 **Ritz, W. (December, 1908). Über die Grundlagen der Elektrodynamik und die Theorie der schwarzen Strahlung. (On the basics of electrodynamics and the theory of black body radiation.)** *Phys. Zeit.*, 9, 903-7; reprinted in (1911). Société Suisse de Physique, ed., *Gesammelte Werke Walther Ritz Œuvres*. Gauthier-Villars, Paris, pp. 493-502; https://ia904708.us.archive.org/3/items/gesammeltewerkeo00ritzuoft/gesammeltewerkeo00ritzuoft.pdf; translation by T. G. Underwood; Ritz notes that the differential equations in the Maxwell-Lorentz formulation of electrodynamics permit infinite solutions, including those with both *retarded* and *advanced* potentials, on which Einstein (1905) relied in deriving the consequences of his two postulates. Ritz shows that *advanced potentials* are inadmissible and how this could be addressed based on *retarded potentials* alone, i.e. by an *emission theory*. He also demonstrates why the role of an ether must be removed from the theory of electrodynamics.

237 **Einstein, A. (March, 1909). Zum gegenwärtigen Stand des Strahlungsproblems. (On the Present Status of the Radiation Problem.)** *Phys. Zeit.*, 10, 6, 185-93; reprinted in John Stachel, ed., *The Collected Papers of Albert Einstein.* Vol. 2. The Swiss Years: Writings, 1900–1909 (Princeton: Princeton University Press, 1989), Doc. 56, pp. 542-50; translation at https://ia904704.us.archive.org/27/items/EinsteinOnPresentStatus_201501/Einstein_On_Present_Status_old.pdf; translation also in A. Beck, P. Havas. (1989). *The Collected Papers of Albert Einstein, Volume 2: The Swiss Years: Writings, 1900-1909*, pp. 357-75; Einstein advocated the use of the Maxwell-Lorentz equations, since they yielded expressions for the energy and momentum of a system at any instant of time, while the exclusive use of *retarded* potentials required knowledge of the earlier states of a system to determine any future state. He denied that *retarded* potentials had some fundamental significance; he viewed them merely as auxiliary mathematical formulations. Ritz responded both in print and in person by visiting Einstein in Zurich.

241 **Ritz, W. (1909). Zum gegenwärtigen Stand des Strahlungsproblems; Erwiderung auf der Aufsatz des Herrn. A Einstein. (On the Present Status of the Radiation Problem; Response to the essay by Mr. A. Einstein.)** *Phys. Zeit.*, 10, 224-5; reprinted in (1911). Société Suisse de Physique, ed., *Gesammelte Werke Walther Ritz Œuvres*. Gauthier-Villars, Paris, pp. 503-6; https://ia904708.us.archive.org/3/items/gesammeltewerkeo00ritzuoft/gesammeltewerkeo00ritzuoft.

pdf; translation by T. G. Underwood; Ritz's response to Einstein (March, 1909).

244 **Ritz, W. & Einstein, A. (1909). Zum gegenwärtigen Stand des Strahlungsproblems. (On the Present Status of the Radiation Problem.)** *Phys. Zeit.*, 10, 323-4; reprinted in (1911). Société Suisse de Physique, ed., *Gesammelte Werke Walther Ritz Œuvres*. Gauthier-Villars, Paris, XXV, pp. 507-8; https://ia904708.us. archive.org/3/items/gesammeltewerkeo00ritzuoft/ gesammeltewerkeo00ritzuoft.pdf; translation in A. Beck, P. Havas. (1989). *The Collected Papers of Albert Einstein, Volume 2: The Swiss Years: Writings, 1900-1909*, Doc. 57, p. 376; translation by T. G. Underwood; this led them to publish a concise joint statement of their main differences of opinion in 1909. Whereas Ritz granted physical meaning only to *retarded potentials* in the interest of obtaining irreversibility, Einstein deemed the apparent irreversibility of radiation phenomena to be grounded solely on probabilistic considerations.

246 **(3) All elementary particles, including electromagnetic waves (photons), are quantized.**

246 **Avoidance of requirement to assume a *point electron*, and address, through a process of *renormalization* the still unresolved *divergencies* in the unsuccessful attempt to introduce *special relativity* into quantum electrodynamics.**

246 **Non-relativistic Quantum Mechanics and Quantum Electrodynamics.**

247 **Heisenberg, W. (July, 1925). Über quantentheoretische Umdeutung kinematischer und mechanischer Beziehungen. (On the quantum-theoretical re-interpretation of kinematic and mechanical relations.)** *Zeit. Phys.*, 33, 879-93; https://doi.org/10.1007/BF01328377; (translation (2014) by Luca Doria, Institute of Theoretical Physics, Gottingen; also translation by D. H. Delphenich; https://neo-classical-physics.info/electromagnetism. html); and translation in van der Waerden, B. L., ed. (1968). *Sources of Quantum Mechanics*, 12, 261-76. Dover, New York; Heisenberg proposes a *quantum mechanics* in which only relationships among observable quantities occur, not possible to assign to the electron a point in space as a function of time, builds on Kramer's dispersion theory and instead assigns to the electron an *emitted radiation*, substitutes *frequencies* and *amplitudes* of Fourier components of emitted radiation of electron, instead of reinterpreting x(t) as a *sum* over transition components represents position by *set* of transition components, assigns *transition frequencies* and *transition amplitudes* as observables, replaces classical component by *transition* component corresponding to the quantum jump from state $n$ to state $n - \alpha$, translates the old *quantum condition*

that fixes the properties of the *states* to a new condition to calculate the amplitude of a *transition* between two states by replacing the differential by a difference, in quantum case *frequencies* do not combine in same way as classical harmonics but in accordance with the *Ritz combination principle* under which spectral lines of any element include frequencies that are either the sum or the difference of the frequencies of two other lines, in quantum case frequencies combine by multiplying *transition amplitudes* (equivalent to matrix multiplication), results in non-commutativity of kinematical quantities, shows simple quantum theoretical connection to Kramers' dispersion theory, the *equation of motion* $\ddot{x} + f(x) = 0$ and the *quantum condition* $h = 4\pi m \sum_{\alpha = -\infty}^{+\infty} \{|a(n, n + \alpha)|^2\omega(n, n + \alpha) - |a(n, n - \alpha)|^2\omega(n, n - \alpha)\}$ together contain if solvable *a complete determination not only of the frequencies and energies but also of the quantum theoretical transition probabilities*.

268 **Dirac, P. A. M. (March, 1927). The quantum theory of the emission and absorption of radiation.** *Roy. Soc. Proc., A*, 114, 767, 243-65; https://doi.org/10.1098/rspa.1927.0039; addresses *non-relativistic quantum electrodynamics*, treats problem of an assembly of similar systems satisfying the Einstein-Bose statistical mechanics which interact with another different system by obtaining a Hamiltonian function to describe the motion, theory of system in which *forces are propagated with velocity of light* instead of instantaneously, time counted as a c-number instead of being treated symmetrically with the space co-ordinates, addition of *interaction term*, production of electromagnetic field (emission of radiation) by moving electron, reaction of radiation field on emitting system, applies to the interaction of an assembly of *light-quanta* with an atom, shows that it leads to *Einstein's laws for the emission and absorption of radiation*, the interaction of an atom with *electromagnetic waves* is then considered, treats *field* of radiation as a dynamical system whose interaction with an ordinary atomic system may be described by a Hamilton function, dynamical variables specifying the *field* are the *energies* and *phases* of the harmonic components of the waves, shows that if one takes the *energies* and *phases* of the waves to be *q-numbers* satisfying the proper quantum conditions instead of *c-numbers* the Hamiltonian function for the interaction of the *field* with an atom takes the same form as that for the interaction of an assembly of *light-quanta* with the atom, provides a complete formal reconciliation between the wave and light-quantum point of view, leads to the correct expressions for Einstein's A's and B's, radiative processes of the more general type considered by Einstein and Ehrenfest in which more than one light-quantum take part simultaneously are not allowed on the present theory, the mathematical development of the theory made possible by Dirac's *general*

*transformation theory* of the quantum matrices [Dirac (January, 1927). The Physical Interpretation of the Quantum Dynamics].

287 **(4) All elementary particles have mass, apart from the photon and gluons.**

287 **Masses of elementary particles in New Physics.**

288 **(5) Elementary and composite particles can have electric charge or be neutral.**

288 **Electric charges of elementary particles in New Physics.**

289 **(6) Elementary particles have a quantum state called spin.**

289 **Spin obeys the mathematical laws of angular momentum quantization.**

290 **Spin of elementary particles in New Physics.**

291 **Non-relativistic theory.**

292 **Compton, A. H. (August, 1921). The Magnetic Electron.** *Journ. Frankl. Inst.*, 192, 2, 145-55; https://www.semanticscholar.org/paper/The-magnetic-electron-Compton/f602176e15e52ed703a67865f4c87eeb7a83048e; Compton's paper on investigations of ferromagnetic substances with X-rays was the first to introduce the idea of *electron spin*. Compton hypothesized that the electron's *magnetic moment* was intrinsically connected to the electron's *spin* and pointed out the possible bearing of this idea on the origin of the natural unit of magnetism.

299 **Uhlenbeck, G. E. & Goudsmit, S. (November, 1925). Ersetzung der Hypothese vom unmechanischen Zwang durch eine Forderung bezuglich des inneren Verhaltens jedes einzelnen Elektrons. (Replacement of the hypothesis of unmechanical coercion by a requirement regarding the internal behavior of each individual electron.)** *Naturw.*, 13, 47, 953-4 (in German); https://doi.org/10.1007/ BF01558878; translation by T. G. Underwood; also in Underwood, T. G. (2023). *Quantum Electrodynamics - annotated sources*, Volume I, pp. 282-6; without being aware of Compton's suggestion Uhlenbeck and Goudsmit noted doublets in the alkali spectra that did not conform to current models of the atom. They proposed applying the model of the *spinning electron* to interpret a number of features of the quantum theory of the *anomalous Zeeman effect*, and applied the classical formula for spherical rotating electron with finite radius and surface charge.

303 **Pauli, W. (February, 1925). Über den Zusammenhang des Abschlusses der Elektronengruppen im Atom mit der Komplexstruktur der Spektren. (On the connection between the completion of electron groups in an atom and the complex structure of spectra.)** *Zeit. Phys*., 31, 1, 765–83; https://doi.org/10.1007/BF02980631; translation at http://www. fisicafundamental. net/relicario/doc/Pauli_1925.pd; Pauli first reviewed the established theories for the energy differences *of the triplet levels of the alkaline earths*, based respectively, on *the anomaly of the relativity correction* of the *optically active electron*, and *the dependence of the interaction between the electron and the atom core on the relative orientation of these two systems*. He noted a serious difficulty with the former is the connection of these ideas with the *correspondence principle*, which was well known to be a necessary means to explain the selection rules for the *quantum numbers* $k_1$, j, and m and the polarization of the Zeeman components, in particular, that *it was necessary that the totality of the stationary states of an atom corresponded to a collection (class) of orbits with a definite type of periodicity properties*. The dynamic explanation of this kind of motion of the *optically active electron*, which was based upon the assumption of deviations of the forces between the *atom* core and the *electron* from central symmetry, *seemed to be incompatible with the possibility to represent the alkali doublet (and thus also the magnitude of the corresponding precession frequency) by relativistic formulae*. Consequently, Pauli, decided to pursue instead the alternative *non-relativistic* theory to the problem of *completion of electron groups in an atom*, in order to draw conclusions only about the *number of possible stationary states* of an *atom* when several equivalent *electrons* are present. But this did not address the position and relative order of the term values. On the basis of these results, Pauli obtained a general classification of every *electron* in the *atom* by the principal quantum number n and two auxiliary quantum numbers $k_1$ and $k_2$ to which he added a further quantum number $m_1$ in the presence of an external field, in agreement with experiments. In particular, his rule explained Stoner's result in a natural way and with it the period lengths 2, 8 18, 32.

318 **Pauli, W. (September, 1927). Zur Quantenmechanik des magnetischen Elektrons. (On the quantum mechanics of magnetic electrons.)** *Zeit. Phys*., 43, 601-23; https://doi.org/10.1007/BF01397326; it will be shown how one can arrive at a formulation of the quantum mechanics of the *magnetic electron* by the Schrödinger method of eigenfunctions, with no use of double-valued functions, when one, on the basis of the Dirac-Jordan general theory of transformations, introduces the components of its *proper impulse moment* in a fixed direction as further independent variables in order to carry out the

computations of its rotational degrees freedom, along with the position coordinates of any *electron*. In contradiction to classical mechanics, these variables can assume only the variables + ½ h/2π and – ½ h/2π, which is completely independent of any sort of external field.

323 **(7) Elementary and composite particles with mass attract each other through the gravitational interaction or gravitational force, according to Newton's law of gravitation.**

323 **Gravity.**

323 **Underwood, T. G. (2023).** ***Gravity***: Newton's universal law of gravitation, pp. 74-7.

324 **Underwood, T. G. (2023).** ***Gravity***: Newton's calculation of Kepler's laws, pp. 71-3.

327 **Newton, I. (July, 1687).** ***Philosophiæ Naturalis Principia Mathematica.*** (The Mathematical Principles of Natural Philosophy.) **Book I: The Motion of Bodies.** 1st Edition, London; 2nd Edition, Cambridge, 1713; 3rd Edition, London, 1726. (In Latin); translation below of 3rd Edition by A. Motte, (1729). London.); https://en.wikisource.org/wiki/The_Mathematical_Principles_of_Natural_Philosophy_(1729); also https://ia601604.us.archive.org/1/items/newtonspmathema00newtrich/newtonspmathema00newtrich_bw.pdf. *Philosophiæ Naturalis Principia Mathematica* (Mathematical Principles of Natural Philosophy) is a work in three books written in Latin, first published July 5, 1687, with encouragement and financial help from Edmond Halley. After annotating and correcting his personal copy of the first edition, Newton published two further editions, during 1713 with errors of the 1687 corrected, and an improved version in 1726. The *Principia* includes *Newton's three laws of motion*, laying the foundation for classical mechanics; *Newton's law of universal gravitation*; and a derivation of *Johannes Kepler's laws of planetary motion* (which Kepler had first obtained empirically). In Book I: The Motion of Bodies, Newton addresses the motion of bodies attracted to each other by centripetal forces.

331 **Newton, I. (July, 1687).** ***Philosophiæ Naturalis Principia Mathematica.*** **Book III: Of the System of the World.** 1st Edition, London; 2nd Edition, Cambridge, 1713; 3rd Edition, London, 1726. (In Latin); translation below of 3rd Edition by A. Motte, (1729). London.); https://en.wikisource.org/ wiki/The_Mathematical_Principles_of_Natural_Philosophy_(1729); also https://ia601604.us.archive.org/1/items/newtonspmathema00newtrich/newtonspmathema00newtrich_bw.pdf. In

Book III, Newton notes that the *centripetal force* which arises between planets is the same as the *gravitational force* attracting matter to the Earth and focusses on gravitational attraction. He then proposes that "all bodies gravitate towards; every Planet and that the Weights of bodies towards any the same Planet, at equal distances from the center of the Planet, are proportional to the quantities of matter which they severally contain; and that there is a power of gravity tending to all bodies, proportional to the several quantities of matter which they contain; and that the force of gravity towards the several equal particles of any body, is reciprocally as the square of the distance of places from the particles". In Proposition VI Newton provides his definition of *gravitational mass*, and in Proposition VII, together with its corollary 2, Newton restates his *universal law of gravitation*.

347 **Quantum entanglement between matter.**

347 **Underwood, T. G. (2023).** ***Gravity***, Application of quantum theory to gravity - quantum entanglement, pp. 199-201.

349 **Introduction of a quantum theory of gravity.**

349 **Dirac, P. A. M. (March, 1927). The quantum theory of the emission and absorption of radiation** (applied to gravity).

367 **(8) Elementary and composite particles with the same electric charge attract each other, and elementary and composite particles with opposite electric charge are repulsed, through the electromagnetic interaction or electromagnetic force, according to Coulomb's law.**

367 **Coulomb's Law; Ampère's Force Law for Magnetism.**

367 **Underwood, T. G. (2023).** ***Electricity & Magnetism.*** Part II: Coulomb's Law (1785), pp. 52-4.

370 **Underwood, T. G. (2023).** ***Electricity & Magnetism.*** Part II: Ampère's Force Law (June, 1822), pp. 124-5.

372 **Coulomb, C-A. (1785). Premier mémoire sur l'électricité et le magnétisme. Construction et usage d'une balance électrique, fondée sur la propriété qu'ont les fils de métal, d'avoir une force de réaction de torsion proportionnelle à l'angle de torsion. (Construction and use of an electric balance based on the property that filaments of metal produce a reactive force in torsion proportional to the angle of twist.)** Experimental determination of the law

according to which the elements of bodies, electrified with the same kind of electricity, are mutually repelled. *Histoire de l'Académie Royale des Sciences avec les mémoires de mathématiques et de physique, partie "Mémoires"*, pp. 569–577; http://www.ampere.cnrs.fr/i-corpuspic/tab/Sources/coulomb/Coulomb_El_1785.pdf; translation by L. L. Bucciarelli, Emeritus Professor of Engineering and Technology Studies (MIT), MIT, 2000 (revised and notes added by Christine Blondel and Bertrand Wolff, 2012); Coulomb describes the construction of a torsion balance and uses this to demonstrate what he describes as the *fundamental law of electricity*, now known as Coulomb's law. *The law states that the magnitude, or absolute value, of the attractive or repulsive electrostatic force between two point-charges is directly proportional to the product of the magnitudes of their charges and inversely proportional to the squared distance between them.*

379 **Ampère, A-M. (1822). Memoire sur la Determination de la formule qui represente l'action mutuelle de deux portions infiniment petites de conducteurs voltaıques.** (Memoir on the Determination of the Formula which Represents the Mutual Action of Two Infinitely Small Portions of Voltaic Conductors.) Annales de Chimie et de Physique, 20, 398–419 (in French); also in Ampère, A-M. (1822), Recueil d'observations électro-dynamiques: contenant divers mémoires, notices, extraits de lettres ou d'ouvrages périodiques sur les sciences, relatifs a l'action mutuelle de deux courans électriques, à celle qui existe entre un courant électrique et un aimant ou le globe terrestre, et à celle de deux aimans l'un sur l'autre (in French), Chez Crochard, Paris; http://www.ampere.cnrs.fr/bibliographies/pdf/1822-P097.pdf, pp. 293-318.

382 **(9) The spin of elementary and composite particles creates an attractive force – the weak interaction or weak force - through exchange interaction resulting from entanglement between two quantum spin states.**

382 **Quantum entanglement between spin states.**

383 **Heitler, W. & London, F. (June, 1927). Wechselwirkung neutraler Atome und homöopolare Bindung nach der Quantenmechanik. (Interaction of neutral atoms and homeopolar bonding according to quantum mechanics.)** *Zeit. Phys.*, 44, 455–72. https://doi.org/10.1007/BF01397394; also at http:// quantum-chemistry-history.com/Heitler_London_Dat/WechselWirk1927/WechselWirk1927.htm (in German); translation by T. G. Underwood; Heitler and London examined the interaction between *neutral atoms* though non-polar bonds, in what is known as valance bonds; and applied quantum mechanics to calculate

the *interaction energy* of the atoms when they move closer together. They found that two neutral atoms could interact with each other in two ways; *the problem was twofold degenerate, corresponding to the two ways of assigning the electrons to the neutral atoms* (known as *quantum entanglement*). Examination of the different cases of two H atoms and two He atoms showed that by applying the *Pauli principle*, the selected eigenfunctions of the system should change or maintain their sign respectively, when two electrons were swapped, if the two electrons compared had the same or different *spin*. They found that in the case of He there was only one solution, which yielded about the right size of the He gas kinetic-radius, *due to the fact that 2 He atoms (and the same applies to all noble gases) cannot differ in their spin* – in contrast to hydrogen (and all atoms with unfinished shells) – so that 2 He atoms have only one possible mode of behaving.

394 **Heisenberg, W. (September, 1928). Zur Theory of Ferromagnetismus. (On the theory of ferromagnetism.)** *Zeit. Phys.*, 49, 619–36; https://doi.org/10.1007/BF01328601; in another brilliant paper, Heisenberg noted that empirical results exhibit *ferromagnetism* as an entirely similar state of affairs to what was previously observed in the spectrum of the helium atom; and it seemed to follow from the levels in the helium atoms that a powerful interaction prevailed between the spin directions of two electrons that led to the splitting of the level structure into systems of singlets and triplets. He also noted that this was closely related to explaining ferromagnetic phenomena as being implied by the *exchange phenomenon* (resulting from *quantum entanglement*). Heisenberg concluded that *an atom in a lattice can only be exchanged with its "neighbors"*; exchanges with atoms that lie further away that the "neighboring atoms" could then be neglected. Then two conditions were necessary for the appearance of *ferromagnetism*: the crystal lattice must be a type such that *any atom has at least 8 neighbors*; and the *principal quantum number* of the electrons that are responsible for magnetism must be $n \geq 3$.

404 **(10) Elementary and composite particles can exist as different quantum states, referred to as isospin states, which creates an attractive force – the strong interaction or strong force - through exchange interaction resulting from entanglement between two quantum isospin states.**

404 **The origin of isospin.**

404 **Isospin.**

406 **Heisenberg, W. (January, 1932). Über den Bau der Atomkerne. I. (About the construction of atomic nuclei. I.); (March, 1932). Über den Bau der Atomkerne. II. (About the construction of atomic nuclei. II.); (September, 1933). Über den Bau der Atomkerne. III.** *Zeit. Phys.*, 77, 1–11; https://doi.org/10.1007/BF01342433; *Ibid.*, 78, 156–64; https://doi.org/10.1007/BF01337585; *Ibid.*, 80, 587–96; https://doi.org/10.1007/BF01335696; three-part paper by Heisenberg, which attempted to treat the protons and neutrons on an equal footing by *considering protons and neutrons as different charge states of the same particle*, which, in 1937, Eugene Wigner referred to as the *isotopic spin parameter.*

408 **Wigner, E. (January, 1937). On the Consequences of the Symmetry of the Nuclear Hamiltonian on the Spectroscopy of Nuclei.** *Phys. Rev.* 51, 2, 106; https://journals.aps.org/pr/abstract/10.1103/PhysRev.51.106; also at https://harvest.aps.org/v2/journals/articles/10.1103/PhysRev.51.106/fulltext; the structure of the *multiplets* of nuclear terms is investigated, using as *first approximation* a Hamiltonian which does not involve the ordinary *spin* and corresponds to equal forces between all nuclear constituents, *protons* and *neutrons*. The *multiplets* turn out to have a rather complicated structure, instead of the S of atomic spectroscopy, one has three quantum numbers S, T, Y. The *second approximation* can either introduce *spin* forces (method 2), or else can discriminate between *protons* and *neutrons* (method 3). The *last approximation* discriminates between *protons* and *neutrons* as in method 2 and takes the spin forces into account as in method 3. The method 2 is worked out schematically and is shown to explain qualitatively the table of *stable nuclei* to about Mo.

## 415 PART III Comparison between the Standard Model and New Physics.

415 **(1) The universe is composed of elementary particles.**

415 The *Standard Model.* **Includes *quarks*, *gluons*, and the $W^+$, $W^-$, $Z^0$, and *Higgs boson*.**

415 **The introduction of quarks**.

415 *New Physics.* **Confined to *elementary particles* that can be *observed.* P*roton* and *neutron* restored to list of *elementary particles*. No quarks.**

416 ***Antiparticles and antimatter***.

416 The *Standard Model.* **Existence of *antiparticles* with *differences in quantum numbers* in additional to electric charge, form *antimatter*, a different form of matter. Unexplained asymmetry of *matter* and *antimatter* in the visible universe.**

417 *New Physics.* **Elimination of notion of *antimatter* and problem of asymmetry of *matter* and *antimatter* in the visible universe. *Antiparticles* are simply the less stable *particles* of similar *mass* but opposite *electric charge***

417 ***Elementary particles.***

417 *Standard Model.* **The universe is composed of 52 *elementary particles* and *antiparticles* of which only the *electron* is stable.**

417 *New Physics.* **The universe is composed of 14 *elementary particles* including the *proton*, *neutron* and *graviton*. *Elementary particles* are confined to those that have the possibility of being *observed.***

418 **(2) The speed of light in vacuum.**

418 *Standard Model.* **The speed of light in a vacuum is constant for all observers, regardless of the motion of light source or observer. Results in *length contraction* and *time dilation* for a moving observer.**

419 *New Physics.* **The speed of light in a vacuum is constant relative to the emitter. Replaces *Einstein's theory of Special Relativity* with *Walter Ritz's emission theory*. Avoids *length contraction* and *time dilation* for a moving observer.**

420 **(3) All elementary particles, including electromagnetic waves (photons), are quantized.**

420 *Standard Model.* **Lack of convergence in current formulations of quantum electrodynamics due to the interaction of the electromagnetic and matter fields with their own vacuum fluctuations. The question is whether all *divergencies* can be isolated in unobservable *renormalization* factors.**

420 *New Physics.* **Only relationships among observable quantities occur. Avoids requirement to assume a *point electron*, and to address through a process of *renormalization* the still unresolved *divergencies*.**

420 ***Non-relativistic Quantum Mechanics and Quantum Electrodynamics.***

421 (4) All elementary particles have mass, apart from the photon and gluons.

421 *Standard Model.* The large *masses* of the $W^+$, $W^-$, $Z^0$ and *Higgs bosons* and the *top quark* (respectively 85.7, 85.7, 97.2, 133.3 and 184.9 times the *mass* of the *proton*) raise questions regarding whether they are really *elementary particles*, in particular in view of how they were created by collisions of high energy *protons* in *proton-antiproton* and *hadron* colliders.

422 *New Physics.* Removal of the $W^+$, $W^-$, $Z^0$ and *Higgs bosons* and the *quarks* avoids the problem of overweight *elementary particles*.

422 (5) Elementary and composite particles can have electric charge or be neutral.

422 *Standard Model. Up*, *charm*, and *top quarks* are assumed to have *electric charges* equal to 2/3 of the charge of the *electron* and *proton*; and *down*, *strange*, and *bottom quarks* to have *electric charges* equal to – 1/3 of the charge of the *electron* and *proton*, but this cannot be observed.

422 *New Physics.* Removal of *quarks* avoids the problem of elementary particles with unobserved fractional *electric charges*.

422 (6) Elementary particles have a quantum state called spin.

423 *Standard Model.* The spin of an elementary particles is a quantum state and consequently a *non-relativistic* concept.

423 *New Physics.* Spin obeys the mathematical laws of angular momentum quantization.

424 *New Physics. Non-relativistic* theory.

425 (7) Elementary and composite particles with mass attract each other through the gravitational interaction or gravitational force.

426 *Standard Model.* Einstein's theory of General Relativity.

427 *New Physics. Quantum entanglement* between matter.

427 *New Physics. Gravity* is explained by a *quantum theory of gravity* based on *quantum entanglement* between *quantum states* of *matter* (*gravitons*).

428 (8) Elementary and composite particles with the same electric charge attract each other, and elementary and composite particles with opposite electric

**charge are repulsed, through the electromagnetic interaction or electromagnetic force, according to Coulomb's law.**

428 **Electromagnetism.**

429 **(9) The spin of elementary and composite particles creates an attractive force – the weak interaction or weak force - through exchange interaction or quantum entanglement between two spin states.**

429 *Standard Model.* **Exchange interaction.**

429 *Standard Model.* **Weak *isospin* and the weak *hypercharge*.**

430 *New Physics.* ***Quantum entanglement* between *spin states*.**

430 *New Physics.* **Exchange phenomenon (*quantum entanglement*).**

431 **(10) Elementary particles, such as protons and neutrons, can exist as different quantum states, referred to as isospin states, which create an attractive force – the strong interaction or strong force - through exchange interaction or quantum entanglement between two isospin states.**

431 **Isospin.**

431 *Standard Model.* ***Exchange interaction*.**

432 *Standard Model.* ***Gauged isospin symmetry*.**

432 *New Physics.* ***Quantum entanglement* between *isospin states*.**

# PREFACE

The exhaustive review of the primary sources of theoretical physics undertaken in my previous books[†] has revealed major problems which persist to the present day.

> [†] (April 2023) "*Quantum Electrodynamics – annotated sources,* Volumes I and II"; (June 2023) "*Special Relativity*"; (November 2023) "*General Relativity*"; (March 2024) "*Gravity*"; (May 2024) "*Electricity & Magnetism*"; (June 2024) "*Quantum Entanglement*"; (September 2024) "*The Standard Model.*"

These can be seen to be largely related to inconsistencies between *Einstein's theories of Special and General Relativity* and *quantum mechanics* and the consequent inability to quantize Einstein's *relativistic field equations*. Of particular concern is the fact that most of the so-called *elementary particles* in the *Standard Model* are largely derived from extremely high energy collisions between *protons*, have very short *half-lives*, between two one millionths and less than one million billion billionth of a second, and have *masses* derived almost entirely from *interaction energy*, making the *Standard Model* appear more like a theory of mass creation in high energy physics than a theory of *elementary particles*. As Dirac noted in his 1933 Nobel Lecture: "To get an interpretation of some modern experimental results one must suppose that particles can be created and annihilated. Thus, *if a particle is observed to come out from another particle, one can no longer be sure that the latter is composite. The former may have been created.*"

These problems include the following:

(1) Einstein's *theory of special relativity*. See Underwood, T. G. (June 2023). *Special Relativity.*

There is no evidence for Einstein's *theory of special relativity*, based directly or indirectly, on the observation of the speed of electromagnetic radiation in a vacuum emitted by an inertial body, or as observed by an inertial observer, moving in a straight line and not involving mirrors. However, by now it may be possible to achieve this in a laboratory experiment in a vacuum without mirrors, using electromagnetic radiation emitted by two sources of the same frequency, one stationary and the other moving at a constant velocity in a straight line; either directly, or by measuring the observed frequency of the radiation.

The Ehrenfest paradox, the *non-relativistic* Doppler red shift and blue shift for light, the known physics of the emission of electromagnetic radiation and of the electron, and the success of *non-relativistic* quantum electrodynamics in explaining the interaction of the electromagnetic field with electrically charged particles, comprise the strongest evidence against Einstein's *second postulate*, the *constancy of the speed of light*.

Quite apart from the enormity of the consequences of Einstein's two postulates taken together, including *length contraction*, *time dilation*, and the requirement to assume a *point electron* in the unsuccessful attempt to introduce special relativity into quantum electrodynamics, the evidence in support of Einstein's *second postulate* on the constancy of the speed of light is far outweighed by the evidence against it.

(2) Einstein's *theory of general relativity*. See Underwood, T. G. (November 2023). *General Relativity.*

Einstein's *theory of general relativity* attempted to extend his *theory of special relativity* beyond space and time, to include *matter* and *gravitational fields*. Whilst this allowed Einstein to construct a *relativistic theory* of the effect of a *gravitational field* on *matter*, it also resulted in him rejecting his *postulate on the constancy of light* in the presence of a gravitational field.

*General relativity* is claimed to generalize *special relativity* and refine Newton's *law of universal gravitation*, providing a unified description of *gravity* as a geometric property of *space and time* or four-dimensional *spacetime*. In particular, the *curvature of spacetime* is directly related to the *energy* and *momentum* of whatever *matter* and *radiation* are present. The relation is specified by the *Einstein field equations*, a system of second-order partial differential equations.

In order to make calculations with his theory, Einstein had to import *Newton's law of gravitation*, which itself is an empirical law with no fundamental foundation. Consequently, the only evidence that Einstein could provide for his *theory of general relativity* was effectively Newtonian.

Einstein, A. (February, 1917). *Kosmologische Betrachtungen zur allgemeinen Relativitätstheorie*. (Cosmological Considerations in the General Theory of Relativity.): describes Einstein's struggles with supplementing the *relativistic differential equations* by *limiting conditions* at *spatial infinity* in order to regard the universe as being of infinite spatial extent. As he noted, "we admittedly had to introduce an extension of the field equations of gravitation which is not justified by our actual knowledge of gravitation".

(3) Gravity. See Underwood, T. G. (March 2024). *Gravity.*

Reconciliation of *general relativity* with the laws of *quantum physics* remains a problem *as there is a lack of a self-consistent theory of quantum gravity*. It is not yet known how gravity can be unified with the three non-gravitational forces: strong, weak and electromagnetic.

(4) Elementary and composite particles with the same electric charge attract each other, and elementary and composite particles with opposite electric charge are repulsed, through the electromagnetic interaction or electromagnetic force, according to Coulomb's law. See Underwood, T. G. (May 2024). *Electricity & Magnetism.*

The Standard Model adds nothing to the classical *non-relativistic* theory.

(5) Quantum field theory (relativistic quantum electrodynamics) and renormalization. See Underwood, T. G. (April 2023). *Quantum Electrodynamics – annotated sources.* Volume II.

The lack of convergence in current formulations of *relativistic quantum electrodynamics* for the *electron*, or *quantum field theory*, due to the interaction of the electromagnetic and matter fields with their own vacuum fluctuations raised the question of whether the still unresolved *divergencies* arising largely, if not entirely, from the assumption of a *point electron*, could be isolated in unobservable *renormalization* factors.

> Underwood, T. G. (April 2023). *Quantum Electrodynamics – annotated sources.* Volume II. Preface, pp. 34-35: "Schwinger, in the Preface of his 1958 book [*Selected Papers on Quantum electrodynamics*], "questioned whether *renormalization* simply corrected a mathematical error that causes the divergencies, or whether *there is a serious flaw in the structure of field theory*". Feynman, in his 1965 Nobel prize speech, described *renormalization* as "simply a way to sweep the difficulties of the divergences of electrodynamics under the rug". Dirac's final judgment on *quantum field theory*, in his last paper published in 1987 [The inadequacies of quantum field theory], was that "These rules of *renormalization* give surprisingly, excessively good agreement with experiments. Most physicists say that these working rules are, therefore, correct. I feel that is not an adequate reason. Just because the results happen to be in agreement with observation does not prove that one's theory is correct."

Despite the claims to the contrary in modern textbooks, there have been no significant developments in the quantum electrodynamics or quantum field theory since 1965 to resolve the underlying occurrence of divergencies.

The *standard model* was established in the 1970s. It was triggered by the development of studies of *gauge theories*. In particular, it was proved that a generalized *gauge theory* is *renormalizable*.

This opened the possibility that all the *interactions* of an *elementary particle* could be described by the *quantum field theory* without the difficulty of *divergence*. Before this time, such description was possible only for *electro-magnetic interaction*.

(6) Elementary particles. See Underwood, T. G. (September 2024). *The Standard Model.*

The *Standard Model* comprises a total of 52 *elementary particles* and their *anti-particles* of which only the *electron* and the *photon* are stable. Most of these have been revealed largely by tracks in cloud chambers from high energy collisions between particles in *proton-antiproton* and *hadron* colliders, or at high altitudes with cosmic rays. The *half-lives* of the other *elementary particles* vary between 2.2 x $10^{-6}$ s for the *muon*, to 3 x $10^{-25}$ s for the $W^+$, $W^-$ and $Z^0$ bosons, 1.6 x $10^{-25}$ s for the Higgs boson, and 5 x $10^{-25}$ s for the top quark, particles with a *very high mass*.

(7) Masses. See Underwood, T. G. (September 2024). *The Standard Model.*

Most of the *mass* of a *proton* or *neutron* is the result of the *strong interaction* energy; the individual *quarks* provide only about 1% of the *mass* of a *proton. Protons* are composed of two *up quarks* (*mass* of each equal to 0.002 of *proton mass*), and one *down quark* (*mass* equal to 0.005 of *proton mass*). *Neutrons* are composed of two *down quarks*, and one *up quark.*

The large *masses* of the $W^+$, $W^-$, $Z^0$ and *Higgs bosons* and the *top quark* (respectively 85.7, 85.7, 97.2, 133.3 and 184.9 times the *mass* of the *proton*) relative to the *masses* of the *proton* and *neutron* raise questions regarding whether they are really *elementary particles*, in particular in view of how they were created.

(8) Symmetries. See Underwood, T. G. (September 2024). *The Standard Model.*

The Standard Model is a *gauge quantum field theory* containing the *internal* (*local*) *symmetries* of the *unitary product group* SU(3) × SU(2) × U(1). Roughly, the three factors of the gauge symmetry give rise to the three fundamental interactions, the *strong, weak* and *electromagnetic interactions*.

The three local symmetries addressed by the Standard Model are:
*C-symmetry* (charge symmetry), a universe where every particle is replaced with its antiparticle; *P-symmetry* (parity symmetry), a universe where everything is mirrored along

the three physical axes. This excludes weak interactions; *T-symmetry* (time reversal symmetry), a universe where the direction of time is reversed.

*CP violation*, the violation of the combination of *C*- and *P-symmetry*, is necessary for the presence of significant amounts of *baryonic matter* in the universe.

(9) Spin. See Underwood, T. G. (September 2024). *Quantum Entanglement.*

The *spin* of an *elementary particle* is a quantum state and consequently a *non-relativistic* concept. We could try to determine the behavior of *spin* under general Lorentz transformations, but we would immediately discover a major obstacle. Unlike SO(3), the group of Lorentz transformations SO(3,1) is *non-compact* and therefore does not have any faithful, unitary, finite-dimensional representations.

(10) Exchange interaction. See Underwood, T. G. (September 2024). *The Standard Model.*

In the *Standard Model of particle physics* four *fundamental interactions* or *forces* are assumed: *gravity*, and the *electromagnetic, weak and strong interactions*, of which the latter three are incorporated in the model.

*Exchange interaction*. According to the *quark formulation* in the *Standard Model*, a *weak interaction occurs when two particles (typically, but not necessarily, half-integer spin fermions) exchange integer-spin, force-carrying bosons*. In the *weak interaction, fermions* can *exchange* three types of force carriers, namely W+, W−, and Z *bosons*. The *weak interaction* is the only fundamental *interaction* that breaks *parity symmetry*, and similarly, but far more rarely, the only *interaction* to break *charge–parity symmetry.* The *weak interaction* is considered unique in that it allows *quarks* to *swap* their flavor for another. The *swapping* of those properties is mediated by the force carrier *bosons*.

*Composite particles*, such as *protons* and *neutrons*, can exist as different *quantum states*, referred to as *isospin states*, which create an attractive force, the *strong interaction*, through *exchange interaction* between two *quantum isospin states*.

*Isospin*. The name of the concept contains the term *spin* because its quantum mechanical description is mathematically similar to that of *angular momentum* (in particular, in the way it *couples*; for example, a *proton–neutron pair* can be *coupled* either in a *state* of *total isospin* 1 or in one of 0. But unlike angular momentum, it is a dimensionless quantity and is not actually any type of spin.

Before the concept of *quarks* was introduced, particles that are affected equally by the *strong force* but had different *electric charges* (e.g. *protons* and *neutrons*) were considered different states of the same particle, but having *isospin* values related to the number of *charge states*. A close examination of *isospin symmetry* ultimately led directly to the discovery and understanding of *quarks* and to the development of *Yang–Mills theory*.

(11) Quarks. See Underwood, T. G. (September 2024). *The Standard Model.*

*Quarks*, which make up composite particles like *neutrons* and *protons*, come in six "*flavors*" – *up*, *down*, *charm*, *strange*, *top* and *bottom* – which give those composite particles their properties.

The *top quark had a mass much larger than expected*, almost as large as that of a gold atom. It has a mass of 172.76 ± 0.3 GeV/c2, (185 times the mass of a proton), which is close to the rhenium atom mass. Because the *top quark* is so massive, *its properties allowed indirect determination of the mass of the Higgs boson*. As such, the *top quark*'s properties are extensively studied as a means to discriminate between competing theories of new physics beyond the *Standard Model*. *The top quark is the only quark that has been directly observed* due to its decay time being shorter than the hadronization time.

*The model was discussed mainly in classical terms*; nothing was proved about the quantized theory. Higgs noted that it should be understood, therefore, that *the conclusions which were presented concerning the masses of particles were conjectures based on the quantization of linearized classical field equations*.

(12) Antimatter. See Underwood, T. G. (September 2024). *The Standard Model.*

Particles with the same *mass* but opposite *electric charge* to an existing *particle* are described as *anti-particles*. In the Standard Model, they are a different form of *matter* known as *antimatter*. The *electric charge* of the *positron* (the *anti-electron*) is – e (i.e. – 1 *electron charge*, positive). *Quarks* have fractional *electric charges*.

According to this theory, there are compelling theoretical reasons to believe that, aside from the fact that *antiparticles* have different signs on all charges (such as *electric* and *baryon charges*), *matter* and *antimatter* have exactly the same properties. This means a *particle* and its corresponding *antiparticle* must have identical *masses* and *decay lifetimes*. It is claimed that the electron's antiparticle, the *positron*, is *stable*, but in condensed matter it typically remains only a short time ($10^{-10}$ sec) before annihilating with an *electron*. Similarly, the *antiproton* is claimed to be *stable* but is short-lived due to collisions with *protons*.

*CP violation* means violation of *symmetry* between *particles* and *anti-particles*. The discovery of *CP violation* implies that there is an essential difference between *particles* and *anti-particles*. *Matter dominance* of the universe seems to require new sources of *CP violation*, because it appears that *CP violation* of the *six-quark model* is too small to explain *matter* dominance.

According to the Standard Model, the asymmetry of *matter* and *antimatter* (*baryon* asymmetry) in the visible universe is one of the great unsolved problems in physics.

(13) Other shortcomings.

The *Standard Model* leaves some physical phenomena unexplained and so falls short of being a complete theory of fundamental interactions. Although the physics of special relativity is included, general relativity is not, and it will fail at energies or distances where the graviton is expected to emerge.

It does not account for the universe's accelerating expansion as possibly described by *dark energy*.

The model does not contain any viable *dark matter* particle that possesses all of the required properties deduced from observational cosmology.

It also does not incorporate *neutrino oscillations* and their *non-zero masses*.

(14) Supersymmetry and String theory. See Underwood, T. G. (September 2024). *The Standard Model.*

*Supersymmetry* could help explain certain phenomena, such as the nature of *dark matter* and the *hierarchy problem* in particle physics. *There is no experimental evidence that either supersymmetry or misaligned supersymmetry holds in our universe*, and *many physicists have moved on from supersymmetry and string theory entirely due to the non-detection of supersymmetry at the Large Hadron Collider* (LHC).

Because *string theory* potentially provides a unified description of *gravity* and *particle physics*, it is a candidate for a theory of everything, a self-contained mathematical model that describes all fundamental forces and forms of matter. Despite much work on these problems, *it is not known to what extent string theory describes the real world* or how much freedom the theory allows in the choice of its details.

**Foundational assumptions.**

This book attempts to highlight these problems by presenting a formulation of the foundations of physics, which I refer to as "*New Physics*", in which they are largely avoided by replacing *Einstein's theory of Special Relativity*, in which the speed of light is constant for all observers *regardless of the motion of light source or observer*, with *Ritz's emission theory*, in which the speed of light is constant *with respect to the emitter*.

Without professing to be a complete theory this is intended to clarify these issues and ideally encourage an experiment to demonstrate directly whether the speed of light travelling in a straight line is in fact independent of the speed of the emitter.

**Part I** examines the foundational assumptions of the Standard Model and **Part II** sets out the corresponding foundational assumptions of *New Physics*. Both include the corresponding annotated primary source documents. In **Part III**, they are brought together under each heading to highlight the differences.

The foundational assumptions of the *Standard Model* are

1) The universe is composed of 52 *elementary particles* and *antiparticles*, of which only the electron and photon are stable. *Antiparticles*, which have *differences in quantum numbers* in additional to electric charge, form *antimatter*, a different form of matter. There is an unexplained asymmetry of *matter* and *antimatter* in the visible universe;
2) The speed of light in a vacuum is constant for all observers, regardless of the motion of light source or observer. This results in *length contraction* and *time dilation* for a moving observer;
3) All elementary particles, including electromagnetic waves (photons), are quantized, but there is a *lack of convergence* in current formulations of quantum electrodynamics due to the interaction of the electromagnetic and matter fields with their own vacuum fluctuations. The question is whether all divergencies can be isolated in unobservable *renormalization* factors;
4) *All elementary* particles have *mass*, apart from the photon and gluons, but the large *masses* of the $W^+$, $W^-$, $Z^0$ and *Higgs bosons* and the *top quark* (respectively 85.7, 85.7, 97.2, 133.3 and 184.9 times the *mass* of the *proton*) raise questions regarding whether they are really *elementary particles*, in particular in view of how they were created by collisions of high energy *protons* in *proton* and *hadron* colliders.
5) *Elementary* and *composite* particles can have *electric charge* or be *neutral*. *Up*, *charm*, and *top quarks* are assumed to have *electric charges* equal to 2/3 of the charge of the *electron* and *proton*; and *down*, *strange*, and *bottom quarks* to have *electric charges* equal to – 1/3 of the charge of the *electron* and *proton*, though this cannot be observed;
6) *Elementary* and *composite* particles have a *quantum state* called *spin*;
7) *Elementary and composite* particles with *mass* attract each other through the *gravitational interaction or gravitational force*, according to Einstein's theory of General Relativity;

8) *Elementary* and *composite* particles with the same *electric charge* attract each other, and *elementary* and *composite* particles with opposite *electric charge* are repulsed, through the *electromagnetic interaction* or *electromagnetic force*, according to Coulomb's law;
9) The *spin* of *elementary* and *composite* particles creates an attractive force between two particles - the *weak interaction* or *weak force* – through *exchange interaction* when *quarks* exchange integer-spin, force carrying *bosons*. This is also ascribed to *weak isospin* and *weak hypercharge*;
10) *Composite* particles, such as *protons* and *neutrons*, exist as *quantum states* with different *baryon numbers*, referred to as *iso-spin states*, which creates an attractive force between two particles – the *strong interaction* or *strong force* – through *exchange interaction* between two *quantum iso-spin states*.

The foundational assumptions of *New Physics* are

1) The universe is composed of 14 *elementary particles*, of which the *electron*, *photon*, *proton*, and *neutron* in the nucleus, are stable. *Elementary particles* are confined to those that have the possibility of being *observed*. No *quarks*, *gluons*, nor $W^+$, $W^-$, $Z^0$, or *Higgs bosons*. Elimination of notion of *antimatter* and problem of asymmetry of *matter* and *antimatter* in the visible universe. *Antiparticles* are simply the less stable *particles* of similar *mass* but of opposite *electric charge*.
2) The speed of light in a vacuum is constant relative to the emitter. Replaces *Einstein's theory of Special Relativity* with *Walter Ritz's emission theory*. Avoids *length contraction* and *time dilation* for a moving observer.
3) All *elementary* particles, including electromagnetic waves (photons), are quantized. Only relationships among observable quantities occur. Avoids requirement to assume a *point electron* and address through a process of *renormalization* the still unresolved *divergencies*;
4) All *elementary* particles have *mass* apart from the *photon*. Removal of $W^+$, $W^-$, $Z^0$ and *Higgs bosons* and the *top quark* avoids problem of overweight particles;
5) *Elementary* and *composite* particles can have *electric charge* or be neutral. Removal of *quarks* avoids the problem of elementary particles with unobserved fractional *electric charges*;
6) *Elementary* and *composite* particles have a *quantum state* called spin. *Spin* obeys the mathematical laws of angular momentum quantization;
7) *Elementary* and *composite* particles with *mass* attract each other through the *gravitational interaction or gravitational force* resulting from *quantum entanglement between matter*. Introduction of a *quantum theory of gravity*;
8) *Elementary* and *composite* particles with the same *electric charge* attract each other, and *elementary* and *composite* particles with opposite *electric charge* are

repulsed, through the *electromagnetic interaction* or *electromagnetic force*, according to Coulomb's law;

9) The *spin* of *elementary* and *composite* particles creates an attractive force between two particles - the *weak interaction* or *weak force* – resulting from *quantum entanglement* between *spin states*;
10) *Elementary* particles, such as *protons* and *neutrons*, and *composite* particles, can exist as *quantum states* referred to as *iso-spin states*, which creates an attractive force between two particles – the *strong interaction* or *strong force* – resulting from *quantum entanglement* between *isospin states*.

I would like to acknowledge Wikipedia, in particular, which provided much of this material, as well as other referenced sources.

Trevor G. Underwood

18 SE 10th Ave
Fort Lauderdale, FL33301.

October 7, 2024.

# PART I Problems with the current version of theoretical physics.

The foundational assumptions of the current version of theoretical physics:

## (1) The universe is composed of elementary particles.

### The Standard Model of Particle Physics.

**Underwood. T. G. (2024). *The Standard Model*,** PART I, The Standard Model of Particle Physics, p. 27: "The *Standard Model of particle physics* is the theory describing three of the four known fundamental *forces* (*electromagnetic, weak and strong interactions* – excluding gravity) in the universe and *classifying all known elementary particles.*

It was developed in stages throughout the latter half of the 20th century, through the work of many scientists worldwide, with the current formulation being finalized in the mid-1970s upon experimental confirmation of the existence of *quarks*. Since then, proof of the *top quark* (1995), the *tau neutrino* (2000), and the *Higgs boson* (2012) have added further credence to the *Standard Model*. In addition, the *Standard Model* has predicted various properties of *weak neutral currents* and the W and Z *bosons* with great accuracy.

Although the *Standard Model* is believed to be theoretically self-consistent*

> * There are mathematical issues regarding quantum field theories still under debate (see e.g. Landau pole), but the predictions extracted from the Standard Model by current methods applicable to current experiments are all self-consistent.

and has demonstrated some success in providing experimental predictions, it leaves some physical phenomena unexplained and so falls short of being a complete theory of fundamental interactions. Although the physics of special relativity is included, general relativity is not, and it will fail at energies or distances where the graviton is expected to emerge. It does not fully explain baryon asymmetry, or account for the universe's accelerating expansion as possibly described by dark energy. The model does not contain any viable dark matter particle that possesses all of the required properties deduced from observational cosmology. It also does not incorporate neutrino oscillations and their non-zero masses.

**The introduction of quarks.**

In 1964, Murray Gell-Mann, and separately George Zweig, proposed that *baryons*, which include *protons* and *neutrons*, and *mesons* were composed of elementary particles.

> [A *meson* is a type of hadronic subatomic *composite particle* composed of *an equal number of quarks and antiquarks, usually one of each, bound together by the strong interaction*. Because *quarks* have a *spin* ½ , the difference in quark number between *mesons* and *baryons* results in *mesons* being *bosons*, whereas *baryons,* the other members of the *hadron* family, composed of *odd numbers of valence quarks* (at least three), are *fermions*.
>
> The existence of *mesons* was predicted by Hideki Yukawa's 1935 *theory of mesons* that postulated the particle as mediating the nuclear force. [Yukawa, H. (1935). On the Interaction of Elementary Particles. *Proceedings of the Physico-Mathematical Society of Japan*, 17, 48; also in (January, 1955). *Progr. Theoret. Phys. Suppl.* 1, 1–10, https://doi.org/10.1143/PTPS.1.1.]

Zweig called the elementary particles "aces" while Gell-Mann called them "*quarks*". The theory came to be called the *quark model* and this became the foundation of the *Standard Model of particle physics*. [Gell-Mann. M. (February, 1964). A Schematic Model of Baryons and Mesons. *Physics Letters*, 8, 3, 214–5] and George Zweig [Zweig, G. (February 21, 1964). An SU(3) Model for Strong Interaction Symmetry and its Breaking: II. *CERN Document Server*. CERN-TH-412].

*Quarks* were introduced as parts of an ordering scheme for *hadrons*, and there was little evidence for their physical existence until deep inelastic scattering experiments at the Stanford Linear Accelerator Center in 1968. Accelerator program experiments have provided evidence for all six *flavors*. The *top quark*, first observed at Fermilab in 1995, was the last to be discovered.

The name "*quark*" was coined by Gell-Mann, and is a reference to the novel Finnegans Wake, by James Joyce ("*Three quarks for Muster Mark!*" book 2, episode 4). Zweig had referred to the particles as "aces", but Gell-Mann's name caught on. *Quarks*, *antiquarks*, and *gluons* were soon established as the underlying elementary objects in the study of the structure of *hadrons*. The 1969 Nobel Prize in Physics was awarded to Gell-Mann "for his contributions and discoveries concerning the classification of elementary particles and their interactions".

In the *quark model*, a *quark* is a type of *elementary particle* and a fundamental constituent of matter. *Quarks* combine to form composite particles called *hadrons*, the most stable of which are *protons* and *neutrons*, the components of atomic nuclei. All commonly observable matter is composed of *up quarks*, *down quarks* and *electrons*.

In the *quark formulation*, *protons* are composite particles, containing three valence *quarks*, and together with *neutrons* are classified as *hadrons*. *Protons* are composed of two *up quarks* of charge +2/3 e (and *mass* equal to 0.002 of the *proton mass*) each, and one *down quark* of charge −1/3 e (and *mass* equal to 0.005 of the *proton mass*). *Neutrons* are composed of two *down quarks* of charge −1/3 e (and *mass* equal to 0.005 of the *proton mass*) each, and one up quark of charge +2/3 e (and *mass* equal to 0.002 of the *proton mass*). The *rest masses* of *quarks* contribute only about 1% of a *proton*'s *mass*. The remainder of a *proton*'s *mass* is due to quantum chromodynamics *binding energy*, which includes the *kinetic energy* of the *quarks* and the *energy* of the *gluon fields* that bind the *quarks* together.

Owing to a phenomenon known as *color confinement*, *quarks* are never found in isolation; they can be found only within *hadrons*, which include *baryons* (such as *protons* and *neutrons*) and *mesons*, or in *quark–gluon* plasmas. For this reason, much of what is known about *quarks* has been drawn from observations of *hadrons*.

*Quarks* have various intrinsic properties, including *electric charge*, *mass*, *color charge*, and *spin*. They are the only elementary particles in the *Standard Model* of particle physics to experience all four fundamental interactions, also known as fundamental forces (*electromagnetism*, *gravitation*, *strong interaction*, and *weak interaction*), as well as the only known particles whose *electric charges* are not integer multiples of the elementary *charge*.

There are six types, known as *flavors*, of *quarks*: *up*, *down*, *charm*, *strange*, *top*, and *bottom*. *Up and down quarks* have the lowest *masses* of all *quarks*. The heavier *quarks* rapidly change into *up and down quarks* through a process of *particle decay*: the transformation from a higher *mass state* to a lower *mass state*. Because of this, *up and down quarks* are generally stable and the most common in the universe, whereas *strange*, *charm*, *bottom*, and *top quarks* can only be produced in high energy collisions (such as those involving cosmic rays and in particle accelerators). For every *quark flavor* there is a corresponding type of *antiparticle*, known as an *antiquark*, that differs from the *quark* only in that some of its properties (such as the *electric charge*) have equal magnitude but opposite sign.

## *The quark model.*

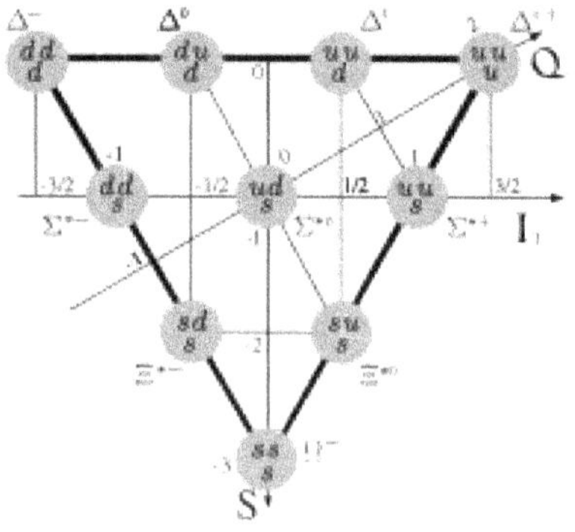


Combinations of three u, d or s-*quarks* forming *baryons* with *spin-*$^3/_2$ form the *baryon decuplet.*

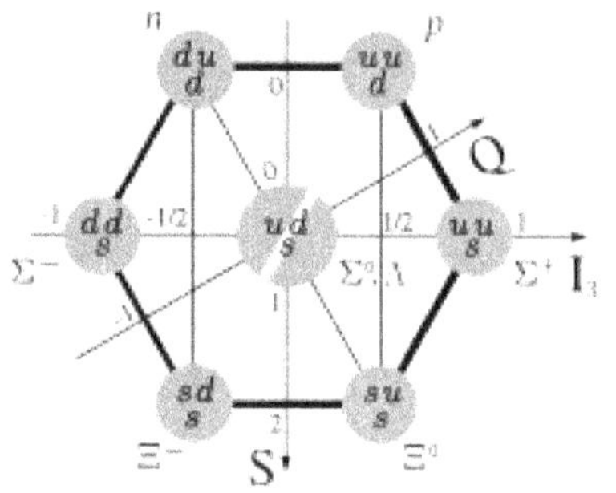


Combinations of three u, d or s-*quarks* forming *baryons* with *spin-*$^1/_2$ form the *baryon octet.*

The discovery and subsequent analysis of additional particles, both *mesons* and *baryons*, made it clear that the concept of *isospin symmetry* could be broadened to an even larger symmetry group, now called *flavor symmetry*. Once the *kaons* and their property of *strangeness* became better understood, it started to become clear that these, too, seemed to be a part of an enlarged symmetry that contained *isospin* as a subgroup. The larger symmetry was named the *Eightfold Way* by Murray Gell-Mann, and was promptly recognized to correspond to the adjoint representation of SU(3). To better understand the origin of this symmetry, Gell-Mann proposed the existence of *up*, *down* and *strange* quarks which would belong to the fundamental representation of the SU(3) *flavor symmetry*.

In the *quark model*, the *isospin projection* ($I_3$) followed from the *up* and *down quark* content of particles; uud for the *proton* and udd for the *neutron*. *Technically, the nucleon doublet states are seen to be linear combinations of products of 3-particle isospin doublet states and spin doublet states*. That is, the (*spin-up*) *proton* wave function, in terms of *quark-flavor eigenstates*, is described by

$$|p\uparrow\rangle = 1/3\sqrt{2}\,(|duu\rangle\ |udu\rangle\ |uud\rangle) \begin{pmatrix} 2 & -1 & -1 \\ -1 & 2 & -1 \\ -1 & -1 & 2 \end{pmatrix} \begin{pmatrix} |\downarrow\uparrow\uparrow\rangle \\ |\uparrow\downarrow\uparrow\rangle \\ |\uparrow\uparrow\downarrow\rangle \end{pmatrix}$$

and the (*spin-up*) *neutron* by

$$|n\uparrow\rangle = 1/3\sqrt{2}\,(|udd\rangle\ |dud\rangle\ |ddu\rangle) \begin{pmatrix} 2 & -1 & -1 \\ -1 & 2 & -1 \\ -1 & -1 & 2 \end{pmatrix} \begin{pmatrix} |\downarrow\uparrow\uparrow\rangle \\ |\uparrow\downarrow\uparrow\rangle \\ |\uparrow\uparrow\downarrow\rangle \end{pmatrix}$$

Here, $|u\rangle$ is the *up quark flavor eigenstate*, and $|d\rangle$ is the *down quark flavor eigenstate*, while $|\uparrow\rangle$ and $|\downarrow\rangle$ are the *eigenstates* of $S_Z$. Although these *superpositions* are the technically correct way of denoting a *proton* and *neutron* in terms of *quark flavor* and *spin eigenstates*, for brevity, they are often simply referred to as "*uud*" and "*udd*". The derivation above assumes exact *isospin symmetry* and is modified by SU(2)-breaking terms.

Similarly, the *isospin symmetry* of the *pions* are given by:

$$|\pi^+\rangle = |u\bar{d}\rangle$$
$$|\pi^0\rangle = 1/\sqrt{2}\,(|u\bar{u}\rangle - |d\bar{d}\rangle)$$
$$|\pi^-\rangle = -\,|d\bar{u}\rangle.$$

Although the discovery of the *quarks* led to reinterpretation of *mesons* as a vector bound *state* of a *quark* and an *antiquark*, it is sometimes still useful to think of them as being the *gauge bosons* of a hidden *local symmetry*.

## **Gell-Mann. M. (February, 1964). A Schematic Model of Baryons and Mesons.***

*Physics Letters*, 8, 3, 214–5; https://doi.org/10.1016/S0031-9163(64)92001-3; also at https://www.nssp.uni-saarland.de/lehre/Vorlesung/Kernphysik_SS19/History/Papers/Gell-Mann.pdf.; also in Underwood, T. G. (2024). *The Standard Model*, Part I, pp. 309-14.

* Work supported in part by the U. S. Atomic Energy Commission.

California Institute of Technology, Pasadena, California.

Received January 4, 1964

In 1964, Murray Gell-Mann, and separately George Zweig, proposed that *baryons*, which include *protons* and *neutrons*, and *mesons* were composed of *elementary particles*. [Zweig, G. (February 21, 1964). An SU(3) Model for Strong Interaction Symmetry and its Breaking: II. *CERN Document Server*. CERN-TH-412; doi:10.17181/CERN-TH-412.] Zweig called the *elementary particles* "aces" while Gell-Mann called them "*quarks*"; the theory came to be called the *quark model*. The bootstrap model for *strongly interacting particles* described in terms of the *broken eightfold way* is discussed to determine algebraic properties of the interactions with scattering amplitudes on the *mass shell*. A mathematical model based on *field theory* is described.

---

If we assume that the *strong interactions* of *baryons* and *mesons* are correctly described in terms of the *broken "eightfold way"*[1,2,3] we are tempted to look for some fundamental explanation of the situation.

[1] Gell-Mann, M. (1961). *California Institute of Technology Synchrotron Laboratory, Report CTSL-20*.
[2] Ne'eman, Y. (1961). *Nuclear Phys*., 26, 222.
[3] Gell-Mann, M. (1962). *Phys. Rev*., 125, 1067.

A highly promised approach is the purely dynamical "bootstrap" model for all the strongly interacting particles within which one may try to derive *isotopic spin* and *strangeness conservation* and *broken eightfold symmetry* from self-consistency alone[4].

[4] E.g.: Capps, R. H. (1963). *Phys. Rev. Lett*., 10, 312; R. E. Cutkosky, R. E., Kalckar, J. & Tarjanne, P. (1962). *Physics Letters*, 1, 93; Abers, E., Zachariasen, F. & Zemaeh, A. C. (1963). *Phys. Rev*., 132, 1831; Glashow, s. (1963). *Phys. Rev*. 130, 2132; R. E. Cutkosky, R. E. & P. Tarjanne, P. (1963). *Phys. Rev*., 132, 1354.

Of course, with only *strong interactions*, the orientation of the asymmetry in the *unitary space* cannot be specified; one hopes that in some way the selection of specific components of the *F-spin* by *electromagnetism* and the *weak interactions* determines the choice of *isotopic spin* and *hypercharge directions*.

[The *strong interaction*, also called the *strong force* or *strong nuclear force*, is a fundamental *interaction* that confines *quarks* into *protons*, neutrons, and other *hadron* particles. The *strong interaction* also binds *neutrons* and *protons* to create *atomic nuclei*, where it is called the *nuclear force*.

Most of the *mass* of a *proton* or *neutron* is the result of the *strong interaction* energy; the individual *quarks* provide only about 1% of the *mass* of a *proton*. Protons are composed of two *up quarks* (*mass* of each equal to 0.002 of *proton mass*), and one *down quark* (*mass* equal to 0.005 of *proton mass*). Neutrons are composed of two *down quarks*, and one *up quark*.

Before 1971, physicists were uncertain as to how the *atomic nucleus* was bound together. It was known that the *nucleus* was composed of *protons* and *neutrons* and that *protons* possessed positive *electric charge*, while *neutrons* were electrically neutral. By the understanding of physics at that time, positive charges would repel one another and the positively charged *protons* should cause the *nucleus* to fly apart. However, this was never observed. A stronger attractive force was postulated to explain how the *atomic nucleus* was bound despite the *protons*' mutual *electromagnetic* repulsion. This hypothesized force was called the *strong force*, which was believed to be a fundamental force that acted on the *protons* and *neutrons* that make up the *nucleus*.

The *strong attraction* between *nucleons* was the side-effect of a more fundamental force that bound the *quarks* together into *protons* and *neutrons*. The theory of *quantum chromodynamics* explains that *quarks* carry what is called a *color charge*, although it has no relation to visible *color*. *Quarks* with unlike *color charge* attract one another as a result of the *strong interaction*, and the particle that mediates this was called the *gluon*.]

Even if we consider the *scattering amplitudes* of *strongly interacting* particles on the *mass shell* only and treat the matrix elements of the *weak*, *electromagnetic*, and *gravitational interactions* by means of *dispersion theory*, there are still meaningful and important questions regarding the algebraic properties of these *interactions* that have so far been discussed only by abstracting the properties from a formal *field theory model* based on fundamental entities[5] from which the *baryons* and *mesons* are built up.

[5] Tarjanne, P. & Teplitz, V. L. (1963). *Phys. Rev. Lett.*, 11, 447.

If these entities were *octets*, we might expect the underlying *symmetry group* to be SU(8) instead of SU(3); it is therefore tempting to try to use *unitary triplets* as fundamental objects. A *unitary triplet* t consists of an *isotopic singlet* s of *electric charge* z (in units of e) and an *isotopic doublet* (u, d) with *charges* z + 1 and z respectively. The *anti-triplet* has, of course, the opposite signs of the *charges*. *Complete symmetry among the members of the triplet gives the exact eightfold way*, while a *mass difference*, for example, between the *isotopic doublet* and *singlet* gives the first-order violation.

For any value of z and of *triplet spin*, we can construct *baryon octets* from a basic neutral *baryon singlet* b by taking combinations ($btt^-$), ($bttt^-t^-$), etc. [**].

[**] This is similar to the treatment in ref. 1. See also ref. 5.

From ($btt^-$), we get the *representations* **1** and **8**, while from ($bttt^-t^-$) we get **1**, **8**, **10**, **10**, and **27**. In a similar way, *meson singlets* and *octets* can be made out of ($tt^-$), ($ttt^-t^-$), etc. The *quantum number* $n_t - n_{t^-}$ would be zero for all known *baryons* and *mesons*. The most interesting example of such a model is one in which the *triplet* has spin ½ and z = – 1, so that the four particles $d^-$, $s^-$, $u^\circ$ and $b^\circ$ exhibit a parallel with the *leptons*.

A simpler and more elegant scheme can be constructed if we allow non-integral values for the *charges*. We can dispense entirely with the basic *baryon* b if we assign to the *triplet* t the following properties: *spin* ½, z = – 1/3 and *baryon number* 1/3. We then refer to the members $u^{2/3}$, $d^{-1/3}$, and $s^{-1/3}$ of the *triplet* as "*quarks*"[6], q and the members of the *anti-triplet* as *anti-quarks* $q^-$.

[6] James Joyce, (1939). *Finnegan's Wake.* Viking Press, New York, p. 383.

*Baryons* can now be constructed from *quarks* by using the combinations (qqq), ($qqqqq^-$), etc., while *mesons* are made out of ($qq^-$), ($qqq^-q^-$), etc. It is assuming that the lowest *baryon* configuration (qqq) gives just the *representations* **1**, **8**, and **10** that have been observed, while the lowest *meson* configuration ($qq^-$) similarly gives just **1** and **8**.

A formal mathematical model based on *field theory* can be built up for the *quarks* exactly as for p, n, Λ in the old Sakata model, for example[3] with all *strong interactions* ascribed to a neutral *vector meson field* interacting symmetrically with the three particles. Within such a framework, the *electromagnetic current* (in units of e) is just

$$i\{2/3\ u^- \gamma_\alpha u - 1/3\ d^- \gamma_\alpha d\ - 1/3\ s^- \gamma_\alpha s\}$$

or $\mathcal{F}_{3\alpha} + \mathcal{F}_{8\alpha}/\sqrt{3}$ in the notation of ref. 3. For the *weak current*, we can take over from the Sakata model the form suggested by Gell-Mann and Levy[7],

[7] Gell-Mann, M. & Levy, M. (1960). *Nuovo Cimento*, 16, 705.

namely i p⁻ γα(1 + γ₅) (n cos θ + Δ sin θ), which gives in the *quark scheme* the expression
***

    i u⁻ γα(1 + γ₅) (d cos θ + s sin θ)

or, in the notation of ref. 3,

$$[\mathcal{F}_{1\alpha} + \mathcal{F}_{1\alpha}^{5} + i\,(\mathcal{F}_{2\alpha} + \mathcal{F}_{2\alpha}^{5})]\cos\theta + [\mathcal{F}_{4\alpha} + \mathcal{F}_{4\alpha}^{5} + i\,(\mathcal{F}_{5\alpha} + \mathcal{F}_{5\alpha}^{5})]\sin\theta.$$

*** The parallel with i v⁻ₑ γα(1 + γ₅) e and i v⁻μ γα(1 + γ₅) μ is obvious. Likewise, in the model with d⁻, s⁻, u°, and b° discussed above, we would take the *weak current* to be i(b⁻° cos θ + u⁻° sin θ) γα(1 + γ₅) s⁻ + i(u⁻° cos θ − b⁻° sin θ) γα(1 + γ₅) d⁻. The part with Δ(nₜ − n⁻ₜ) = 0 is just i u⁻° γα(1 + γ₅) (d⁻ cos θ − s⁻° sin θ).

We thus obtain all the features of Cabibbo's picture[8]

[8] Cabibbo, N. (1963). *Phys. Rev. Lett.*, 10, 531.

of the *weak current*, namely the rules $|\Delta I| = 1$, $\Delta Y = 0$ and $|\Delta I| = \frac{1}{2}$, $|\Delta Y/\Delta Q = +1$, the conserved $\Delta Y = 0$ *current* with coefficient cos θ, the *vector current* in general as a component of the *current* of the F-spin, and the *axial vector current* transforming under SU(3) as the same component of another *octet*. Furthermore, we have[3] the equal-time commutation rules for the fourth components of the *currents*:

$$[\mathcal{F}_{j4}(x) \pm \mathcal{F}_{j4}^{5}(x), \mathcal{F}_{k4}(x') \pm \mathcal{F}_{k4}^{5}(x')] = -2f_{jkl}\,[\mathcal{F}_{k4}(x) \pm \mathcal{F}_{k4}^{5}(x)]\,\delta(x - x'),$$
$$[\mathcal{F}_{j4}(x) \pm \mathcal{F}_{j4}^{5}(x), \mathcal{F}_{k4}(x') \pm \mathcal{F}_{k4}^{5}(x')] = 0,$$

i = 1, ... 8, yielding the group SU(3) × SU(3). We can also look at the behavior of the *energy density* $\theta_{44}(x)$ (in the gravitational interaction) under equal-time commutation with the *operators* $\mathcal{F}_{j4}(x') \pm \mathcal{F}_{j4}^{5}(x')$. That part which is non-invariant under the group will transform like particular *representations* of SU(3) × SU(3), for example like (3, 3⁻) and (3⁻, 3) if it comes just from the *masses* of the *quarks*.

All these relations can now be abstracted from the field theory model and used in a *dispersion theory* treatment. The *scattering amplitudes* for *strongly interacting particles* on the *mass shell* are assumed known; there is then a system of linear *dispersion relations* for the matrix elements of the *weak currents* (and also the *electromagnetic* and *gravitational interactions*) to lowest order in these *interactions*. These *dispersion relations*, un-subtracted and supplemented by the non-linear *commutation rules* abstracted from the *field theory*, may be powerful enough to determine all the matrix elements of the *weak currents*, including the effective strengths of the *axial vector current* matrix elements compared with those of the *vector current*.

It is fun to speculate about the way *quarks* would behave if they were *physical particles* of finite *mass* (instead of purely mathematical entities as they would be in the limit of infinite mass). Since *charge* and *baryon number* are exactly conserved, one of the *quarks* (presumably $u^{2/3}$ or $d^{-1/3}$) would be absolutely stable* while the other member of the *doublet* would go into the first member very slowly by β-*decay* or K-*capture*.

* There is the alternative possibility that the quarks are unstable under decay into *baryon* plus *anti-di-quark* or *anti-baryon* plus *quadri-quark*. In any case, some particle of fractional *charge* would have to be absolutely stable.

The *isotopic singlet quark* would presumably decay into the *doublet* by *weak interactions*, much as Δ goes into N. Ordinary *matter* near the earth's surface would be contaminated by stable *quarks* as a result of high energy *cosmic ray* events throughout the earth's history, but the contamination is estimated to be so small that it would never have been detected. A search for stable *quarks* of charge – 1/3 or + 2/3 and/or stable *di-quarks* of charge – 2/3 or + 1/3 or + 4/3 at the highest energy accelerators would help to reassure us of the non-existence of real *quarks*.

These ideas were developed during a visit to Columbia University in March 1963; the author would like to thank Professor Robert Serber for stimulating them.

## Makoto Kobayashi – Nobel Lecture, December 8, 2008. CP Violation and Flavor Mixing.

Makoto Kobayashi – Nobel Lecture. NobelPrize.org. https://www.nobelprize.org/prizes/physics/2008/kobayashi/lecture/

In his Nobel Prize lecture Makoto Kobayashi provided a brief history of the development of the *six-quark model.*

---

### *1. INTRODUCTION.*

We know that ordinary *matter* is made of *atoms*. An *atom* consists of the *atomic nucleus* and *electrons*. The *atomic nucleus* is made of a number of *protons* and *neutrons*. A *proton* and a *neutron* are further made of two kinds of *quarks*, u and d. Therefore, the fundamental building blocks of ordinary matter are the *electron* and two kinds of *quarks*, u and d.

The *standard model*, which gives a comprehensive description of current understanding of the *elementary particle* phenomena, however, tells us that the number of species of *quarks* is six. The additional *quarks* are called s, c, b and t. The reason why we do not find them in ordinary *matter* is that they are unstable in the usual environment. Similarly, the *electron* belongs to a family of six members called *leptons*. Three types of *neutrinos* are included among these six.

Another important ingredient of the *standard model* is *fundamental interactions*. Three kinds of *interactions* act on the *quarks* and *leptons*. The *strong interaction* is described by *quantum chromodynamics* (QCD) and the *electro-magnetic* and *weak interactions* by the *Weinberg-Salam-Glashow theory* in a unified manner. All of them belong to a special type of *field theory* called *gauge theory*.

...

Figure 1. The Standard Model.

The *standard model* was established in the 1970s. It was triggered by the development of studies of *gauge theories*. In particular, it was proved that generalized *gauge theory* is *renormalizable*[1].

[1] 't Hooft, G., (December, 1971). Renormalizable Lagrangians for massive Yang-Mills fields. *Nucl. Phys. B*, 35, 167-88; see above; 't Hooft, G. & Veltman, M. J. G. (1972). Regularization and Renormalization of Gauge Fields. *Nucl. Phys.*, B 44, 189.

This opened the possibility that all the *interactions* of an *elementary particle* can be described by the *quantum field theory* without the difficulty of *divergence*. Before this time, such description was possible only for *electro-magnetic interaction.*

The discovery of the new *flavors* made in 1970s played an important role in the establishment of the *standard model*. In particular, the τ-*lepton* and c- and b-*quarks* were found in the 1970s. When we proposed the *six-quark model* to explain *CP violation* with Dr. Toshihide Maskawa in 1973[2], only three *quarks* were widely accepted, and a slight hint of the fourth *quark* was there, but no one thought there would be six *quarks*.

> [2] Kobayashi, M. & Maskawa, T. (February, 1973). CP-Violation in the Renormalizable Theory of Weak Interaction. *Progress of Theoretical Physics*, 49, 2, 652–7. See above.

In the following, I will describe the development of the studies on *CP violation* and the *quark* and *lepton flavors*, putting some emphasis on contributions from Japan. The next section will be devoted to the pioneering works of the Sakata School, from which I learned many things. The work on *CP violation* will be discussed in Section 3. I will explain what we thought and what we found at that time. Section 4 will describe subsequent development related to our work. Experimental verification of the proposed model has been done by using accelerators called *B-factories*. A brief outline of those experiments will be given. Finally in Section 5, I will look briefly at *flavor mixing* in the *lepton* sector, because this is a phenomenon parallel to the *flavor mixing* in the *quark* sector, and Japan has made unique and important contributions in this field.

## *2. SAKATA SCHOOL.*

Both Dr. Maskawa and I graduated from and obtained our PhD's from Nagoya University. When I entered the graduate program, the theoretical particle physics group of Nagoya University was known for its unique research activity and was led by Professor Shoichi Sakata.

In the early 1950s, a number of *strange* particles were discovered, with the first evidence having been found in the *cosmic ray* events of 1947. In the current terminology, *strange* particles contain an *s-quark* or *anti-s-quark* as a constituent, while non-strange particles do not. But what we are about to consider is the era before the *quark model* appeared.

In 1956, Sakata[3] proposed a model which is known as the *Sakata model.*

> [3] Sakata, S. (September, 1956). On a Composite Model for the New Particles. *Progr. Theor. Phys.*, 16, 6, 686-8. See above.

In this model, all *hadrons*, strange and non-strange, are supposed to be composite *states* of the *triplet* of *baryons*, the *proton* (p), the *neutron* (n), and the *lambda* (Λ). In other words, three *baryons*, p, n, and Λ are the fundamental building blocks of the *hadrons* in the model. Eventually, the *Sakata model* was replaced by the *quark model*, where the *triplet* of *quarks*, u, d, and s replace p, n, and Λ. But the root of the idea of fundamental *triplet* is in the *Sakata model*.

In the following, we focus on the *weak interactions* in the *Sakata model*. Usual *beta-decays* of the *atomic nucleus* are caused by the transition of a *neutron* into a *proton*. Similarly, we can consider the transition of a *lambda* into a *proton*. In the *Sakata model*, all the *weak interaction* of the *hadrons* can be explained by these two kinds of transitions among the fundamental *triplet*.

This pattern of the *weak interaction* is quite similar to the *weak interaction* of the *leptons*;

...

It should be noted that at that time, the *neutrino* was thought of as a single species. This similarity of the *weak interaction* between the *baryons* and the *leptons* was pointed out by Gamba, Marshak and Okubo.

In 1960, Maki, Nakagawa, Ohnuki and Sakata[5] developed the idea of *baryon-lepton* or *B-L symmetry* further and proposed the so-called *Nagoya model*.

[5] Maki, Z., Nakagawa, M., Ohnuki, Y. & Sakata, S. (June, 1960). A Unified Model for Elementary Particles. *Progr. Theor. Phys*., 23, 6, 1174-80; https://doi.org/10.1143/PTP.23.1174.

They considered that the *triplet baryons*, p, n, and Λ are composite states of a hypothetical object called *B-matter* and the *neutrino*, the *electron*, and the *muon*, respectively;

$$p = \langle B^{+}v \rangle, \quad n = \langle B^{+}e \rangle, \quad \Lambda = \langle B^{+}\mu \rangle,$$

where B-matter is denoted as $B^{+}$.

Although the composite picture of the *Nagoya model* did not lead to remarkable progress, some ideas in the *Nagoya model* developed in an interesting way. In 1962, it was discovered that there exist two kinds of *neutrinos*, corresponding to the *electron* and the *muon*, respectively. When the results of this discovery at Brookhaven National Laboratory were to come out, two interesting papers were published, one written by Maki, Nakagawa and Sakata[7] and the other by Katayama, Matsumoto, Tanaka and Yamada[8].

[7] Maki, Z., Nakagawa, M., & Sakata, S. (1962). Remarks on the Unified Model of Elementary Particles. *Progr. Theor. Phys.*, 28, 870.
[8] Katayama, Y., Matumoto, K., Tanaka, S. & Yamada, E. (1962.) Possible Unified Models of Elementary Particles with Two Neutrinos. *Progr. Theor. Phys.*, 28 (1962) 675.

Both papers discussed the modification of the *Nagoya model* to accommodate two *neutrinos* in the model.

In the course of the argument to associate *leptons* and *baryons*, Maki et al. discussed the *masses* of *neutrinos* and derived the relation describing the mixing of the *neutrino states*;

$$\nu_1 = \cos\vartheta\, \nu_e + \sin\vartheta\, \nu_\mu,$$
$$\nu_2 = \sin\vartheta\, \nu_e + \cos\vartheta\, \nu_\mu,$$

where $\nu_1$ and $\nu_2$ are the *mass eigenstates* of *neutrinos*, and they assumed that the *proton* is the composite state of the *B-matter* and $\nu_1$. Although the last assumption is not compatible with the current experimental evidence, it is remarkable that they did present the correct formulation of *lepton flavor mixing*. To recognize their contribution, the *lepton flavor mixing matrix* is called the *MNS matrix* today.

*Lepton flavor mixing* gives rise to the phenomenon called *neutrino oscillation*. Many years later, *neutrino oscillation* was discovered in an unexpected manner. We will come back to this point later.

Another important outcome of this argument is the possible existence of the fourth fundamental particle associated with $\nu_2$. This was discussed by Katayama et al. in some detail. At the time, the fundamental particles were still considered *baryons* but the structure of the *weak interaction* discussed here is the same as that of the *Glashow-Illiopoulos-Miani*[9] *scheme*.

[9] Glashow, S. L., Iliopoulos, J. & Maiani, L. (October, 1970). Weak Interactions with Lepton-Hadron Symmetry. *Phys. Rev. D*, 2, 7, 1285; https://doi.org/10.1103/PhysRevD.2.1285.

These works were revived in 1971, when Niu and his collaborators found new kind of events in emulsion chambers exposed to *cosmic rays*. One of the events they found is shown in Figure 3. In this event, we see kinks on two tracks, which indicate the decay of new particles produced in pairs. The estimated *mass* of the new particle was 2~3 GeV and the *life* was a few times $10^{-14}$ sec. under some reasonable assumptions.

When this result came to his attention, Shuzo Ogawa, a member of the Sakata group, immediately pointed out that this new particle might be related to the fourth element expected in the extended version of the *Nagoya model*. By that time, the Sakata model had already been replaced by the *quark model*, so that what he meant was that those new particles might be *charmed* particles, in the current terminology. Following this suggestion, several Japanese groups, including mine, began to investigate the *four-quark model*. At that time, I was a graduate student at Nagoya University.

...

Figure 3. A cosmic ray event.

So far, I have explained about the unique activities of the Sakata School. I mentioned the *four-quark model* in some detail. But I do not mean to imply that the *six-quark model* we proposed is a simple extension of the *four-quark model*. What was most important for me was the atmosphere of the particle physics group of Nagoya University. Although most of the work we discussed in this section had been done by Sakata and his group before I entered the graduate course, the spirit created by this work was still there. I learned the importance of capturing the entire picture, which is necessary for this kind of work.

### *3. SIX-QUARK MODEL.*

In 1971, the *renormalizability* of the non-Abelian gauge theory was proved[1]. This enabled a description of the *weak interactions* with the *quantum field theory* in a consistent manner, and the *Weinberg-Salam-Glashow*[13] *theory* began to attract attention.

> [13] Weinberg, S. (November, 1967). Model of Leptons. *Phys. Rev. Lett*., 19, 21, 1264–6, see above; Salam, A. (1968). Weak and Electromagnetic Interactions, originally printed in "*N. Svartholm: Elementary Particle Theory, Proceedings of the Nobel Symposium Held 1968 at Lerum, Sweden*", pp. 367–377; Glashow, S. L. (February, 1961). Partial-symmetries of weak interactions. *Nuclear Physics*, 22, 4, 579–88, see above.

In 1972, I obtained my PhD from Nagoya University and moved to Kyoto University. Then my work on *CP violation* started.

*CP violation* was first found in 1964 by Cronin, Fitch et al.[14] in the decay of the *neutral K-meson* [*Kaon*].

> [14] Christenson, J. H., Cronin, J. W., Fitch, V. L. & Turlay, R. (July 1964). Evidence for the $2\pi$ Decay of the $K_2^0$ Meson. *Phys. Rev. Lett*., 13, 4, 138–40; https://doi.org/10.1103/PhysRevLett.13.13.

> [*Kaons* have played a distinguished role in our understanding of fundamental conservation laws: *CP violation*, a phenomenon generating the observed *matter–*

*antimatter asymmetry* of the universe, was discovered in the *kaon* system in 1964. This was acknowledged by the award of the 1980 Nobel Prize in Physics jointly to James Watson Cronin and Val Logsdon Fitch "for the discovery of violations of fundamental symmetry principles in the decay of neutral K-mesons".

*CP violation* is a violation of *CP-symmetry* (or *charge conjugation parity symmetry*): the combination of *C-symmetry* (*charge conjugation symmetry*) and *P-symmetry* (*parity symmetry*). *CP-symmetry* states that the laws of physics should be the same if a particle is interchanged with its *antiparticle* (*C-symmetry*) while its *spatial coordinates* are inverted ("mirror" or *P-symmetry*).

*A parity transformation* (also called *parity inversion)* is the flip in the sign of one spatial coordinate.

*Charge conjugation* is a transformation that switches all particles with their corresponding *antiparticles*, thus changing the sign of all *charges*: not only *electric charge* but also the *charges* relevant to other forces.]

*CP violation* means violation of *symmetry* between *particles* and *anti-particles*. The discovery of *CP violation* implies that there is an essential difference between *particles* and *anti-particles*.

We thought that if the *gauge theory* can describe the *interactions* of particles consistently, *CP violating interaction* should also be included in it. It was rather straightforward to solve the problem. We simply investigated conditions for *CP violation* in the *renormalizable gauge theory*. What we found then is summarized as follows[2].

At that time only three *quarks* were widely accepted, but the *three-quark model* had some flaws in the *gauge theory*. Therefore, from a theoretical point of view, the *four-quark* model of the *Glashow-Illiopoulos-Miani* (GIM) type was considered preferable. However, it is impossible to accommodate *CP violation* in a model of the GIM type. We found that even if we relax the conditions for the GIM type, we cannot make any realistic model of *CP violation* with four *quarks*. This implies that there must be some unknown particles besides the fourth *quark*. I thought that this was quite strong and an important conclusion of our argument.

Then we considered a few possible mechanisms of *CP violation* by introducing new particles. We proposed the *six-quark model* as one such possible mechanism.

Below we will discuss the *quark flavor mixing* in some detail, in order to understand why four *quarks* are not enough and six *quarks* are needed to accommodate *CP violation*.

In the frame work of the *gauge theory*, *flavor mixing* arises from a mismatch between *gauge symmetry* and *particle states*. *Gauge symmetry* lumps a certain number of particles into a group called a *multiplet*.

> [A *multiplet* is the *state space* for 'internal' degrees of freedom of a particle, that is, degrees of freedom associated to a particle itself, as opposed to 'external' degrees of freedom such as the particle's position in space. Examples of such degrees of freedom are the *spin state* of a particle in *quantum mechanics*, or the *color*, *isospin* and *hypercharge state* of particles in the *Standard Model* of particle physics. Formally, this state space is described by a *vector space* which carries the *action* of a group of *continuous symmetries*.]

However, each *multiplet* member is not necessarily identical to a single species of particles, but sometimes it is a *superposition of particles*. The *flavor mixing* is nothing but this *superposition*. In the present case, the relevant *gauge group* is SU(2) of the *Weinberg-Salam-Glashow theory* and the *multiplet* is a *doublet*.

Assuming that four *quarks* consist of two *doublets* of the SU(2) *group*, we can denote the most general form of them as

$$\begin{pmatrix} u \\ d' \end{pmatrix}, \quad \begin{pmatrix} c \\ s' \end{pmatrix},$$

where d' and s' are the *superposition* of real *quark states* d and s, described in a matrix form as

$$\begin{pmatrix} d' \\ s' \end{pmatrix} = \begin{pmatrix} V_{ud} & V_{us} \\ V_{cd} & V_{cs} \end{pmatrix} \begin{pmatrix} d \\ s \end{pmatrix},$$

where the matrix describing the mixing should be what is called a *unitary matrix* in mathematics.

The next problem is what the condition for *CP violation* is. In *quantum field theory*, *CP violation* is related to *complex coupling constants*. To be more concrete in the present formulation, *CP violation* will occur if *irreducible* complex numbers appear in the elements of the *mixing matrix*. The matrix elements of a *unitary matrix* are complex numbers in general, but some of them can be made real by adjusting the *phase factor* of the *particle state* without changing the physics results. In such case, those complex numbers are called *reducible*, and otherwise, *irreducible*. Therefore, *one condition of CP violation is that there remain complex numbers which cannot be removed by the phase adjustment of the particle states*.

In the *four-quark model*, adjustment factors are described by two diagonal matrices whose elements are mere *phase factors*. It is easy to see that, if we choose them properly, we can make any 2 x 2 unitary matrix into a real matrix:

… .

Therefore, in this case, we cannot accommodate *CP violation*.

How does this argument change in the *six-quark model*? In this case, we can express the *flavor mixing* as follows:

(u), (c), (t),
(d') (s') (b'),

| | | | | | |
|---|---|---|---|---|---|
| (d') | = | ($V_{ud}$ | $V_{us}$ | $V_{ub}$) | (d), |
| (s') | | ($V_{cd}$ | $V_{cs}$ | $V_{cb}$) | (s) |
| (b') | | ($V_{td}$ | $V_{ts}$ | $V_{tb}$) | (b) |

This time the mixing matrix is a 3 x 3 *unitary matrix*. In this case, however, we cannot remove all the *phase factors* of the matrix elements by adjusting the *phases* of the *quark states*. The best we can do by adjusting the *phases* is to express them by a certain standard form with four parameters. A popular parameterization is

V = … .

where $c_{ij} = \cos \vartheta_{ij}$ and $s_{ij} = \sin \vartheta_{ij}$ with i, j = 1, 2, 3. *We note that, unless $\delta = 0$, the imaginary part remains in the matrix elements and therefore CP symmetry is violated.*

Taking into account the hierarchy of the actual values of the parameters, the following approximate parameterization is frequently used in phenomenological analyses.

V ≈ … .

In this parameterization, *if $\eta$ is not zero, the system is violating CP symmetry.*

We thought that this mechanism of *CP violation* is very interesting and elegant, *but we had no further reason to single out the six-quark model from the other possibilities*. The model was not so special, because if the system has sufficiently many particles, it is not difficult to violate *CP symmetry*. However, *the subsequent experimental development pushed up the six-quark model to a special position.*

In 1974, the *J/ψ particle* was discovered, and soon it turned out that it is the *bound state* of the fourth *quark* c and its *anti-particle*. The discovery had a great impact on particle physics, but it had little effect on the *six-quark model*.

[In 1974, the *J/ψ (J/psi) meson* (a composite particle which is a *vector meson* (*quark* and *antiquark*) was discovered by groups headed by Burton Richter and Samuel Ting, both American physicists, demonstrating the existence of the *charm quark and charm anti-quark*, for which they shared the 1976 Nobel Prize for Physics. They discovered that they had found the same particle, and both announced their discoveries on November 11, 1974. [Aubert, J. J. et al. (1974). Experimental Observation of a Heavy Particle J. *Phys. Rev. Let*., 33, 23, 1404–6; https://doi.org/10.1103/PhysRevLett.33.1404; Augustin, J. -E. et al. (1974). Discovery of a Narrow Resonance in e+e− Annihilation. *Phys. Rev. Let*., 33, 23, 1406–8; https://doi.org/10.1103/PhysRevLett.33.1406.]
Burton Richter (March 22, 1931–July 18, 2018) led the Stanford Linear Accelerator Center (SLAC) team, and Samuel Ting (born January 27, 1936) led the Brookhaven National Laboratory (BNL) team. The importance of this discovery is highlighted by the fact that the subsequent, rapid changes in high-energy physics at the time have become collectively known as the "*November Revolution*".

The *J/ψ meson* is a subatomic particle, a *flavor-neutral meson* consisting of a *charm quark* and a *charm antiquark*. *Mesons* formed by a bound state of a *charm quark* and a *charm anti-quark* are generally known as "*charmonium*" or *psions*. The J/ψ is the most common form of *charmonium*, due to its *spin* of 1 and its low *rest mass*. The J/ψ has a *rest mass* of 3.0969 GeV/$c^2$, just above that of the $\eta_c$ (2.9836 GeV/$c^2$), and a *mean lifetime* of $7.2\times10^{-21}$ s. *This lifetime was about a thousand times longer than expected.*

The *J/ψ meson* had been proposed by James Bjorken and Sheldon Glashow in 1964) [Bjørken, B. J. & Glashow, S. L. (1964). Elementary Particles and SU(4). *Phys. Let*., 11, 3, 255–7; https://www.sciencedirect.com/science/article/abs/pii/0031916364904330.]

[A *vector meson* is a *meson* with *total spin* 1 and odd parity. *Vector mesons* have been seen in experiments since the 1960s, and are well known for their spectroscopic pattern of masses.

The *vector mesons* contrast with the *pseudovector mesons*, which also have a *total spin* 1 but instead have *even parity*. The *vector* and *pseudovector mesons* are also dissimilar in that the spectroscopy of *vector mesons* tends to show nearly pure states of constituent *quark flavors*, whereas *pseudovector mesons* and *scalar mesons* tend to be expressed as composites of mixed states.]

In 1975, the *τ-lepton* was discovered. This discovery had a significant effect on our model. The *τ-lepton* is the fifth member of *leptons*. Although it is a *lepton*, it suggested the existence of a third family in the *quark* sector, too. That was when people began to pay attention to our model. Early works which discussed the *six-quark model* include … .

[In 1975, the *tau* discovered by a group headed by Martin Perl. [Perl, M. L. (1975). Evidence for Anomalous Lepton Production in e+–e− Annihilation. *Phys. Rev. Let.*, 35, 22, 1489–92; https://doi.org/10.1103/PhysRevLett.35.1489.] Their equipment consisted of SLAC's then-new electron–positron colliding ring, called SPEAR, and the LBL magnetic detector. They could detect and distinguish between *leptons*, *hadrons*, and *photons*. They did not detect the *tau* directly, but rather discovered anomalous events.

The search for *tau* started in 1960 at CERN by the Bologna-CERN-Frascati (BCF) group led by Antonino Zichichi. Zichichi came up with the idea of a new sequential *heavy lepton*, now called *tau*, and invented a method of search. He performed the experiment at the ADONE facility in 1969 once its accelerator became operational; however, the accelerator he used did not have enough energy to search for the *tau* particle.

The *tau* was independently anticipated in a 1971 article by Yung-su Tsai. [Tsai, Y-S. (November, 1971). Decay correlations of heavy leptons in e+ + e− → ℓ+ + ℓ−. *Phys. Rev. D.*, 4, 9, 2821; https://doi.org/10.1103/PhysRevD.4.2821.] Providing the theory for this discovery, the *tau* was detected in a series of experiments between 1974 and 1977 by Perl with his and Tsai's colleagues at the Stanford Linear Accelerator Center (SLAC) and Lawrence Berkeley National Laboratory (LBL) group.

The *tau*, also called the *tau lepton*, *tau particle*, *tauon* or *tau electron*, is an elementary particle similar to the *electron*, with negative *electric charge* and a *spin* of ½. Like the *electron*, the *muon*, and the three *neutrinos*, the *tau* is a *lepton*, and like all elementary particles with *half-integer spin*, the *tau* has a corresponding *antiparticle* of *opposite charge but equal mass and spin*. In the *tau*'s case, this is the "*anti-tau*" (also called the *positive tau*). *Tau leptons* have a *lifetime* of $2.9 \times 10^{-13}$ s and a *mass* of 1776.9 MeV/$c^2$ (compared to 105.66 MeV/$c^2$ for *muons* and 0.511 MeV/$c^2$ for *electrons*). Since their *interactions* are very similar to those of the *electron*, a *tau* can be thought of as a much heavier version of the *electron*. Because of their greater *mass, tau* particles do not emit as much *bremsstrahlung* (braking radiation) as *electrons;* consequently, they are potentially much more highly penetrating than *electrons*.

Because of its *short lifetime*, the *range* of the *tau* is mainly set by its *decay length*, which is too small for *bremsstrahlung* to be noticeable. *Its penetrating power appears only at ultra-high velocity and energy (above peta-electronvolt energies)*, when time dilation extends its otherwise very short path-length. As with the case of the other *charged leptons*, the *tau* has an associated *tau neutrino*.]

In 1977, the *upsilon particle* was discovered, and it turned out that it is a *bound state* of the fifth *quark*, the *bottom quark* and the *anti-bottom quark*.

[In 1977, the *upsilon meson* was observed by a team at Fermilab led by Leon Lederman, demonstrating the existence of the *bottom quark*. This was a strong indicator of the *top quark*'s existence: without the *top quark*, the *bottom quark* would have been without a partner. [Herb, S. W. et al. (1977). Observation of a Dimuon Resonance at 9.5 GeV in 400-GeV Proton-Nucleus Collisions. *Phys. Rev. Let*., 39, 5, 252–5; https://doi.org/10.1103/PhysRevLett.39.252.] The evidence for the *bottom quark* was first obtained by the Fermilab E288 experiment team led by Leon M. Lederman, when *proton-nucleon* collisions produced bottomonium decaying to pairs of *muons*. The discovery was confirmed about a year later by the PLUTO and DASP2 Collaborations at the electron-positron collider DORIS at DESY.

The *upsilon meson* is a quarkonium state (i.e. flavorless *meson*) formed from a *bottom quark* and its *antiparticle*. It was the first particle containing a *bottom quark* to be discovered because it is the lightest that can be produced without additional massive particles. It has a *lifetime* of $1.21\times10^{-20}$ s and a *mass* about 9.46 GeV/$c^2$ in the ground state.

The *bottom quark* is an elementary particle of the *third generation*. It is a *heavy quark* with a *charge* of − 1/3 e. All *quarks* are described in a similar way by *electroweak interaction* and *quantum chromodynamics*, but the *bottom quark* has exceptionally low rates of transition to lower-mass quarks. The *bottom quark* is also notable because *it is a product in almost all top quark decays*, and is a frequent decay product of the *Higgs boson*.

The *bottom quark* was proposed by Kobayashi and Maskawa in 1973, for which they shared half of the 2008 Nobel Prize in Physics "for [their 1973] discovery of *the origin of the broken symmetry which predicts the existence of at least three families of quarks in nature*", with the other half going to Yoichiro Nambu "for the discovery of *the mechanism of spontaneous broken symmetry* in subatomic physics". [Kobayashi, M. & Maskawa, T. (February, 1973). CP-Violation in the Renormalizable Theory of Weak Interaction. *Progress of Theoretical Physics*, 49, 2, 652–7; https://doi.org/10.1143/PTP.49.652. See above.]]

The discovery of the last *quark*, the *top quark*, occurred as recently as in 1995, but before that time the *six-quark model* became a standard one.

[It was not until 1995 that the *top quark* was finally observed at Fermilab [Abe, F. et al. (CDF collaboration) (1995). Observation of Top quark production in p–p Collisions with the Collider Detector at Fermilab. *Phys. Rev. Let.*, 74, 14, 2626–31; arXiv:hep-ex/9503002; doi:10.1103/PhysRevLett.74.2626; S. Arabuchi et al. (D0 collaboration) (1995). Observation of the Top Quark. *Phys. Rev. Let.*, 74, 14, 2632–7; arXiv:hep-ex/9503003; doi:10.1103/ PhysRevLett.74.2632.]

[The DØ experiment (sometimes written D0 experiment) was a worldwide collaboration of scientists conducting research on the fundamental nature of matter. DØ was one of two major experiments (the other was the CDF experiment) both located at the Tevatron Collider at Fermilab in Batavia, Illinois. The Tevatron was the world's highest-energy accelerator from 1983 until 2009, when its energy was surpassed by the Large Hadron Collider. The DØ experiment stopped taking data in 2011, when the Tevatron shut down, but data analysis is still ongoing.]

*It had a mass much larger than expected*, almost as large as that of a gold atom.

The *top quark* is *the most massive of all observed elementary particles*. It derives its mass from its *coupling to the Higgs boson*. This coupling is very close to unity; in the *Standard Model* of particle physics, *it is the largest (strongest) coupling at the scale of the weak interactions and above*. Like all other quarks, the *top quark* is a fermion with *spin ½* and *participates in all four fundamental interactions: gravitation, electromagnetism, weak interactions, and strong interactions*. It has an *electric charge* of + 2/3 e. It has a mass of 172.76 ± 0.3 GeV/c2, which is close to the rhenium atom mass. The *antiparticle* of the *top quark* is the *top antiquark* (sometimes called *antitop quark* or simply *antitop*), which differs from it only in that some of its properties have equal magnitude but opposite sign.

The *top quark* interacts with *gluons* of the *strong interaction* and is typically produced in *hadron* colliders via this interaction. However, once produced, the *top* (or antitop) can decay only through the *weak force*. It decays to a *W boson* and either a *bottom quark* (most frequently), a *strange quark*, or, on the rarest of occasions, a *down quark*.

The *Standard Model* determines the *top quark*'s mean *lifetime* to be roughly $5\times10^{-25}$ s. This is about a twentieth of the timescale for *strong interactions*, and therefore *it does not form hadrons*, giving physicists a unique opportunity to study a "bare" *quark* (all other *quarks* hadronize, meaning that they combine with other *quarks* to form *hadrons* and can only be observed as such).

Because the *top quark* is so massive, *its properties allowed indirect determination of the mass of the Higgs boson*. As such, the *top quark*'s properties are extensively studied as a means to discriminate between competing theories of new physics beyond the *Standard Model*. *The top quark is the only quark that has been directly observed* due to its decay time being shorter than the hadronization time.]

Meanwhile, it was pointed out that we could expect large *CP asymmetry* in the *B-meson* system[21].

[21] Carter, A. B. & Sanda, A. I. (1981). CP Violation in B Meson Decays. *Phys. Rev. D*, 23, 1567; Bigi, I. I. Y. & Sanda, A. I. (1981). Notes on the Observability of CP Violations in B Decays. *Nucl. Phys. B*, 193, 85.

This opened the possibility to test the model with *B-factories*. *B-meson* implies a meson containing b- or *anti*-b as a constituent, and *B-factory* means an accelerator, which produces a lot of *B-mesons* like a factory.

### *4. EXPERIMENTAL VERIFICATION AT B-FACTORIES.*

In order to verify the *six-quark model* experimentally, two B-factories, KEKB at KEK in Japan and PEPII at SLAC in the US, were built. Those accelerators are unusual ones. Colliding *electrons* and *positrons* have different energies, so that the *B-mesons* produced are boosted. Both experimental groups, Belle (KEKB) and BaBar (PEPII), are large international teams organized with participation from many countries.

They were approved and started experiments almost at the same time. PEPII/BaBar ceased operation this year, while KEKB/Belle is still running. They achieved luminosities more than $10^{34}$ $cm^{-2}s^{-1}$, which are record high. Luminosity is a key parameter representing the performance of the colliding accelerator.

…

In the light of the B-factory results, the present status of *CP violation* may be summarized as follows.

- B-factory results show that *quark mixing* of the *six-quark model* is the dominant source of the observed *CP violation.*
- B-factory results, however, allow small room for additional source from new physics beyond the *standard model.*
- And *matter dominance* of the universe seems to require new sources of *CP violation*, because it appears that *CP violation* of the *six-quark model* is too small to explain matter dominance.

It has been proposed that the last point may be related to *lepton flavor mixing*, which is the counterpart of *quark mixing*. In regard to *lepton flavor mixing*, very important contributions have been made in Japan, which will be discussed in the next section.

## *5. LEPTON FLAVOUR MIXING.*

The most important achievement is the discovery of *neutrino oscillation* at Super Kamiokande, which is a huge water tank detector built in the Kamioka mine in central Japan. They were observing *neutrinos* produced by *cosmic rays* in the atmosphere surrounding the earth. Since *neutrinos* penetrate the earth, those *neutrinos* come to the detector also from below. The *neutrino oscillation* implies the species of *neutrino* changes during its flight. So, if the *neutrino oscillation* takes place while *neutrinos* are travelling the distance from the other side of the earth, the observed number of the particular kind of *neutrino* will be reduced …

… The results show a clear deficit of observed *neutrinos* and are completely consistent with *neutrino oscillation.*

This great discovery was led by Yoji Totsuka (1942-2008). To our deep regret, he passed away in this last July. The *neutrino oscillation* was further confirmed by two experiments using man-made *neutrinos*. One is the K2K experiment. In this experiment, *neutrinos* were produced by the *proton synchrotron* in the KEK laboratory and those *neutrinos* were observed by the Super-Kamiokande. … Data show a clear oscillation pattern. …

## Antimatter and antiparticles.

The term *antimatter* was first used by Arthur Schuster in two rather whimsical letters to Nature in 1898, in which he coined the term. He hypothesized *antiatoms*, as well as whole *antimatter* solar systems, and discussed the possibility of *matter* and *antimatter* annihilating each other. Schuster's ideas were not a serious theoretical proposal, merely speculation, and differed from the modern concept of *antimatter* in that it possessed negative gravity.

*Antimatter* is defined as *matter* composed of the *antiparticles* of the corresponding particles in "ordinary" matter, and in the *Standard Model* can be thought of as *matter with reversed electric charge, parity, and time, known as CPT reversal. Antimatter* particles carry the same *electric charge* as *matter* particles, but of opposite sign. That is, an *antiproton* is negatively charged and an *antielectron* (*positron*) is positively charged. *Neutrons* do not carry a net charge, but *according to quark formulation of the Standard Model* their constituent *quarks* do. In this theory, a *particle* and its *antiparticle* (for example, a *proton* and an *antiproton*) have the same *mass*, but opposite *electric charge*, and, *according to the Standard Model, other differences in quantum numbers. Protons* and *neutrons* have a *baryon number* of +1, while *antiprotons* and *antineutrons* have a *baryon number* of –1. Similarly, *electrons* have a *lepton number* of +1, while that of *positrons* is –1.

[The *baryon number* is a strictly conserved additive *quantum number* of a system. It is defined as

$$B = 1/3\,(n_q - n_q^-),$$

where $n_q$ is the number of *quarks*, and $n_q^-$ is the number of *antiquarks*. *Baryons* (*three quarks*) have a *baryon number* of +1, *mesons* (*one quark, one antiquark*) have a *baryon number* of 0, and *antibaryons* (*three antiquarks*) have a *baryon number* of −1.

The *lepton number* is a conserved *quantum number* representing the difference between the number of *leptons* and the number of *antileptons* in an *elementary particle* reaction. *Lepton number* is an additive *quantum number*, so its sum is preserved in interactions (as opposed to multiplicative *quantum numbers* such as *parity*, where the product is preserved instead). The *lepton number* L is defined by

$$L = n_\ell - n_\ell^-,$$

where

$n_\ell$ is the number of *leptons* and
$n_\ell^-$ is the number of *antileptons*.]

When a *particle* and its corresponding *antiparticle* collide, they are both converted into energy. In this view, a collision between any *particle* and its *anti-particle* partner is seen to lead to their mutual *annihilation* [???] giving rise to various proportions of intense *photons* (*gamma rays*), *neutrinos*, and sometimes less-massive *particle–antiparticle* pairs. The majority of the total energy of annihilation emerges in the form of ionizing radiation. If surrounding *matter* is present, the energy content of this radiation will be absorbed and converted into other forms of energy, such as heat or light. The amount of energy released is usually proportional to the total mass of the collided *matter* and *antimatter*, in accordance with the notable mass–energy equivalence equation, $E=mc^2$.

*Antiparticles* bind with each other to form *antimatter*; just as ordinary particles bind to form normal *matter*. For example, a *positron* (the *antiparticle* of the *electron*) and an *antiproton* (the *antiparticle* of the *proton*) can form an *antihydrogen* atom. The nuclei of *antihelium* have been artificially produced, albeit with difficulty, and are the most complex *anti-nuclei* so far observed. Physical principles indicate that complex *antimatter* atomic *nuclei* are possible, as well as *anti-atoms* corresponding to the known chemical elements.

There is no difference in the gravitational behavior of *matter* and *antimatter*. In other words, *antimatter* falls down when dropped, not up. There are compelling theoretical reasons to believe that, aside from the fact that *antiparticles* have different signs on all charges (such as *electric* and *baryon charges*), *matter* and *antimatter* have exactly the same properties. This means a *particle* and its corresponding *antiparticle* must have identical *masses* and *decay lifetimes* (if unstable) [???].

*Antimatter* occurs in natural processes like cosmic ray collisions and some types of radioactive decay, but only a tiny fraction of these have successfully been bound together in experiments to form *antiatoms*. Minuscule numbers of *antiparticles* can be generated at particle accelerators; however, total artificial production has been only a few nanograms. No macroscopic amount of *antimatter* has ever been assembled due to the extreme cost and difficulty of production and handling. Nonetheless, *antimatter* is an essential component of widely available applications related to beta decay, such as *positron* emission tomography, radiation therapy, and industrial imaging.

There is strong evidence that the observable universe is composed almost entirely of ordinary *matter*, as opposed to an equal mixture of *matter* and *antimatter*. According to this theory, this asymmetry of *matter* and *antimatter* in the visible universe is one of the great unsolved problems in physics. The process by which this inequality between *matter* and *antimatter* particles developed is called *baryogenesis*.

## The four "fundamental interactions or forces".

**Underwood. T. G. (2024).** ***The Standard Model***, PART I. The four "fundamental interactions or forces", pp. 47-51: "In the *Standard Model of particle physics* four *fundamental interactions* or *forces* are assumed: *gravity*, and the *electromagnetic, weak and strong interactions*, of which the latter three are incorporated in the model. ... The *Standard Model* … provides a uniform framework for understanding *electromagnetic*, *weak*, and *strong interactions*. …

**Gravity (the *gravitational interaction* or *gravitational force*).**

*Gravity* (from Latin gravitas 'weight') is a *fundamental interaction* primarily observed as mutual attraction between all things that have *mass*. *Gravity* is, by far, the weakest of the four *fundamental interactions*, approximately $10^{38}$ times weaker than the *strong interaction*, $10^{36}$ times weaker than the *electromagnetic force* and $10^{29}$ times weaker than the *weak interaction*. As a result, it has no significant influence at the level of subatomic particles. …

**The *electromagnetic interaction* or *electromagnetic force*.**

The *electromagnetic interaction* or *electromagnetic force* occurs between particles with *electric charge* via *electromagnetic fields*. It is the dominant *force* in the *interactions of atoms and molecules*. *Electromagnetism* can be thought of as a combination of *electrostatics* and *magnetism*, which are distinct but closely intertwined phenomena. *Electromagnetic forces* occur between any two *charged particles*. *Electric forces* cause an attraction between particles with opposite *charges* and repulsion between particles with the same *charge*, while *magnetism* is an *interaction* that occurs between *charged particles* in *relative motion*. These two forces are described in terms of *electromagnetic fields*. Macroscopic charged objects are described in terms of *Coulomb's law for electricity* and *Ampère's force law for magnetism*; the *Lorentz force* describes microscopic *charged particles*. …

**The *weak interaction* or *weak force*.**

The *weak interaction*, also called the *weak force*, is the mechanism of *interaction* between subatomic particles that is responsible for the *radioactive decay* of atoms: the *weak interaction* participates in nuclear fission and nuclear fusion. The effective range of the *weak force* is limited to subatomic distances and is less than the diameter of a *proton*.

In 1933, Enrico Fermi proposed the first theory of the *weak interaction*, known as *Fermi's interaction*. He suggested that *beta decay* could be explained by a four-*fermion* interaction, involving a contact force with no range. He proposed a quantitative theory of *β decay*, in which the existence of the *neutrino* was assumed, and the emission of *electrons* and *neutrinos* from a nucleus in the β case was treated with a method similar to that of the emission of a quantum of light from an excited atom in radiation theory.

In the 1960s, Sheldon Glashow, Abdus Salam and Steven Weinberg unified the *electromagnetic force* and the *weak interaction* by showing them to be two aspects of a single force, now termed the *electroweak force. A weak interaction occurs when two particles (typically, but not necessarily, half-integer spin fermions) exchange integer-spin, force-carrying bosons.*

...

In the *weak interaction, fermions* can *exchange* three types of force carriers, namely W+, W−, and Z *bosons*. The *masses* of these *bosons* are far greater than the *mass* of a *proton* or *neutron*, which is consistent with the short range of the *weak force*. In fact, the force is termed *weak* because its field strength over any set distance is typically several orders of magnitude less than that of the *electromagnetic force*, which itself is further orders of magnitude less than the *strong nuclear force. ...*

The *weak interaction* is the only fundamental *interaction* that breaks *parity symmetry*, and similarly, but far more rarely, the only *interaction* to break *charge–parity symmetry.*

The *weak interaction* is unique in that it allows *quarks* to *swap* their flavor for another. *Quarks*, which make up composite particles like *neutrons* and *protons*, come in six "*flavors*" – *up*, *down*, *charm*, *strange*, *top* and *bottom* – which give those composite particles their properties. The *swapping* of those properties is mediated by the force carrier *bosons*. For example, during *beta-minus decay*, a *down quark* within a *neutron* is changed into an *up quark*, thus converting the *neutron* to a *proton* and resulting in the emission of an *electron* and an *electron antineutrino. ...*

**The *strong interaction* or *strong force.***

The *strong interaction* between *nucleons*, is the side-effect of a more fundamental force that binds *quarks* together into *protons* and *neutrons* to create *atomic nuclei*, where it is called the *strong force* or *strong nuclear force*, The theory of *quantum chromodynamics* explains that *quarks* carry what is called a *color charge*, although it has no relation to visible *color*. *Quarks* with unlike *color charge* attract one another as a result of the *strong interaction*, and the particle that mediates this was called the *gluon*.

Most of the *mass* of a *proton* or *neutron* is the result of the *strong interaction* energy; the individual *quarks* provide only about 1% of the *mass* of a *proton. ...*"

## Symmetry in the Standard Model.

**Underwood. T. G. (2024).** ***The Standard Model***, PART I. Symmetry in the Standard Model, p. 53-55: "The Standard Model of particle physics is a *gauge quantum field theory* containing the *internal* (*local*) *symmetries* of the *unitary product group* SU(3) × SU(2) × U(1). Roughly, *the three factors of the gauge symmetry give rise to the three fundamental interactions*, the *strong, weak* and *electromagnetic interactions*.

> [*Gauge theories* are a type of *quantum field theory*. In a *gauge theory*, there is a group of transformations of the *field variables* (*gauge transformations*) that leaves the basic physics of the *quantum field* unchanged. This condition, called *gauge invariance*, gives the theory a certain *symmetry*, which governs its equations. *Gauge* theories constrain the laws of physics, because all the changes induced by a *gauge transformation* have to cancel each other out when written in terms of observable quantities. ...]

The *unitary group* of degree n, denoted by U(n), is the group of n × n *unitary matrices*, with the *group operation of matrix multiplication*. In the simple case n = 1, the *group U(1)* corresponds to the one dimensional *circle group*, consisting of all complex numbers with absolute value 1, under multiplication. All the unitary groups contain copies of this group.

The *special unitary group* of degree n, denoted SU(n), is the Lie group of n × n *unitary matrices* with *determinant* 1. The SU(n) groups find wide application in the *Standard Model* of particle physics, especially SU(2) in the *electroweak interaction*, and SU(3) in the *strong interaction*, in *quantum chromodynamics*. The group SU(2) is a simple Lie group consisting on all 2 x 2 matrices of *determinant* 1. The group SU(3) is a simple Lie group consisting on all 3 x 3 matrices of *determinant* 1.

The simplest case, SU(1), is the trivial group, with the 1 x 1 matrix having only a single element. Representations of SU(2) describe *non-relativistic spin*, due to being a double covering of the rotation group of Euclidean 3-space. SU(2) symmetry also supports concepts of *isobaric spin* and *weak isospin*, collectively known as *isospin*. When an element of SU(2) is written as a complex 2 × 2 matrix, it is simply a multiplication of column 2-vectors. It is known in physics as the spin-1/2.

The group SU(3) is an 8-dimensional simple Lie group consisting of all 3 × 3 unitary matrices with determinant 1. The Gell-Mann matrices, developed by Murray Gell-Mann, are a set of eight linearly independent 3 × 3 traceless Hermitian matrices used in the study of the *strong interaction* in particle physics. These matrices are traceless, Hermitian, and obey the extra trace orthonormality relation, so they can generate unitary matrix group elements of SU(3) through exponentiation. These properties were chosen by Gell-Mann because they then naturally generalize the Pauli matrices for SU(2) to SU(3), which formed the basis for Gell-Mann's *quark model.*

*Symmetries* may be broadly classified as *global* or *local.* A *global symmetry* is one that keeps a property invariant for a transformation that is applied simultaneously at all points of spacetime, whereas a *local symmetry* is one that keeps a property invariant when a possibly different symmetry transformation is applied at each point of spacetime; specifically, a *local symmetry transformation* is parameterized by the spacetime co-ordinates, whereas a *global symmetry* is not. This implies that a *global symmetry* is also a *local symmetry. Local symmetries* play an important role in physics as they form the basis for *gauge theories.*

According to the *Standard Model*, there are three *local symmetries* in the universe in which we live, which should be indistinguishable where a certain type of change is introduced.

The three *local symmetries* addressed by the *Standard Mode*l are:

> *C-symmetry* (*charge symmetry*), a universe where every particle is replaced with its *antiparticle.*
>
> *P-symmetry* (*parity symmetry*), a universe where everything is mirrored along the three physical axes. This excludes *weak interactions.*
>
> *T-symmetry* (*time reversal symmetry*), a universe where the direction of time is reversed. *T-symmetry* is counterintuitive (the future and the past are not symmetrical) but explained by the fact that the *Standard Model* describes *local* properties, not *global* ones like entropy. To properly reverse the direction of time, one would have to put the Big Bang and the resulting low-entropy state in the "future". Since we perceive the "past" ("future") as having lower (higher) entropy than the present, the inhabitants of this hypothetical time-reversed universe would perceive the future in the same way as we perceive the past, and vice versa.

These *symmetries* are near-symmetries because each is broken in the present-day universe. However, the *Standard Model* predicts that the combination of the three (that is, the simultaneous application of all three transformations) must be a symmetry, called *CPT symmetry. CP violation,* the violation of the combination of *C-* and *P-symmetry,* is necessary for the presence of significant amounts of *baryonic matter* in the universe.

*Symmetry breaking* is a phenomenon where a disordered but symmetric state collapses into an ordered, but less symmetric state. This collapse is often one of many possible bifurcations that a particle can take as it approaches a lower energy state. Due to the many possibilities, an observer may assume the result of the collapse to be arbitrary. This phenomenon is fundamental to *quantum field theory*. Specifically, it plays a central role in the *Glashow–Weinberg–Salam model* which forms part of the *Standard Model* modelling the *electroweak* sector. …"

## Supersymmetry.

**Underwood. T. G. (2024).** ***The Standard Model***, PART II, Supersymmetry, p. 458-9:
"*Supersymmetry* is a theoretical framework in physics that suggests the existence of a *symmetry* between particles with *integer spin* (*bosons*) and particles with *half-integer spin* (*fermions*). It proposes that for every known particle, there exists a partner particle with different *spin* properties. *There have been multiple experiments on supersymmetry that have failed to provide evidence that it exists in nature*. If evidence is found, *supersymmetry* could help explain certain phenomena, such as the nature of *dark matter* and the *hierarchy problem* in particle physics.

> [*There is no experimental evidence that either supersymmetry or misaligned supersymmetry holds in our universe*, and *many physicists have moved on from supersymmetry and string theory entirely due to the non-detection of supersymmetry at the Large Hadron Collider* (LHC).]

A *supersymmetric theory* is a theory in which the *equations for force* and the *equations for matter* are identical. In theoretical and mathematical physics, any theory with this property has the *principle of supersymmetry*. Dozens of supersymmetric theories exist. In theory, *supersymmetry* is a type of *spacetime symmetry* between two basic classes of particles: *bosons*, which have an integer-valued *spin* and follow Bose–Einstein statistics, and *fermions*, which have a half-integer-valued *spin* and follow Fermi–Dirac statistics. [Haber, H. *Supersymmetry. Part I (Theory).* Reviews, Tables and Plots. Particle Data Group.] *The names of bosonic partners of fermions are prefixed with s-, because they are scalar particles*.

In *supersymmetry*, each particle from the class of *fermions* would have an associated particle in the class of *bosons*, and vice versa, known as a *superpartner*. *The spin of a particle's superpartner is different by a half-integer.* For example, if the *electron* exists in a *supersymmetric theory*, then there would be a particle called a *selectron* (superpartner electron), a *bosonic* partner of the *electron*. In the simplest supersymmetry theories, with perfectly *"unbroken" supersymmetry*, each pair of *superpartners* would share the same *mass* and *internal quantum numbers* besides *spin*. *More complex supersymmetry theories have a spontaneously broken symmetry*, allowing *superpartners* to differ in *mass*.

… However, *no supersymmetric extensions of the Standard Model have been experimentally verified.*"

## String Theory.

**Underwood. T. G. (2024). The Standard Model**, PART III. String Theory, p. 468; p. 477: "*String theory is a theoretical framework in which the point-like particles of particle physics are replaced by one-dimensional objects called strings. String theory* describes how these *strings* propagate through space and *interact* with each other. On distance scales larger than the string scale, a *string* looks just like an ordinary particle, *with its mass, charge, and other properties determined by the vibrational state of the string*. In *string theory*, one of the many *vibrational states* of the string corresponds to the *graviton*, a *quantum mechanical particle* that carries the *gravitational force*. Thus, *string theory is a theory of quantum gravity*.

*String theory* is a broad and varied subject that attempts to address a number of deep questions of fundamental physics. *String theory* has contributed a number of advances to mathematical physics, which have been applied to a variety of problems in black hole physics, early universe cosmology, nuclear physics, and condensed matter physics, and it has stimulated a number of major developments in pure mathematics. Because *string theory* potentially provides a unified description of *gravity* and *particle physics*, it is a candidate for a theory of everything, a self-contained mathematical model that describes all fundamental forces and forms of matter. Despite much work on these problems, *it is not known to what extent string theory describes the real world* or how much freedom the theory allows in the choice of its details.
…

*One of the challenges of string theory is that the full theory does not have a satisfactory definition in all circumstances*. Another issue is that the theory is thought to describe an enormous landscape of possible universes, which has complicated efforts to develop theories of particle physics based on *string theory*. These issues have led some in the community to criticize these approaches to physics, and to question the value of continued research on *string theory unification*."

## List of elementary particles in the Standard Model.

The *Standard Model* of particle physics evolved from the Bohr model of the atom in 1913, based on what were believed to be 3 stable particles, *electrons* in orbit around a *nucleus* comprised of *protons* and *neutrons*, to its emergence in 1973 as the *six-quark model*, comprising 26 *elementary particles* or a total of 52 *elementary particles* and their *anti-particles*, of which only the *electron* and the *photon* are stable. Most of these have been revealed largely by tracks in cloud chambers from high energy collisions between particles, or at high altitudes with cosmic rays. See Underwood, T. G. (2024). *The Standard Model.*

**List of elementary particles in the Standard Model.**

| | **Fermions:** | **Half-life** |
|---|---|---|
| Leptons: | | |
| 1 | Electron | $6.6 \times 10^{28}$ years |
| 2 | Electron neutrino | stable? |
| 3 | Muon | $2.2 \times 10^{-6}$ s |
| 4 | Muon neutrino | stable? |
| 5 | Tau | $2.9 \times 10^{-13}$ s |
| 6 | Tau neutrino | stable? |
| Quarks: | | |
| 7 | Up quark | |
| 8 | Down quark | |
| 9 | Charm quark | |
| 10 | Strange quark | |
| 11 | Top quark | $5 \times 10^{-25}$ s |
| 12 | Bottom quark | |
| **Bosons:** | | |
| Gauge bosons: | | |
| 13 | Photon | stable |
| 14 | $W^+$ boson | $3 \times 10^{-25}$ s |
| 15 | $W^-$ boson | $3 \times 10^{-25}$ s |
| 16 | $Z^0$ boson | $3 \times 10^{-25}$ s |
| 17-24 | Gluons (8) | |
| 25 | Graviton (hypothetical) | |
| Scalar boson: | | |
| 26 | Higgs boson | $1.6 \times 10^{-22}$ s |

## (2) The speed of light in vacuum is constant for all observers, regardless of the motion of light source or observer.

**Einstein's theory of Special Relativity.**

In 1905 Albert Einstein published his *Theory of Special Relativity*, a scientific theory regarding the relationship between space and time. In Einstein's original treatment, the theory is based on two postulates:
(1) The laws of physics are invariant (that is, identical) in all inertial frames of reference (that is, frames of reference with no acceleration), known as the *principle of relativity.*
(2) The *speed of light in vacuum is the same for all observers*, regardless of the motion of the light source or observer.

Many modern treatments of *special relativity* base it on the single postulate of universal Lorentz covariance, or, equivalently, on the single postulate of Minkowski spacetime.

However, Einstein's *theory of general relativity* attempted to extend his *theory of special relativity* beyond space and time, to include *matter* and *gravitational fields*. Whilst this allowed Einstein to construct a *relativistic theory* of the effect of a *gravitational field* on *matter*, it also resulted in him rejecting his *postulate on the constancy of light* in the presence of a gravitational field.

**Underwood, T. G. (2023). *Special Relativity*,** Conclusion, p. 381: "There is no evidence, based directly or indirectly, on the observation of the speed of electromagnetic radiation in a vacuum emitted by an inertial body, or as observed by an inertial observer, moving in a straight line and not involving mirrors. However, by now it may be possible to achieve this in a laboratory experiment in a vacuum without mirrors, using electromagnetic radiation emitted by two sources of the same frequency, one stationary and the other moving at a constant velocity in a straight line; either directly, or by measuring the observed frequency of the radiation.

The Ehrenfest paradox, the *non-relativistic* Doppler red shift and blue shift for light, the known physics of the emission of electromagnetic radiation and of the electron, and the success of *non-relativistic* quantum electrodynamics in explaining the interaction of the electromagnetic field with electrically charged particles, comprise the strongest evidence against Einstein's *second postulate*, the *constancy of the speed of light*.

Recognizing that evidence based on *celestial observations*, experiments with *light passing through a medium*, and observations on *rotating platforms* and other *accelerated* systems, is suspect, the evidence in support of the *theory of special relativity* reduces to extremely slim pickings, of which the Beckman & Mandics (1965) experiment seems most plausible, though this may suffer from a potential problem with the use of a mirror.

Quite apart from the enormity of the consequences of Einstein's two postulates taken together, including *length contraction*, *time dilation*, and the requirement to assume a *point electron* in the unsuccessful attempt to introduce special relativity into quantum electrodynamics, the evidence in support of Einstein's *second postulate* on the constancy of the speed of light is far outweighed by the evidence against it.

For this reason, until more satisfactory evidence in support of Einstein's *second postulate*, a refutation of the Ehrenfest paradox, and an explanation for the observed Doppler red shift and blue shift consistent with Einstein's two postulates, is provided, under any normal measure of a theory in physics, *Einstein's second postulate, and consequently his theory of special relativity, must be rejected. …*"

**Underwood, T. G. (2023).** ***General Relativity***, Conclusion, p.474: "… A detailed examination of Einstein's *theory of general relativity* reveals that it is not a *theory of gravity*; it is a *relativistic* theory about the *effects* of *gravitation*, or more strictly, of a *uniformly accelerated reference frame*. There is nothing in any version of this theory that represents or explains or provides any connection to the weak attractive *gravitational force* between matter. We are no further forward in understanding the origin of this fundamental force. Whilst Einstein's and others' objectives in removing a preferred reference frame and the existence of an ether from physics were admirable intentions, Einstein's subsequent fixation on the constancy of the speed of light, or some form of invariant space-time, in the face of reasonable alternatives, such as *Ritz's emission theory* on which *quantum electrodynamics* is founded, was not.

Einstein's *theory of general relativity* attempted to extend his *theory of special relativity* beyond space and time, to include *matter* and *gravitational fields*. *Gravitation* was introduced through the "*equivalence principle*", the equivalence of the *outcome* of the force of *gravity* and the acceleration of *matter*, first recognized in Newton's *Principia*. This allowed Einstein to construct a *relativistic theory* of the effect of a *gravitational field* on *matter*, but it also resulted in him rejecting his *postulate on the constancy of light* in the presence of a gravitational field, and provided no connection to or explanation of the weak attractive *gravitational force* between *matter*. In order to make calculations with his theory, Einstein had to import *Newton's law of gravitation*, which itself is an empirical law with no fundamental foundation. Consequently, the only evidence that Einstein could provide for his *theory of general relativity* was effectively Newtonian.

In the light of the continued failure of Einstein's efforts to overcome the main objections to his *theory of special relativity* - the Ehrenfest paradox, and its failure to explain the observed Doppler redshift and blueshift of light – or to provide any evidence for it, and in the absence of any supportive evidence for his *theory of general relativity*, both theories must be rejected until such objections are overcome and such evidence is provided."

**Einstein, A. (September, 1905). Zur Elektrodynamik bewegter Körper. (On the electrodynamics of moving bodies.)**

*Ann. Phys.*, 322, 10, 891-921; http://www.physik.uni-augsburg.de/annalen/history/einstein-papers/1905_17_891-921.pdf; translated by W. Perrett & G. B. Jeffery; in (1923). *The Principle of Relativity*, Methuen and Company, Ltd., London; https://www.fourmilab.ch/etexts/einstein/specrel/specrel.pdf.; translation also in A. Beck (translator), P. Havas (consultant). (1989). *The Collected Papers of Albert Einstein, Volume 2: The Swiss Years: Writings, 1900-1909.* English translation), pp. 140-71; also in Underwood, T. G. (2023). *Special Relativity*, Part I, pp. 176-98.

Received June 30, 1905.

Berne, Switzerland.

In this paper, Einstein introduces the two postulates of his theory of (special) relativity, notes that the introduction of a "luminiferous ether" will prove to be superfluous, and describes some of the consequences of the two postulates. His first *postulate*, the *principle of relativity*, that *"... not only in mechanics, but also in electrodynamics, no properties of observed facts correspond to a concept of absolute rest; but that for all coordinate systems for which the mechanical equations hold, the equivalent electrodynamical and optical equations hold also ... "* is unexceptional, and had been the main driver behind the search for an alternative to current theories of electromagnetic waves based on an ether in a rest frame. In order to explain the Michelson-Morley result, he added his second postulate, that *"...light is propagated in vacant space, with a velocity c which is independent of the nature of motion of the emitting body"*, which is not. It is important to note that Einstein's second postulate refers to the state of motion of the *emitting body*, whereas Lorentz referred to the state of *observers* in different *inertial reference frames*. After stating the postulates, Einstein's *annus mirabilis* paper proceeds to work through the consequences of these assumptions on kinematics and electrodynamics. These consequences, which all refer to observations by an *inertial observer*, are consequences of the *second postulate*, include relativity of simultaneity, length contraction of a moving body, time dilation of the body in the moving system, the relativistic velocity addition formula, and the relativistic Doppler effect. However, Einstein's *first postulate* states that the *same physical laws* which apply to observations of body in a stationary frame by an observer in a frame moving with constant velocity relative to the stationary frame *also apply to observations of the body moving with the same velocity relative to the stationary frame*. Although this is not referred to by Einstein, it implies that *these consequences apply to any inertial body, without reference to an observer*, and, as such, result in *real consequences for moving bodies*.

[Einstein, A. (1949). Autobiographical Notes: "The assumptions *relativity* and *light speed invariance* are compatible if relations of a new type ("Lorentz transformation") are postulated for the conversion of coordinates and times of events ... The universal principle of the *special theory of relativity* is contained in the postulate: *The laws of physics are invariant with respect to Lorentz transformations* (for the transition from one inertial system to any other arbitrarily chosen inertial system). This is a restricting principle for natural laws ..."

In this paper, Einstein notes: "...These two postulates suffice for the attainment of a simple and consistent theory of the electrodynamics of moving bodies based on Maxwell's theory for stationary bodies. The introduction of a "luminiferous ether" will prove to be superfluous inasmuch as the view here to be developed will not require an "absolutely stationary space" provided with special properties, nor assign a velocity-vector to a point of the empty space in which electromagnetic processes take place".]

---

It is known that Maxwell's electrodynamics—as usually understood at the present time—when applied to *moving bodies*, leads to asymmetries which do not appear to be inherent in the phenomena. Take, for example, the reciprocal electrodynamic action of a magnet and a conductor. The observable phenomenon here depends only on the relative motion of the conductor and the magnet, whereas the customary view draws a sharp distinction between the two cases in which either the one or the other of these bodies is in motion. For if the magnet is in motion and the conductor at rest, there arises in the neighborhood of the magnet an electric field with a certain definite energy, producing a current at the places where parts of the conductor are situated. But if the magnet is stationary and the conductor in motion, no electric field arises in the neighborhood of the magnet. In the conductor, however, we find an electromotive force, to which in itself there is no corresponding energy, but which gives rise—assuming equality of relative motion in the two cases discussed—to electric currents of the same path and intensity as those produced by the electric forces in the former case.

[This is a poor example, as one would not expect absolute symmetry in this situation in the stationary frame of the magnet and the stationary frame conductor. The point is that in both frames the same electric current is observed.]

Examples of this sort, together with the unsuccessful attempts to discover any motion of the earth relatively to the "light medium," suggest that *the phenomena of electrodynamics as well as of mechanics possess no properties corresponding to the idea of absolute rest*. They suggest rather that, as has already been shown to the first order of small quantities, *the same laws of electrodynamics and optics will be valid for all frames of reference for which the equations of mechanics hold good.*[1]

[1] The preceding memoir by Lorentz was not at this time known to the author.

[Einstein not only omits all references to prior work by other authors, but even fails to properly identify this memoir, presumably Lorentz, H. A. (May, 1904). Electromagnetic phenomena in a system moving with any velocity smaller than that of light. *Proc. Royal Acad. Amsterdam*, 6, 809-31, which was published more than a year before Einstein's paper was submitted on June 30, 1905.]

We will raise this conjecture (the purport of which will hereafter be called the "*Principle of Relativity*") to the status of a postulate, *and also introduce another postulate*, which is only apparently irreconcilable with the former, namely, *that light is always propagated in empty space with a definite velocity c which is independent of the state of motion of the emitting body*. These two postulates suffice for the attainment of a simple and consistent theory of the electrodynamics of moving bodies based on Maxwell's theory for stationary bodies. *The introduction of a "luminiferous ether" will prove to be superfluous inasmuch as the view here to be developed will not require an "absolutely stationary space" provided with special properties*, nor assign a velocity-vector to a point of the empty space in which electromagnetic processes take place.

The theory to be developed is based—like all electrodynamics—on the kinematics of the rigid body, since the assertions of any such theory have to do with the relationships between rigid bodies (systems of co-ordinates), clocks, and electromagnetic processes. Insufficient consideration of this circumstance lies at the root of the difficulties which the electrodynamics of moving bodies at present encounters.

## *I. KINEMATICAL PART.*

### *§ 1. Definition of Simultaneity.*

Let us take a system of co-ordinates in which the equations of Newtonian mechanics hold good.[2]

[2] i.e. to the first approximation.

In order to render our presentation more precise and to distinguish this system of co-ordinates verbally from others which will be introduced hereafter, we call it the "*stationary system.*"

If a material point is at rest relatively to this system of co-ordinates, its position can be defined relatively thereto by the employment of rigid standards of measurement and the methods of Euclidean geometry, and can be expressed in Cartesian co-ordinates.

If we wish to describe the *motion* of a material point, we give the values of its co-ordinates as functions of the time. Now we must bear carefully in mind that a mathematical description of this kind has no physical meaning unless we are quite clear as to what we understand by "time." We have to take into account that all our judgments in which time plays a part are always judgments of *simultaneous events*. If, for instance, I say, "That train arrives here at 7 o'clock," I mean something like this: "The pointing of the small hand of my watch to 7 and the arrival of the train are simultaneous events."[3]

[3] We shall not here discuss the inexactitude which lurks in the concept of simultaneity of two events at approximately the same place, which can only be removed by an abstraction.

It might appear possible to overcome all the difficulties attending the definition of "time" by substituting "the position of the small hand of my watch" for "time." And in fact, such a definition is satisfactory when we are concerned with defining a time exclusively for the place where the watch is located; but it is no longer satisfactory when we have to connect in time series of events occurring at different places, or—what comes to the same thing—to evaluate the times of events occurring at places remote from the watch.

We might, of course, content ourselves with time values determined by an observer stationed together with the watch at the origin of the co-ordinates, and coordinating the corresponding positions of the hands with light signals, given out by every event to be timed, and reaching him through empty space. But this co-ordination has the disadvantage that it is not independent of the standpoint of the observer with the watch or clock, as we know from experience. We arrive at a much more practical determination along the following line of thought.

If at the point A of space there is a clock, an observer at A can determine the time values of events in the immediate proximity of A by finding the positions of the hands which are simultaneous with these events. If there is at the point B of space another clock in all respects resembling the one at A, it is possible for an observer at B to determine the time values of events in the immediate neighborhood of B. But it is not possible without further assumption to compare, in respect of time, an event at A with an event at B. We have so far defined only an "A time" and a "B time." We have not defined a common "time" for A and B, for the latter cannot be defined at all unless we establish *by definition* that the "time" required by light to travel from A to B equals the "time" it requires to travel from B to A. Let a *ray of light* start at the "A time" $t_A$ from A towards B, let it at the "B time" $t_B$ be reflected at B in the direction of A, and arrive again at A at the "A time" $t'_A$.

In accordance with definition the two clocks synchronize if

$$t_B - t_A = t'_A - t_B.$$

We assume that this definition of synchronism is free from contradictions, and possible for any number of points; and that the following relations are universally valid: —

1. If the clock at B synchronizes with the clock at A, the clock at A synchronizes with the clock at B.

2. If the clock at A synchronizes with the clock at B and also with the clock at C, the clocks at B and C also synchronize with each other.

Thus, with the help of certain imaginary physical experiments we have settled what is to be understood by synchronous stationary clocks located at different places, and have evidently obtained a definition of "simultaneous," or "synchronous," and of "time." The "time" of an event is that which is given simultaneously with the event by a stationary clock located at the place of the event, this clock being synchronous, and indeed synchronous for all time determinations, with a specified stationary clock.

In agreement with experience, we further assume the quantity

$$\frac{2\mathrm{AB}}{t'_A - t_A} = c,$$

to be a universal constant—the velocity of light in empty space.

It is essential to have time defined by means of stationary clocks in the stationary system, and the time now defined being appropriate to the stationary system we call it "the time of the stationary system."

***§ 2. On the Relativity of Lengths and Times.***

The following reflections are based on the *principle of relativity* and on the *principle of the constancy of the velocity of light*. These two principles we define as follows: —

1. The laws by which the states of physical systems undergo change are not affected, whether these changes of state be referred to the one or the other of two systems of co-ordinates in *uniform translatory motion*.

2. Any ray of light moves in the "stationary" system of co-ordinates with the determined velocity $c$, whether the ray be emitted by a stationary or by a moving body. Hence

$$\text{velocity} = \frac{\text{light path}}{\text{time interval}}$$

where time interval is to be taken in the sense of the definition in § 1.

Let there be given a stationary rigid rod; and let its length be $l$ as measured by a measuring-rod which is also stationary. We now imagine the axis of the rod lying along the axis of x of the stationary system of co-ordinates, and that a *uniform motion of parallel translation* with velocity $\upsilon$ along the axis of x in the direction of increasing x is then imparted to the rod. We now inquire as to the length of the moving rod, and imagine its length to be ascertained by the following two operations: —

(a) The observer moves together with the given measuring-rod and the rod to be measured, and measures the length of the rod directly by superposing the measuring-rod, in just the same way as if all three were at rest.

(b) By means of stationary clocks set up in the stationary system and synchronizing in accordance with § 1, the observer ascertains at what points of the stationary system the two ends of the rod to be measured are located at a definite time. The distance between these two points, measured by the measuring-rod already employed, which in this case is at rest, is also a length which may be designated "the length of the rod."

In accordance with the *principle of relativity* the length to be discovered by the operation (a)—we will call it "the length of the rod in the moving system"—must be equal to the length $l$ of the stationary rod.

The length to be discovered by the operation (b) we will call "the length of the (moving) rod in the stationary system." This we shall determine on the basis of our two principles, and we shall find that it differs from $l$.

Current kinematics tacitly assumes that the lengths determined by these two operations are precisely equal, or in other words, that a moving rigid body at the epoch $t$ may in geometrical respects be perfectly represented by *the same* body *at rest* in a definite position.

We imagine further that at the two ends A and B of the rod, clocks are placed which synchronize with the clocks of the stationary system, that is to say that their indications correspond at any instant to the "time of the stationary system" at the places where they happen to be. These clocks are therefore "synchronous in the stationary system."

We imagine further that with each clock *there is a moving observer*, and that these observers apply to both clocks the criterion established in § 1 for the synchronization of two clocks. *Let a ray of light* depart from A at the time[4] $t_A$, let it be reflected at B at the time $t_B$, and reach A again at the time $t'_A$.

[4] "Time" here denotes "time of the stationary system" and also "position of hands of the moving clock situated at the place under discussion."

Taking into consideration the *principle of the constancy of the velocity of light* we find that

$$t_{\mathrm{B}} - t_{\mathrm{A}} = \frac{r_{\mathrm{AB}}}{c - v} \text{ and } t'_{\mathrm{A}} - t_{\mathrm{B}} = \frac{r_{\mathrm{AB}}}{c + v}$$

where $r_{AB}$ denotes the length of the moving rod—measured in the stationary system. Observers moving with the moving rod would thus find that the two clocks were not synchronous, while observers in the stationary system would declare the clocks to be synchronous.

So, we see that we cannot attach any *absolute* signification to the concept of simultaneity, but that two events which, viewed from a system of co-ordinates, are simultaneous, can no longer be looked upon as simultaneous events when envisaged from a system which is in motion relatively to that system.

### *§ 3. Theory of the Transformation of Co-ordinates and Times from a Stationary System to another System in Uniform Motion of Translation Relatively to the Former.*

Let us in "stationary" space take two systems of co-ordinates, i.e., two systems, each of three rigid material lines, perpendicular to one another, and issuing from a point. Let the axes of X of the two systems coincide, and their axes of Y and Z respectively be parallel. Let each system be provided with a rigid measuring-rod and a number of clocks, and let the two measuring-rods, and likewise all the clocks of the two systems, be in all respects alike.

Now to the origin of one of the two systems (k) *let a constant velocity* $\upsilon$ be imparted in the direction of the increasing x of the other stationary system (K), and let this velocity be communicated to the axes of the co-ordinates, the relevant measuring-rod, and the clocks. To any time of the stationary system K there then will correspond a definite position of the axes of the moving system, and from reasons of symmetry we are entitled to assume that the motion of k may be such that the axes of the moving system are at the time t (this "t" always denotes a time of the stationary system) parallel to the axes of the stationary system.

We now imagine space to be measured from the stationary system K by means of the stationary measuring-rod, and also from the moving system k by means of the measuring-rod moving with it; and that we thus obtain the co-ordinates x, y, z, and $\xi$, $\eta$, $\zeta$ respectively. Further, let the time t of the stationary system be determined for all points thereof at which there are clocks *by means of light signals* in the manner indicated in § 1; similarly let the time t' of the moving system be determined for all points of the moving system at which there are clocks at rest relatively to that system by applying the *method of light signals* between the points at which the latter clocks are located.

To any system of values x, y, z, t, which completely defines the place and time of an event in the stationary system, there belongs a system of values $\xi$, $\eta$, $\zeta$, $\tau$, determining that event relatively to the system k, and our task is now to find the system of equations connecting these quantities.

In the first place it is clear that *the equations must be linear on account of the properties of homogeneity which we attribute to space and time.*

If we place $x' = x - \upsilon t$, it is clear that a point at rest in the system k must have a system of values x', y, z, independent of time. We first define $\tau$ as a function of x', y, z, and t. To do

this we have to express in equations that $\tau$ is nothing else than the summary of the data of clocks at rest in k, which have been synchronized according to the rule given in § 1.
From the origin of system k *let a ray be emitted* at the time $\tau_0$ along the X-axis to x', and at the time $\tau_1$ *be reflected* thence to the origin of the co-ordinates, arriving there at the time $\tau_2$; we then must have $(\tau_0 + \tau_2)/2 = \tau_1$, or, by inserting the arguments of the function $\tau$ and *applying the principle of the constancy of the velocity of light* in the stationary system: —

$$[\tau(0, 0, 0, t) + \tau(0, 0, 0, t + x'/(c - \upsilon) + x'(c + \upsilon)]/2 = \tau(x', 0, 0, t + x'/(c - \upsilon)),$$

Hence, *if x' be chosen infinitesimally small,*

$$1/2\ [1/(c - v) + 1/(c + \upsilon)]\delta\tau/\delta t = \delta\tau/\delta x' + [1/(c - \upsilon)]\delta\tau/\delta t,$$

or

$$\delta\tau/\delta x' + \upsilon/(c^2 - \upsilon^2)]\delta\tau/\delta t = 0$$

It is to be noted that instead of the origin of the co-ordinates we might have chosen any other point for the point of origin of the ray, and the equation just obtained is therefore valid for all values of x', y, z.

An analogous consideration—applied to the axes of Y and Z—it being borne in mind that light is always propagated along these axes, when viewed from the stationary system, with the velocity $\sqrt{(c^2 - \upsilon^2)}$ gives us

$$\delta\tau/\delta y = 0, \qquad \delta\tau/\delta z = 0.$$

Since $\tau$ is a linear function, it follows from these equations that

$$\tau = a[t - \{\upsilon/(c^2 - \upsilon^2)\}x']$$

where $a$ is a function $\phi(\upsilon)$ at present unknown, and where for brevity it is assumed that at the origin of k, $\tau = 0$, when $t = 0$.

With the help of this result, we easily determine the quantities $\xi$, $\eta$, $\zeta$ by expressing in equations that *light* (as required by the *principle of the constancy of the velocity of light*, in combination with the *principle of relativity*) *is also propagated with velocity c when measured in the moving system*. For a ray of light *emitted* at the time $\tau = 0$ in the direction of the increasing $\xi$

$$\xi = c\tau \text{ or } \xi = ac\left(t - \frac{v}{c^2 - v^2}x'\right).$$

But the ray moves relatively to the initial point of *k*, when measured in the stationary system, with the velocity $c - \upsilon$, so that

$$\frac{x'}{c-v}=t.$$

If we insert this value of *t* in the equation for ξ, we obtain

$$\xi=a\frac{c^2}{c^2-v^2}x'.$$

In an analogous manner we find, by considering rays moving along the two other axes, that

$$\eta=c\tau=ac\left(t-\frac{v}{c^2-v^2}x'\right)$$

when

$$\frac{y}{\sqrt{c^2-v^2}}=t,\ x'=0.$$

Thus

$$\eta=a\frac{c}{\sqrt{c^2-v^2}}y \text{ and } \zeta=a\frac{c}{\sqrt{c^2-v^2}}z.$$

Substituting for *x*' its value, we obtain

$$\begin{aligned}\tau &= \phi(v)\beta(t-vx/c^2),\\ \xi &= \phi(v)\beta(x-vt),\\ \eta &= \phi(v)y,\\ \zeta &= \phi(v)z,\end{aligned}$$

where

$$\beta=\frac{1}{\sqrt{1-v^2/c^2}},$$

and ϕ is an as yet unknown function of υ. If no assumption whatever be made as to the initial position of the moving system and as to the zero point of $\tau$, an additive constant is to be placed on the right side of each of these equations.

*We now have to prove that any ray of light, measured in the moving system, is propagated with the velocity c, if, as we have assumed, this is the case in the stationary system*; for we have not as yet furnished the proof that *the principle of the constancy of the velocity of light is compatible with the principle of relativity.*

At the time $t=\tau=0$, when the origin of the co-ordinates is common to the two systems, *let a spherical wave be emitted* therefrom, and be propagated with the velocity *c* in system K. If (*x*, *y*, *z*) be a point just attained by this wave, then

$x^2 + y^2 + z^2 = c^2t^2$.

Transforming this equation with the aid of our equations of transformation we obtain after a simple calculation

$$\xi^2 + \eta^2 + \zeta^2 = c^2\tau^2.$$

The wave under consideration is therefore no less a spherical wave with velocity of propagation *c* when viewed in the moving system. *This shows that our two fundamental principles are compatible.*[5]

[5] *The equations of the Lorentz transformation may be more simply deduced directly from the condition that in virtue of those equations the relation $x^2 + y^2 + z^2 = c^2t^2$ shall have as its consequence the second relation.*

In the equations of transformation which have been developed there enters an unknown function ϕ of υ, which we will now determine.

For this purpose we introduce a third system of co-ordinates $K'$, which relatively to the system *k* is *in a state of parallel translatory motion* parallel to the axis of Ξ,[*1] such that the origin of co-ordinates of system $K'$, moves with velocity – υ on the axis of Ξ.

[*1] In Einstein's original paper, the symbols (Ξ, H, Z) for the co-ordinates of the moving system *k* were introduced without explicitly defining them. In the 1923 English translation, (X, Y, Z) were used, creating an ambiguity between X co-ordinates in the fixed system K and the parallel axis in moving system *k*. Here and in subsequent references *we use* Ξ *when referring to the axis of system k along which the system is translating with respect to K.* In addition, the reference to system $K'$, later in this sentence was incorrectly given as "*k*" in the 1923 English translation.

At the time t = 0 let all three origins coincide, and when t = x = y = z = 0 let the time t' of the system $K'$ be zero. We call the co-ordinates, measured in the system $K'$, x', y', z', and by a twofold application of our equations of transformation we obtain

$$\begin{aligned} t' &= \phi(-v)\beta(-v)(\tau + v\xi/c^2) &&= \phi(v)\phi(-v)t, \\ x' &= \phi(-v)\beta(-v)(\xi + v\tau) &&= \phi(v)\phi(-v)x, \\ y' &= \phi(-v)\eta &&= \phi(v)\phi(-v)y, \\ z' &= \phi(-v)\zeta &&= \phi(v)\phi(-v)z. \end{aligned}$$

Since the relations between x', y', z' and x, y, z do not contain the time t, the systems K and $K'$ are at rest with respect to one another, and it is clear that the transformation from K to $K'$ must be the identical transformation. Thus

$$\phi(v)\phi(-v) = 1.$$

We now inquire into the signification of $\phi(v)$. We give our attention to that part of the axis of Y of system *k* which lies between $\xi = 0, \eta = 0, \zeta = 0$ and $\xi = 0, \eta = l, \zeta = 0$. This part of the axis of Y is a rod moving perpendicularly to its axis with velocity $v$ relatively to system K. Its ends possess in K the co-ordinates

$$x_1 = vt,\ y_1 = \frac{l}{\phi(v)},\ z_1 = 0$$

and

$$x_2 = vt,\ y_2 = 0,\ z_2 = 0.$$

The length of the rod measured in K is therefore $l/\phi(v)$; and this gives us the meaning of the function $\phi(v)$. From reasons of symmetry, it is now evident that the length of a given rod moving perpendicularly to its axis, measured in the stationary system, must depend only on the velocity and not on the direction and the sense of the motion. The length of the moving rod measured in the stationary system does not change, therefore, if $v$ and $-v$ are interchanged. Hence follows that

$$l/\phi(v) = l/\phi(-v)$$

or

$$\phi(v) = \phi(-v).$$

It follows from this relation and the one previously found that $\phi(v) = 1$, *so that the transformation equations which have been found become*

$$\begin{aligned} \tau &= \beta(t - vx/c^2), \\ \xi &= \beta(x - vt), \\ \eta &= y, \\ \zeta &= z, \end{aligned}$$

where

$$\beta = 1/\sqrt{1 - v^2/c^2}.$$

### *§ 4. Physical Meaning of the Equations Obtained in Respect to Moving Rigid Bodies and Moving Clocks.*

We envisage a *rigid sphere*[6] of radius R, at rest relatively to the moving system *k*, and with its center at the origin of co-ordinates of *k*.

[6] That is, a body possessing spherical form when examined at rest.

The equation of the surface of this sphere moving relatively to the system K with velocity $v$ is

$$\xi^2 + \eta^2 + \zeta^2 = R^2.$$

The equation of this surface expressed in x, y, z at the time t = 0 is

$$\frac{x^2}{(\sqrt{1 - v^2/c^2})^2} + y^2 + z^2 = R^2.$$

A rigid body which, measured in a state of rest, has the form of a *sphere*, therefore has in a state of motion—viewed from the stationary system—the form of an *ellipsoid* of revolution with the axes

$$R\sqrt{1 - v^2/c^2},\ R,\ R.$$

Thus, whereas the Y and Z dimensions of the sphere (and therefore of every rigid body of no matter what form) do not appear modified by the motion, the X dimension appears *shortened in the ratio 1 : √(1 − $v^2/c^2$)*, i.e. the greater the value of υ, the greater the shortening. *For υ = c all moving objects—viewed from the "stationary" system—shrivel up into plane figures.*[*2]

[*2] In the original 1923 English edition, this phrase was erroneously translated as "plain figures". I have used the correct "plane figures" in this document.

For velocities greater than that of light our deliberations become meaningless; we shall, however, find in what follows, that the velocity of light in our theory plays the part, physically, of an infinitely great velocity.

It is clear that the same results hold good of bodies at rest in the "stationary" system, viewed from a system in uniform motion.

Further, we imagine one of the clocks which are qualified to mark the time t when at rest relatively to the stationary system, and the time $\tau$ when at rest relatively to the moving system, to be located at the origin of the co-ordinates of *k*, and so adjusted that it marks the time $\tau$. What is the rate of this clock, when viewed from the stationary system?

Between the quantities x, t, and $\tau$, which refer to the position of the clock, we have, evidently, x = υt and

$$\tau = \frac{1}{\sqrt{1 - v^2/c^2}}(t - vx/c^2).$$

Therefore,

$$\tau = t\sqrt{1 - v^2/c^2} = t - (1 - \sqrt{1 - v^2/c^2})t$$

whence it follows that the time marked by the clock (viewed in the stationary system) is slow by $1 - \sqrt{(1 - \upsilon^2/c^2)}$ seconds per second, or—neglecting magnitudes of fourth and higher order—by $\frac{1}{2} \upsilon^2/c^2$.

*From this there ensues the following peculiar consequence.* If at the points A and B of K there are stationary clocks which, viewed in the stationary system, are synchronous; and if the clock at A is moved with the velocity υ along the line AB to B, then on its arrival at B the two clocks no longer synchronize, but *the clock moved from A to B lags behind the other which has remained at B* by $\frac{1}{2} t\upsilon^2/c^2$ (up to magnitudes of fourth and higher order), t being the time occupied in the journey from A to B.

It is at once apparent that this result still holds good if the clock moves from A to B in any polygonal line, and also when the points A and B coincide.

If we assume that the result proved for a polygonal line is also valid for a continuously curved line, we arrive at this result: *If one of two synchronous clocks at A is moved in a closed curve with constant velocity until it returns to A, the journey lasting t seconds, then by the clock which has remained at rest the travelled clock on its arrival at A will be* $\frac{1}{2} t\upsilon^2/c^2$ *second slow*. Thence we conclude that a balance-clock[7] at the equator must go more slowly, by a very small amount, than a precisely similar clock situated at one of the poles under otherwise identical conditions.

[7] Not a pendulum-clock, which is physically a system to which the Earth belongs. This case had to be excluded.

## *§ 5. The Composition of Velocities.*

In the system *k* moving along the axis of X of the system K with velocity υ, let a *point* move in accordance with the equations

$$\xi = w_\xi \tau, \eta = w_\eta \tau, \zeta = 0,$$

where $\omega_\xi$ and $\omega_\eta$ denote constants.

Required: the motion of the *point* relatively to the system K. If with the help of the *equations of transformation* developed in § 3 we introduce the quantities x, y, z, t into the *equations of motion* of the *point*, we obtain

$$x = \frac{w_\xi + v}{1 + vw_\xi/c^2}t,$$
$$y = \frac{\sqrt{1 - v^2/c^2}}{1 + vw_\xi/c^2}w_\eta t,$$
$$z = 0.$$

Thus, *the law of the parallelogram of velocities is valid according to our theory only to a first approximation.* We set

$$V^2 = \left(\frac{dx}{dt}\right)^2 + \left(\frac{dy}{dt}\right)^2,$$
$$w^2 = w_\xi^2 + w_\eta^2,$$
$$a = \tan^{-1} w_\eta/w_\xi; \qquad *3$$

*3 This equation was incorrectly given in Einstein's original paper and the 1923 English translation as $a = \tan^{-1} \omega_y/\omega_x$.

*a* is then to be looked upon as the angle between the velocities υ and ω. After a simple calculation we obtain*4

*4 The exponent of *c* in the denominator of the sine term of this equation was erroneously given as 2 in the 1923 edition of this paper. It has been corrected to unity here.

$$V = \frac{\sqrt{(v^2 + w^2 + 2vw\cos a) - (vw\sin a/c)^2}}{1 + vw\cos a/c^2}.$$

It is worthy of remark that υ and ω enter into the expression for the resultant velocity in a symmetrical manner. If *w* also has the direction of the axis of X, we get

$$V = \frac{v + w}{1 + vw/c^2}.$$

It follows from this equation that *from a composition of two velocities which are less than c, there always results a velocity less than c*. For if we set υ = c – κ, and ω = c – λ, κ and λ being positive and less than c, then

$$V = c\frac{2c - \kappa - \lambda}{2c - \kappa - \lambda + \kappa\lambda/c} < c.$$

It follows, further, that *the velocity of light c cannot be altered by composition with a velocity less than that of light*. For this case we obtain

$$V = \frac{c + w}{1 + w/c} = c.$$

We might also have obtained the formula for V, for the case when υ and ω have the same direction, by compounding two transformations in accordance with § 3. If in addition to the systems K and *k* figuring in § 3 we introduce still another system of co-ordinates k' moving parallel to *k*, its initial point moving on the axis of Ξ *5

*5 "X" in the 1923 English translation.

with the velocity ω, we obtain equations between the quantities x, y, z, t and the corresponding quantities of k', which differ from the equations found in § 3 only in that the place of "υ" is taken by the quantity

$$\frac{v + w}{1 + vw/c^2};$$

from which we see that such parallel transformations—necessarily—form a group.

*We have now deduced the requisite laws of the theory of kinematics corresponding to our two principles*, and we proceed to show their application to electrodynamics.

## *II. ELECTRODYNAMICAL PART.*

### *§ 6. Transformation of the Maxwell-Hertz Equations for Empty Space. On the Nature of the Electromotive Forces Occurring in a Magnetic Field During Motion.*

Let the Maxwell-Hertz equations for empty space hold good for the stationary system K, so that we have

$$\begin{aligned}
\frac{1}{c}\frac{\partial X}{\partial t} &= \frac{\partial N}{\partial y} - \frac{\partial M}{\partial z}, & \frac{1}{c}\frac{\partial L}{\partial t} &= \frac{\partial Y}{\partial z} - \frac{\partial Z}{\partial y}, \\
\frac{1}{c}\frac{\partial Y}{\partial t} &= \frac{\partial L}{\partial z} - \frac{\partial N}{\partial x}, & \frac{1}{c}\frac{\partial M}{\partial t} &= \frac{\partial Z}{\partial x} - \frac{\partial X}{\partial z}, \\
\frac{1}{c}\frac{\partial Z}{\partial t} &= \frac{\partial M}{\partial x} - \frac{\partial L}{\partial y}, & \frac{1}{c}\frac{\partial N}{\partial t} &= \frac{\partial X}{\partial y} - \frac{\partial Y}{\partial x},
\end{aligned}$$

where (X, Y, Z) denotes the vector of the *electric force*, and (L, M, N) that of the *magnetic force*.

If we apply to these equations the *transformation* developed in § 3, by referring the electromagnetic processes *to the system of co-ordinates* there introduced, *moving with the velocity υ*, we obtain the equations*6

*6 In the 1923 English translation, the quantities "ζ" and "ξ" were interchanged in the second equation. They were given correctly in the original 1905 paper.

$$
\begin{aligned}
\frac{1}{c}\frac{\partial \mathrm{X}}{\partial \tau} &= \frac{\partial}{\partial \eta}\left\{\beta\left(\mathrm{N}-\frac{v}{c}\mathrm{Y}\right)\right\} - \frac{\partial}{\partial \zeta}\left\{\beta\left(\mathrm{M}+\frac{v}{c}\mathrm{Z}\right)\right\}, \\
\frac{1}{c}\frac{\partial}{\partial \tau}\left\{\beta\left(\mathrm{Y}-\frac{v}{c}\mathrm{N}\right)\right\} &= \frac{\partial \mathrm{L}}{\partial \zeta} - \frac{\partial}{\partial \xi}\left\{\beta\left(\mathrm{N}-\frac{v}{c}\mathrm{Y}\right)\right\}, \\
\frac{1}{c}\frac{\partial}{\partial \tau}\left\{\beta\left(\mathrm{Z}+\frac{v}{c}\mathrm{M}\right)\right\} &= \frac{\partial}{\partial \xi}\left\{\beta\left(\mathrm{M}+\frac{v}{c}\mathrm{Z}\right)\right\} - \frac{\partial \mathrm{L}}{\partial \eta}, \\
\frac{1}{c}\frac{\partial \mathrm{L}}{\partial \tau} &= \frac{\partial}{\partial \zeta}\left\{\beta\left(\mathrm{Y}-\frac{v}{c}\mathrm{N}\right)\right\} - \frac{\partial}{\partial \eta}\left\{\beta\left(\mathrm{Z}+\frac{v}{c}\mathrm{M}\right)\right\}, \\
\frac{1}{c}\frac{\partial}{\partial \tau}\left\{\beta\left(\mathrm{M}+\frac{v}{c}\mathrm{Z}\right)\right\} &= \frac{\partial}{\partial \xi}\left\{\beta\left(\mathrm{Z}+\frac{v}{c}\mathrm{M}\right)\right\} - \frac{\partial \mathrm{X}}{\partial \zeta}, \\
\frac{1}{c}\frac{\partial}{\partial \tau}\left\{\beta\left(\mathrm{N}-\frac{v}{c}\mathrm{Y}\right)\right\} &= \frac{\partial \mathrm{X}}{\partial \eta} - \frac{\partial}{\partial \xi}\left\{\beta\left(\mathrm{Y}-\frac{v}{c}\mathrm{N}\right)\right\},
\end{aligned}
$$

where

$$\beta = 1/\sqrt{1-v^2/c^2}.$$

Now the *principle of relativity* requires that if the Maxwell-Hertz equations for empty space hold good in system K, they also hold good in system *k*; that is to say that the *vectors of the electric and the magnetic force* — $(\mathrm{X}', \mathrm{Y}', \mathrm{Z}')$ and $(\mathrm{L}', \mathrm{M}', \mathrm{N}')$ — of the moving system *k*, *which are defined by their ponderomotive effects on electric or magnetic masses respectively*, satisfy the following equations: —

$$
\begin{aligned}
\frac{1}{c}\frac{\partial \mathrm{X}'}{\partial \tau} &= \frac{\partial \mathrm{N}'}{\partial \eta} - \frac{\partial \mathrm{M}'}{\partial \zeta}, & \frac{1}{c}\frac{\partial \mathrm{L}'}{\partial \tau} &= \frac{\partial \mathrm{Y}'}{\partial \zeta} - \frac{\partial \mathrm{Z}'}{\partial \eta}, \\
\frac{1}{c}\frac{\partial \mathrm{Y}'}{\partial \tau} &= \frac{\partial \mathrm{L}'}{\partial \zeta} - \frac{\partial \mathrm{N}'}{\partial \xi}, & \frac{1}{c}\frac{\partial \mathrm{M}'}{\partial \tau} &= \frac{\partial \mathrm{Z}'}{\partial \xi} - \frac{\partial \mathrm{X}'}{\partial \zeta}, \\
\frac{1}{c}\frac{\partial \mathrm{Z}'}{\partial \tau} &= \frac{\partial \mathrm{M}'}{\partial \xi} - \frac{\partial \mathrm{L}'}{\partial \eta}, & \frac{1}{c}\frac{\partial \mathrm{N}'}{\partial \tau} &= \frac{\partial \mathrm{X}'}{\partial \eta} - \frac{\partial \mathrm{Y}'}{\partial \xi}.
\end{aligned}
$$

Evidently the two systems of equations found for system *k* must express exactly the same thing, since both systems of equations are equivalent to the Maxwell-Hertz equations for system K. Since, further, the equations of the two systems agree, with the exception of the symbols for the vectors, it follows that the functions occurring in the systems of equations at corresponding places must agree, with the exception of a factor $\psi(v)$, which is common for all functions of the one system of equations, and is independent of $\xi$, $\eta$, $\zeta$ and $\tau$ but depends upon $v$. Thus, we have the relations

$$
\begin{aligned}
\mathrm{X}' &= \psi(v)\mathrm{X}, & \mathrm{L}' &= \psi(v)\mathrm{L}, \\
\mathrm{Y}' &= \psi(v)\beta\left(\mathrm{Y}-\frac{v}{c}\mathrm{N}\right), & \mathrm{M}' &= \psi(v)\beta\left(\mathrm{M}+\frac{v}{c}\mathrm{Z}\right), \\
\mathrm{Z}' &= \psi(v)\beta\left(\mathrm{Z}+\frac{v}{c}\mathrm{M}\right), & \mathrm{N}' &= \psi(v)\beta\left(\mathrm{N}-\frac{v}{c}\mathrm{Y}\right).
\end{aligned}
$$

If we now form the reciprocal of this system of equations, firstly by solving the equations just obtained, and secondly by applying the equations to the *inverse transformation* (from *k* to K), which is characterized by the velocity $-v$, it follows, when we consider that the two systems of equations thus obtained must be identical, that $\psi(v)\psi(-v) = 1$. Further, from reasons of symmetry and therefore

$$\psi(v) = 1,$$

and our equations assume the form

$$\begin{aligned} X' &= X, & L' &= L, \\ Y' &= \beta\left(Y - \frac{v}{c}N\right), & M' &= \beta\left(M + \frac{v}{c}Z\right), \\ Z' &= \beta\left(Z + \frac{v}{c}M\right), & N' &= \beta\left(N - \frac{v}{c}Y\right). \end{aligned}$$

[8] If, for example, X = Y = Z = L = M = 0, and N ≠ 0, then from reasons of symmetry it is clear that when υ changes sign without changing its numerical value, $Y'$ must also change sign without changing its numerical value.

As to the interpretation of these equations we make the following remarks: Let a point charge of electricity have the magnitude "one" when measured in the stationary system K, i.e. let it when at rest in the stationary system exert a force of one dyne upon an equal quantity of electricity at a distance of one cm. By the *principle of relativity* this electric charge is also of the magnitude "one" when measured in the moving system. If this quantity of electricity is at rest relatively to the stationary system, then by definition the vector (X, Y, Z) is equal to the force acting upon it. If the quantity of electricity is at rest relatively to the moving system (at least at the relevant instant), then the force acting upon it, measured in the moving system, is equal to the vector ($X'$, $Y'$, $Z'$). Consequently, the first three equations above allow themselves to be clothed in words in the two following ways: —

1. *If a unit electric point charge is in motion in an electromagnetic field, there acts upon it, in addition to the electric force, an "electromotive force"* which, if we neglect the terms multiplied by the second and higher powers of *v*/*c*, is *equal to the vector-product of the velocity of the charge and the magnetic force, divided by the velocity of light*. (Old manner of expression.)

2. *If a unit electric point charge is in motion in an electromagnetic field, the force acting upon it is equal to the electric force which is present at the locality of the charge, and which we ascertain by transformation of the field to a system of co-ordinates at rest relatively to the electrical charge*. (New manner of expression.)

The analogy holds with "magnetomotive forces." We see that *electromotive force plays in the developed theory merely the part of an auxiliary concept*, which owes its introduction to the circumstance that *electric and magnetic forces do not exist independently of the state of motion of the system of co-ordinates*.

Furthermore, it is clear that *the asymmetry mentioned in the introduction* as arising when we consider the currents produced by the relative motion of a magnet and a conductor, *now disappears*. Moreover, questions as to the "seat" of electrodynamic electromotive forces (unipolar machines) now have no point.

### *§ 7. Theory of Doppler's Principle and of Aberration.*

In the system K, very far from the origin of co-ordinates, let there be a source of electrodynamic waves, which in a part of space containing the origin of co-ordinates may be represented to a sufficient degree of approximation by the equations

$$\begin{array}{ll} X = X_0 \sin\Phi, & L = L_0 \sin\Phi, \\ Y = Y_0 \sin\Phi, & M = M_0 \sin\Phi, \\ Z = Z_0 \sin\Phi, & N = N_0 \sin\Phi, \end{array}$$

where

$$\Phi = \omega\left\{t - \frac{1}{c}(lx + my + nz)\right\}.$$

Here (**X₀**, **Y₀**, **Z₀**) and (**L₀**, **M₀**, **N₀**) are the vectors defining the *amplitude* of the *wave-train*, and *l*, *m*, *n* the direction-cosines of the wave-normals. We wish to know the constitution of these waves, when they are examined by an observer at rest in the moving system *k*.

Applying the *equations of transformation* found in § 6 for electric and magnetic forces, and those found in § 3 for the co-ordinates and the time, we obtain directly

$$\begin{array}{ll} X' = X_0 \sin\Phi', & L' = L_0 \sin\Phi', \\ Y' = \beta(Y_0 - vN_0/c)\sin\Phi', & M' = \beta(M_0 + vZ_0/c)\sin\Phi', \\ Z' = \beta(Z_0 + vM_0/c)\sin\Phi', & N' = \beta(N_0 - vY_0/c)\sin\Phi', \end{array}$$
$$\Phi' = \omega'\left\{\tau - \tfrac{1}{c}(l'\xi + m'\eta + n'\zeta)\right\}$$

where

$$\begin{aligned} \omega' &= \omega\beta(1 - lv/c), \\ l' &= \frac{l - v/c}{1 - lv/c}, \\ m' &= \frac{m}{\beta(1 - lv/c)}, \\ n' &= \frac{n}{\beta(1 - lv/c)}. \end{aligned}$$

From the equation for $\omega'$ it follows that if an observer is moving with velocity υ relatively to an infinitely distant source of light of frequency $\nu$, in such a way that the connecting line "source-observer" makes the angle ϕ with the velocity of the observer referred to a system of co-ordinates which is at rest relatively to the source of light, the frequency $\nu'$ of the light perceived by the observer is given by the equation

$$\nu' = \nu \frac{1 - \cos\phi \cdot v/c}{\sqrt{1 - v^2/c^2}}.$$

This is Doppler's principle for any velocities whatever. When $\phi = 0$ the equation assumes the perspicuous form

$$\nu' = \nu \sqrt{\frac{1 - v/c}{1 + v/c}}.$$

*We see that, in contrast with the customary view, when* $\upsilon = -c$, $\nu' = \infty$.

[This calculation assumes *non-relativistic* addition and subtraction of the velocity of the inertial observer (or the inertial emitter), and the speed of light, and is inconsistent with Einstein's *second postulate*. As noted by Einstein in § 5 above "It follows, further, that *the velocity of light c cannot be altered by composition with a velocity less than that of light*". See "Relativistic longitudinal Doppler effect" in Part II, below.].

If we call the angle between the wave-normal (direction of the ray) in the moving system and the connecting line "source-observer" $\phi'$, the equation for $\phi'$ [*7] assumes the form

$$\cos\phi' = \frac{\cos\phi - v/c}{1 - \cos\phi \cdot v/c}.$$

[*7] Erroneously given as *l'* in the 1923 English translation, propagating an error, despite a change in symbols, from the original 1905 paper.

This equation expresses the law of aberration in its most general form. If $\phi = \frac{1}{2}\pi$, the equation becomes simply

$$\cos\phi' = -v/c.$$

We still have to find the *amplitude* of the waves, as it appears in the moving system. If we call the *amplitude* of the electric or magnetic force A or $\mathrm{A}'$ respectively, accordingly as it is measured in the stationary system or in the moving system, we obtain

$$\mathrm{A}'^2 = \mathrm{A}^2 \frac{(1 - \cos\phi \cdot v/c)^2}{1 - v^2/c^2}$$

which equation, if $\phi = 0$, simplifies into

$$A'^2 = A^2 \frac{1 - v/c}{1 + v/c}.$$

It follows from these results that to an observer approaching a source of light with the velocity $c$, this source of light must appear of infinite intensity.

### *§ 8. Transformation of the Energy of Light Rays. Theory of the Pressure of Radiation Exerted on Perfect Reflectors.*

…

### *§ 9. Transformation of the Maxwell-Hertz Equations when Convection-Currents are Taken into Account.*

…

### *§ 10. Dynamics of the Slowly Accelerated Electron.*

Let there be in motion in an electromagnetic field an electrically charged particle (in the sequel called an "electron"), for the law of motion of which we assume as follows: —

If the electron is at rest at a given epoch, the motion of the electron ensues in the next instant of time according to the equations

$$\begin{aligned} m\frac{d^2x}{dt^2} &= \epsilon X \\ m\frac{d^2y}{dt^2} &= \epsilon Y \\ m\frac{d^2z}{dt^2} &= \epsilon Z \end{aligned}$$

where $x$, $y$, $z$ denote the co-ordinates of the electron, and $m$ the mass of the electron, as long as its motion is slow.

Now, secondly, let the velocity of the electron at a given epoch be $\upsilon$. We seek the law of motion of the electron in the immediately ensuing instants of time.

Without affecting the general character of our considerations, we may and will assume that the electron, at the moment when we give it our attention, is at the origin of the co-ordinates, and moves with the velocity $\upsilon$ along the axis of X of the system K. It is then clear that at the given moment ($t = 0$) the electron is at rest relatively to a system of co-ordinates which is in parallel motion with velocity $\upsilon$ along the axis of X.

From the above assumption, in combination with the principle of relativity, it is clear that in the immediately ensuing time (for small values of $t$) the electron, viewed from the system $k$, moves in accordance with the equations

$$
\begin{aligned}
m\frac{d^2\xi}{d\tau^2} &= \epsilon \mathrm{X}',\\
m\frac{d^2\eta}{d\tau^2} &= \epsilon \mathrm{Y}',\\
m\frac{d^2\zeta}{d\tau^2} &= \epsilon \mathrm{Z}',
\end{aligned}
$$

in which the symbols $\xi$, $\eta$, $\zeta$, $\mathrm{X}'$, $\mathrm{Y}'$, $\mathrm{Z}'$ refer to the system $k$. If, further, we decide that when $t = x = y = z = 0$ then $\tau = \xi = \eta = \zeta = 0$, the transformation equations §§ 3 and 6 hold good, so that we have

$$
\begin{aligned}
&\xi = \beta(x - vt), \eta = y, \zeta = z, \tau = \beta(t - vx/c^2),\\
&\mathrm{X}' = \mathrm{X}, \mathrm{Y}' = \beta(\mathrm{Y} - v\mathrm{N}/c), \mathrm{Z}' = \beta(\mathrm{Z} + v\mathrm{M}/c).
\end{aligned}
$$

With the help of these equations, we transform the above equations of motion from system $k$ to system K, and obtain

$$
\left.
\begin{aligned}
\frac{d^2x}{dt^2} &= \frac{\epsilon}{m\beta^3}\mathrm{X}\\
\frac{d^2y}{dt^2} &= \frac{\epsilon}{m\beta}\left(\mathrm{Y} - \frac{v}{c}\mathrm{N}\right)\\
\frac{d^2z}{dt^2} &= \frac{\epsilon}{m\beta}\left(\mathrm{Z} + \frac{v}{c}\mathrm{M}\right)
\end{aligned}
\right\} \qquad \ldots \quad \textbf{(A)}
$$

Taking the ordinary point of view, we now inquire as to the "longitudinal" and the "transverse" mass of the moving electron. We write the equations **(A)** in the form

$$
\begin{aligned}
m\beta^3\frac{d^2x}{dt^2} &= \epsilon \mathrm{X} &&= \epsilon \mathrm{X}',\\
m\beta^2\frac{d^2y}{dt^2} &= \epsilon\beta\left(\mathrm{Y} - \frac{v}{c}\mathrm{N}\right) &&= \epsilon \mathrm{Y}',\\
m\beta^2\frac{d^2z}{dt^2} &= \epsilon\beta\left(\mathrm{Z} + \frac{v}{c}\mathrm{M}\right) &&= \epsilon \mathrm{Z}',
\end{aligned}
$$

and remark firstly that $\epsilon\mathrm{X}'$, $\epsilon\mathrm{Y}'$, $\epsilon\mathrm{Z}'$ are the components of the ponderomotive force acting upon the electron, and are so indeed as viewed in a system moving at the moment with the electron, with the same velocity as the electron. (This force might be measured, for example, by a spring balance at rest in the last-mentioned system.) Now if we call this force simply "the force acting upon the electron,"[9]

[9] The definition of force here given is not advantageous, as was first shown by M. Planck. It is more to the point to define force in such a way that the laws of momentum and energy assume the simplest form.

and maintain the equation—mass × acceleration = force—and if we also decide that the accelerations are to be measured in the stationary system K, we derive from the above equations

$$\text{Longitudinal mass} = \frac{m}{(\sqrt{1-v^2/c^2})^3}.$$

$$\text{Transverse mass} = \frac{m}{1-v^2/c^2}.$$

With a different definition of force and acceleration we should naturally obtain other values for the masses. This shows us that in comparing different theories of the motion of the electron we must proceed very cautiously.

We remark that these results as to the mass are also valid for ponderable material points, because a ponderable material point can be made into an electron (in our sense of the word) by the addition of an electric charge, *no matter how small.*

We will now determine the kinetic energy of the electron. If an electron moves from rest at the origin of co-ordinates of the system K along the axis of X under the action of an electrostatic force X, it is clear that the energy withdrawn from the electrostatic field has the value $\int \varepsilon X\, dx$. As the electron is to be slowly accelerated, and consequently may not give off any energy in the form of radiation, the energy withdrawn from the electrostatic field must be put down as equal to the energy of motion W of the electron. Bearing in mind that during the whole process of motion which we are considering, the first of the equations **(A)** applies, we therefore obtain

$$\begin{aligned} W &= \int \epsilon X\, dx = m\int_0^v \beta^3 v\, dv \\ &= mc^2\left\{\frac{1}{\sqrt{1-v^2/c^2}}-1\right\}. \end{aligned}$$

Thus, when $v = c$, W becomes infinite. Velocities greater than that of light have—as in our previous results—no possibility of existence.

This expression for the kinetic energy must also, by virtue of the argument stated above, apply to ponderable masses as well.

We will now enumerate the properties of the motion of the electron which result from the system of equations **(A)**, and are accessible to experiment.

1. From the second equation of the system **(A)** it follows that an electric force Y and a magnetic force N have an equally strong deflective action on an electron moving with the velocity $v$, when $Y = Nv/c$. Thus, we see that it is possible by our theory to determine

the velocity of the electron from the ratio of the magnetic power of deflection $A_m$ to the electric power of deflection $A_e$, for any velocity, by applying the law

$$\frac{A_m}{A_e} = \frac{v}{c}.$$

This relationship may be tested experimentally, since the velocity of the electron can be directly measured, e.g. by means of rapidly oscillating electric and magnetic fields.

2. From the deduction for the kinetic energy of the electron it follows that between the potential difference, P, traversed and the acquired velocity $v$ of the electron there must be the relationship

$$P = \int X dx = \frac{m}{\epsilon} c^2 \left\{ \frac{1}{\sqrt{1 - v^2/c^2}} - 1 \right\}.$$

3. We calculate the radius of curvature of the path of the electron when a magnetic force N is present (as the only deflective force), acting perpendicularly to the velocity of the electron. From the second of the equations **(A)** we obtain

$$-\frac{d^2 y}{dt^2} = \frac{v^2}{R} = \frac{\epsilon}{m} \frac{v}{c} N \sqrt{1 - \frac{v^2}{c^2}}$$

or

$$R = \frac{mc^2}{\epsilon} \cdot \frac{v/c}{\sqrt{1 - v^2/c^2}} \cdot \frac{1}{N}.$$

These three relationships are a complete expression for the laws according to which, by the theory here advanced, the electron must move.

In conclusion I wish to say that in working at the problem here dealt with I have had the loyal assistance of my friend and colleague M. Besso, and that I am indebted to him for several valuable suggestions.

**Einstein, A. (May, 1912). Lichtgeschwindigkeit und Statik des Gravitionsfeldes. (The Speed of Light and the Statics of the Gravitational Field.)**

*Ann. Phys.*, 38, 355-69; translation in A. Beck (translator), D. Howard (consultant). (1996). *The Collected Papers of Albert Einstein,* Volume 4: The Swiss Years: Writings, 1912-1914. (English translation), Princeton University Press, Princeton, Vol. 4, Doc. 3, 95-106; https://einsteinpapers.press.princeton.edu/vol4-doc/151; translation below by D. H. Delphenich at http://www.neo-classical-physics.info/uploads/3/4/3/6/34363841/einstein_-_speed_of_light_and_grav.pdf.; also in Underwood, T. G. (2023). *General Relativity*, Part II, pp. 161-74.

Prague

Received February 26, 1912.

Einstein (1911) showed that the validity of one of the fundamental laws of his theory of special relativity, namely, the law of the *constancy of the speed of light*, could claim to be valid only for space-time domains of constant gravitational potential. Einstein noted that despite the fact that this result *excluded the general applicability of the Lorentz transformation*, it should not deter us from pursuing the consequences of that path. Here he took that further by demonstrating that the Lorentz transformation could not be established for infinitely-small space-time regions either *as soon as one abandons the universal constancy of c*.

> [Bacelar Valente, M. (2018). Einstein's redshift derivations: its history from 1907 to 1921. *Circumscribere: International Journal for the History of Science*, 22, 1-16: "*In his subsequent work on a theory of a static gravitational field, Einstein did not mention explicitly the redshift*, but mentioned the effect of the field on clocks, which according to his previous treatment causes the redshift:
>
> > "A clock runs faster the greater the [gravitational potential] of the location to which we bring it." [Einstein, A. (February, 1912). Lichtgeschwindigkeit und Statik des Gravitionsfeldes. (The Speed of Light and the Statics of the Gravitational Field.) *Ann. Phys.*, 38, 355-69; translation in A. Beck (translator), D. Howard (consultant). (1996). *The Collected Papers of Albert Einstein,* Volume 4: The Swiss Years: Writings, 1912-1914. (English translation), Princeton University Press, Princeton, Vol. 4, Doc. 3, 95-106, on p. 104].
>
> We see that in the context of an application of the equivalence principle or the scalar theory of gravitation, the redshift is due to the effect of the gravitational field on

clocks (e.g. atoms) 'at rest' in the field. The main aspects of these derivations are then the following:

a) The *gravitational field* affects the rate of clocks (which leads to the redshift).

b) Clocks are taken to be at 'rest' in the *gravitational field* (i.e., at rest in an inertial reference frame with a homogeneous *gravitational field*).

c) Atoms are an example of clocks affected by the *gravitational field*."]

---

In a paper that appeared last year[1],

[1] Einstein, A. (1911). Über den Einfluss der Schwerkraft auf die Ausbreitung des Lichtes. (On the Influence of Gravitation on the Propagation of Light.) *Ann. Phys*., 4, 35, 898-908.

starting from the hypothesis that the *gravitational field* and the state of *acceleration* of a coordinate system are physically equivalent, I inferred some consequences that are in very good agreement with the results of the theory of relativity (viz., the theory of relativity of uniform motion). *However, it was shown that the validity of one of the fundamental laws of that theory, namely, the law of the constancy of the speed of light, can claim to be valid only for space-time domains of constant gravitational potential*. Despite the fact that this result excluded the general applicability of the Lorentz transformation, it should not deter us from pursuing consequences of choosing that path. At the very least, my opinion in regard to the hypothesis that the "acceleration field" is a special case of the gravitational field seems so likely to be true, especially when one recalls the consequences in regard to the gravitational mass of the energy content that were inferred before in the latter paper, that a more precise analysis of the consequences of that equivalence hypothesis would seem to be in order.

Since then, Abraham has presented a theory of gravitation[2] that includes the consequences of my first paper as special cases.

[2] Abraham, M. (1912). Zur Theorie der Gravitation. (On the New theory of Gravitation.) *Phys. Zeit*., 13, 19, 1-4.

However, in what follows we will see that Abraham's system of equations cannot be brought into agreement with the *equivalence hypothesis* and that its concept of space and time cannot be maintained, even from a purely-mathematical standpoint.

*1. – Space and time in the acceleration field.*

The reference system K (coordinates x, y, z) is found to be in a state of uniform *acceleration* in the direction of its x-coordinate. Let that acceleration be a uniform in the Born sense, i.e., let the acceleration of its origin relative to a system that is unaccelerated, relative to which the points of K possess no velocity at all (an infinitely-small velocity, resp.), be a constant quantity. According to the *equivalence hypothesis*, such a system K is equivalent to a system at rest in which one finds a mass-free static *gravitational field*[3] of a certain kind.

[3] One must imagine that the masses that produce that field are at infinity.

The spatial measurement of K happens by means of yardsticks that possess equal lengths (when compared to each other in the rest state at the aforementioned location for K). The laws of geometry shall be valid for lengths that are measured in that way, so they will also be valid for the relationships between the coordinates x, y, z and other lengths. It is not obvious that this convention is legitimate since *it includes physical assumptions that might possibly prove to be incorrect. For example, it does not seem likely that it is true in a uniformly-rotating system, in which the Lorentz contraction would imply that the ratio of the circumference to the diameter would have to be different from π when one applies our definition of length.*

> [This is a reference to the *Ehrenfest paradox*, which is one of the three unresolved problems with Einstein's *theory of special relativity*, and resulted in the present author coming to the conclusion "For this reason, until more satisfactory evidence in support of Einstein's *second postulate*, a refutation of the Ehrenfest paradox, and an explanation for the observed Doppler redshift and blueshift consistent with Einstein's two postulates, is provided, under any normal measure of a theory in physics, *Einstein's second postulate, and consequently his theory of special relativity, must be rejected.*" Underwood, T. G. (1923). *Special Relativity*, p. 381. The failure to resolve this paradox may have finally persuade Einstein that it was time to move on from his *theory of special relativity*.]
>
>> [Underwood, T. G. (1923). *Special Relativity*, p. 313: "The *Ehrenfest paradox* concerns the rotation of a "rigid" disc in the theory of special relativity. In its original 1909 formulation as presented by Paul Ehrenfest in relation to the concept of Born rigidity within special relativity, it discusses an ideally rigid cylinder that is made to rotate about its axis of symmetry rotating with constant angular velocity ω. The reference frame is fixed to the stationary center of the disk. The radius R as seen in the laboratory frame is always perpendicular to its motion and should therefore be equal to its value $R_0$ when stationary. Then the magnitude of the relative velocity of any

point in the circumference of the disk is ωR. However, the circumference (2πR) should appear Lorentz-contracted to a smaller value than at rest, by the usual factor of √{1 – (ωR)2/c2}. However, since the radius is aways perpendicular to the direction of motion, it will not undergo any contraction. This leads to the contradiction that R = R0 and R < R0. Thus, Ehrenfest argued by reductio ad absurdum that Born rigidity is not generally compatible with special relativity.
The paradox was deepened further by Albert Einstein, who showed that since measuring rods aligned along the periphery and moving with it should appear contracted, more would fit around the circumference, which would thus measure greater than 2πR. This leads to the paradox that the rigid measuring rods would have to separate from one another due to Lorentz contraction."]

The yardstick, as well as the coordinate axes, are imagined to be rigid bodies. *That is permissible, despite the fact that according to the theory of relativity, rigid bodies cannot exist in reality*. One can then think of the rigid measuring devices as being replaced with a large number of small non-rigid bodies that are arranged with respect to each other in such a way that they exert no forces of repulsion on each other, which will keep each of them in place. We imagine that the time t in the system K is measured by clocks that have such a nature and such a fixed arrangement at spatial points of the system K that the time interval (as measured by those clocks) that a light ray needs in order to arrive at a point B in the system K from a point A does not depend on the time-point of the emission of the light ray at A. It will be shown further that one can make a consistent definition of simultaneity such that all light rays that pass through a point A in K will possess the same speed of propagation independently of the direction at A relative to the readings on the clocks that one gets by continuation.

We now think of the reference frame K (x, y, z, t) from an *acceleration*-free reference frame (*of constant gravitational potential*) Σ (ξ, η, ζ, τ). We assume that the x-axis is permanently parallel to the ξ-axis and that the y-axis is permanently parallel to the η-axis, and the z-axis is permanently parallel to the ζ-axis. *This determination is possible on the assumption that the state of acceleration has no influence on the shape of K with respect to E*. We take this physical assumption as a basis. It implies that for arbitrary τ, we must have:

$$\eta = y, \qquad (1)$$
$$\zeta = z,$$

such that all we still need to look for is the relationship that exists between $\xi$ and $\tau$, on the one hand, and between x and t, on the other. Both reference systems might coincide at time $\tau = 0$. In any event, the desired equations of the substitution must have the form:

$$\xi = \lambda + \alpha t^2 + \dots \qquad (2)$$
$$\tau = \beta + \gamma t^2 + \delta t^2 + \dots$$

The coefficients of these series, which are valid for sufficiently-small positive and negative values of $\tau$, are regarded as unknown functions of x, for the time being. When we restrict ourselves to the terms that were written down, we will get:

$$d\xi = (\lambda' + \alpha' t^2)\, dx + 2\alpha t\, dt, \qquad (3)$$
$$d\tau = (\beta' + \gamma' t^2 + \delta' t^2)\, dx + (\gamma + 2\delta t)\, dt$$

upon differentiation.

In the system $\Sigma$, we think of time as being measured in such a way that the speed of light will be equal to 1. We can then write the equation of a shell that propagates with the speed of light from an arbitrary space-time point, *when we restrict ourselves to an infinitely-small neighborhood of that space-time point*, in the form:

$$d\xi^2 + d\eta^2 + d\zeta^2 - d\tau^2 = 0.$$

The same shell must have the equation:

$$dx^2 + dy^2 + dz^2 - c^2\, dt^2 = 0.$$

in the system K. The equations of the substitution (2) must be such that those two equations are equivalent. Due to (1), that requires the identity:

$$d\xi^2 - d\tau^2 = dx^2 - c^2\, dt^2. \qquad (4)$$

If one sets the expressions for dx and dt in the left-hand side of this equation equal to unity and sets the coefficients of $dx^2$, $dt^2$, and dx dt equal to each other on the left-hand and right-hand sides then one will get the equations:

$$1 = (\lambda' + \alpha' t^2)^2 - (\beta' + \gamma' t^2 + \delta' t^2)^2,$$
$$-c^2 = 4\alpha^2 t^2 - (\gamma + 2\delta t)^2,$$
$$0 = (\lambda' + \alpha' t^2) \,.\, 2\alpha t - (\beta' + \gamma' t^2 + \delta' t^2)(\gamma + 2\delta t).$$

Those equations are valid at t identically, up to higher powers of t, in such a way that the terms that were omitted from (2) can have no influence, so the first equation is valid up to the second power of t, and the second and third ones are valid up to the first power of t. That will imply the equations:

$$1 = \lambda'^2 \beta'^2, \qquad 0 = \beta'\gamma', \qquad 2\,\lambda\alpha' - \gamma'^2 - 2\beta'\delta' = 0,$$
$$-c^2 = -\gamma^2, \qquad 0 = \gamma\delta,$$
$$0 = \beta'\gamma, \qquad 0 = 2\,\alpha\lambda' - 2\beta'\delta - \gamma\gamma'.$$

Since $\gamma$ cannot vanish, it will follow from the first equation in the third row that $\beta' = 0$. $\beta$ is a constant then that we can set equal to zero by a suitable choice of time origin. Furthermore, the coefficient $\gamma$ must be positive, so from the first equation in the second row:

$$\gamma = c.$$

From the second equation in the second row:

$$\delta = 0.$$

Since $\beta'$ vanishes and one can assume that x increases with $\xi$, it will follow from the first equation in the first row that:

$$\lambda' = 1,$$

so if one is to have $x = 0$, $\xi = 0$ for $t = 0$ then:

$$\lambda = x.$$

Finally, when one employs the relations that were found above, the third equation in the first row and the second equation in the third row will imply the differential equations:

$$2\alpha' - c'^2 = 0,$$
$$2\alpha - cc' = 0.$$

When we denote integration constants by $c_0$ and $a$, it will follow from the latter equations that:

$$c = c_0 + ax,$$
$$2\alpha = a\,(c_0 + ax) = ac.$$

The desired substitution is ascertained by that for sufficiently-small values of t. When one neglects third and higher powers of t, one will have the equations:

$$\xi = x + ac/2\ t^2, \qquad (4)$$
$$\eta = y,$$
$$\zeta = z,$$
$$\tau = ct,$$

by which, the speed of light c in the system K, which can depend upon only x, but not t, will be given by the relation that was just derived as:

$$c = c_0 + a x. \tag{5}$$

The constant $c_0$ depends upon the rate at which the clock that one measures time with ticks at the origin of K. One gets the meaning of the constant *a* in the following way: When one recalls (5), the first and fourth of equations (4) yield the *equation of motion*:

$$\xi = a/c_0\ \tau^2$$

for the origin (x = 0) of K. $a/c_0$ is then the acceleration of the origin of K relative to Σ when measured in time units in which the speed of light is equal to 1.

***§ 2. – Differential equation of the static gravitational field.***
***Equation of motion of a material point in a static gravitational field.***

It already emerges from the previous paper that a relationship exists between c and the *gravitational potential* of a static *gravitational field*, or in other words, that the field is determined by c. In those *gravitational fields* that correspond to the *acceleration field* that was considered in § 1, from (5) and the *equivalence principle*, the equation:

$$\Delta c = \partial^2 c/\partial x^2 + \partial^2 c/\partial y^2 + \partial^2 c/\partial z^2 = 0 \tag{5.a}$$

is fulfilled, and that suggests that we have assumed that this equation is valid in every mass-free static *gravitational field*[1].

[1] In a paper that will follow shortly, it will be shown that equations (5.a) and (5.b) cannot be correct exactly. However, in this article, they will be employed provisionally.

In any event, that equation is the simplest one that is compatible with (5).

It is easy to exhibit the presumably-valid equation that would correspond to Poisson's. Namely, it follows immediately from the meaning of c that c is determined only up to a constant factor that depends upon how one measures t at the origin of K with a suitable clock. The equation that corresponds to Poisson's must then be homogeneous in c. The simplest equation of that kind is the linear equation:

$$\Delta c = k c \rho, \tag{5.b}$$

when k is understood to mean the (universal) *constant of gravitation* and ρ is the *density* of matter. The latter must be defined such that it is already given by the mass distribution, i.e., it is independent of c for given matter in the spatial element. We can achieve that when we set the mass of a cubic centimeter of water equal to 1, which might also be found to be in a *gravitational potential*. ρ will then be the ratio of the mass that is found in a cubic centimeter to that unit.

*We now seek to ascertain the law of motion for a material point in a static gravitational field.* To that end, we shall seek the *law of motion* of a force-free material point that moves in the *acceleration field* that was considered in § 1. That *law of motion* in the system Σ is:

$$\xi = A_1\,\tau + B_1,$$
$$\eta = A_2\,\tau + B_2,$$
$$\zeta = A_3\,\tau + B_3,$$

in which the A and B are constant. By means of (4), those equations will go to the equations:

$$x = A_1\,c\,t + B_1 - ac/2\;t^2,$$
$$y = A_2\,c\,t + B_2,$$
$$z = A_3\,c\,t + B_3,$$

which are true for sufficiently small t. Upon repeatedly differentiating the first equation, when one sets t = 0 in it, one will get the two equations[1]:

[1] The terms in (2) that were dropped have no effect on the result of that double differentiation and subsequent setting of t to zero.

$$\dot{x} = A_1\,c,$$
$$\ddot{x} = 2A_1\,\dot{c} - ac.$$

When one eliminates $A_1$ from those two equations, it will follow that:

$$c\ddot{x} - 2\dot{c}\dot{x} = -\,ac^2,$$

or the equation:

$$d/dt\,(\dot{x}/c^2) = -\,a/c^2.$$

In an analogous way, it results that the other two components satisfy the equations:

$$d/dt\,(\dot{y}/c^2) = 0,$$
$$d/dt\,(\dot{z}/c^2) = 0.$$

Initially, those three equations are true at the instant t = 0. However, they are true in general, since that time-point is not distinguished from any other one by anything except for the fact that we have made it the starting point for our series development. *The equations that are found in that way are the desired equations of motion of the force-free moving point in a constant acceleration field.* If we consider that $a = \partial c/\partial x$ and that $(\partial c/\partial y) = (\partial c/\partial z) = 0$ then we can also write those equations in the form:

$$d/dt\,(\dot{x}/c^2) = -\,1/c\;\partial c/\partial x, \qquad (6)$$
$$d/dt\,(\dot{y}/c^2) = -\,1/c\;\partial c/\partial x,$$

$$d/dt\,(\dot{z}/c^2) = -\,1/c\ \partial c/\partial x.$$

The x-axis is no longer distinguished in this form for the equations; both sides have a vector character. For that reason, we must probably also regard those equations as the equations of motion of a material point in a static gravitational field when the point is subject to only the influence of gravity.

The relationship of the constant k that appears in (5.b) to the gravitational constant K in the usual sense next follows from (6). Namely, *in the case of a speed that is less than c*, one has from (6) that:

$$\ddot{x} = -\,c\ \partial c/\partial x = -\,\partial\Phi/\partial x,$$

such that (5.b) will go to:

$$\Delta\Phi = kc^2\ \rho$$

when one neglects certain terms. One then has:

$$K = kc^2.$$

The *gravitational constant* is not a constant then, but only the quotient $K/c^2$ is constant.

If we multiply equations (6) by $\dot{x}/c^2$, $\dot{y}/c^2$, $\dot{z}/c^2$, in succession and add them then when we set:

$$q^2 = \dot{x}^2 + \dot{y}^2 + \dot{z}^2,$$

we will get:

$$d/dt\,(\tfrac{1}{2}\,q^2/c^4) = -\,\dot{c}/c^3 = d/dt\,(1/2c^2)$$

or

$$d/dt\,[1/c^2\,(1 - q^2/c^2)] = 0,$$

or

$$c/\surd\,(1 - q^2/c^2) = \text{const.} \qquad (7)$$

*That equation includes the law of energy for the material point that moves in a stationary gravitational field.* The left-hand side of that equation depends upon q in precisely the same way that the energy of the material point depends upon q in the usual *theory of relativity*. We must then regard the left-hand side of the equation as the energy E of the point, up to a factor (that depends upon only the mass-point itself). Obviously, that factor is equal to the *mass* m, in the sense that was established above, *because that definition of mass was established independently of the gravitational potential*. One then has:

$$E = mc/\surd\,(1 - q^2/c^2), \qquad (8)$$

or approximately:

$$E = mc + m/2c\ q^2. \tag{8.a}$$

It next emerges from the second terms of that development that the quantity that we have deferred to as *energy* possesses a dimension that deviates from the more familiar one. Correspondingly, the unit of the individual energy quantities will also be different, namely, it will be c times smaller than it is in the system that is familiar to us. Furthermore, the "*kinetic energy*", which generally cannot be separated from the *gravitational energy* using (8), taken rigorously, depends upon not only m and q, but also on c, i.e., on the gravitational potential. *(8) further implies the important result that the energy of the point at rest in the gravitational field is mc.* If we would like to preserve the relation:

force · path length = energy supplied

then the *force* $\mathfrak{K}$ that is exerted on the material point at rest will be:

$$\mathfrak{K} = - m \text{ grad } c.$$

*We would now like to derive the equations of motion for a material point in an arbitrary gravitational field for the case in which other forces act on the point besides gravity.* We remark that *equations (6) are not similar to the equations of motion that are true in relativistic mechanics.* However, if we multiply them by the left-hand side of (7) then we will get the equations

$$d/dt\ \{(\dot{x}/c)/\sqrt{(1 - q^2/c^2)}\} = - (\partial c/\partial x)/\sqrt{(1 - q^2/c^2)}, \text{ etc.,} \tag{6.a}$$

which are equivalent to equations (6). *Except for the factor 1/c in the numerator, which is irrelevant in the ordinary theory of relativity, the left-hand side has precisely the same form that it has in the ordinary theory of relativity.* For that reason, we will have to refer to the quantity in brackets as the x-component of the quantity of motion (for a point of mass 1). Furthermore, we have just shown that $- \partial c/\partial x$ must be regarded as the x-component of the force that is exerted by the *gravitational field* on an arbitrary moving mass-point of mass 1. The force that is exerted by the *gravitational field* on an arbitrary moving mass-point differ from it by only a factor that vanishes with q. The equation that was just presented then leads one to set that force $\mathfrak{K}_g$ equal to $- (\partial c/\partial x)/\sqrt{(1 - q^2/c^2)}$. The right-hand side of the equation presented will then be $\mathfrak{K}g$. The time derivative of the impulse will then be equal to the applied force. If another force $\mathfrak{K}$ acts on the point then one will have to add a term $\mathfrak{K}/m$ to the right-hand side of the equation, such that the *equation of motion* of a point of mass m will assume the form:

$$d/dt\ \{(m\dot{x}/c)/\sqrt{(1 - q^2/c^2)}\} = - (m\ \partial c/\partial x)/\sqrt{(1 - q^2/c^2)} + \mathfrak{K}_x, \text{ etc.,} \tag{6.b}$$

However, that equation is permissible only when the *law of energy* is fulfilled in the form:

$$\mathfrak{K}q = \dot{E}.$$

That can be accomplished in the following way:

If one writes (6.b) in the form:

$$d/dt \{(x^{\bullet}/c)\, E\} + 1/c\; \partial c/\partial x\; E = Ꝁ_x, \text{ etc.,}$$

and multiplies those equations by $x^{\bullet}/c^2$, …, in succession, then one will find that:

$$\tfrac{1}{2}\, q^2/c^4\, E^{\bullet} + \tfrac{1}{2}\, E\; d/dt\, (q^2/c^4) + E\, c^{\bullet}/c^3 = Ꝁq/c^2.$$

That will imply the desired relation when one considers the fact that, from (8), one will have:

$$q^2/c^4 = 1/c^2 - m^2/E^2$$

and

$$d/dt\, (q^2/c^4) = -\, c^{\bullet}/c^3 + m^2 E/E^3.$$

The relationship between force and the law of energy-impulse then remains preserved.

### *§ 3. – Remarks on the physical meaning of the static gravitational potential.*

If we measure the speed of light in a space with an almost constant *gravitational potential* when we measure time by means of a certain clock *that makes light traverse a closed path with a well-defined length* then we will always obtain the same number for the speed of light, independently of how large the gravitational potential is in the space where we perform that measurement[1].

[1] The clock that is employed in order to measure time is therefore always the same one. It is always brought to the position where c is to be ascertained.

That follows immediately from the *equivalence principle*. When we say that the speed of light at a point is $c/c_0$ greater than it is at a point $P_0$ then that will mean that we must appeal to a clock that runs $c/c_0$ slower at P, where we measure time[1], than the clock that is employed to measure time at $P_0$ in the event that the ways that both clocks would work at the same location are comparable to each other.

[1] Namely, we measure the time that was denoted by t in the equations.

In other words: *A clock will run faster when we bring it to a location where c is greater*. That dependency of the rate of passage of time on the gravitational potential (c) is true for the rate at which arbitrary processes proceed. That was explained already in the previous article.

Similarly, the tension in a spring that is stretched in a certain way, and above all, the *force* (energy, resp.) in an arbitrary system, *always depends upon how large c is found to be at a location in the system*. That emerges easily from the following elementary argument: When we successively experiment in several small spatial regions of varying c and continually appeal to the same clock, the same yardstick, etc., we will find the same regularities with the same constants everywhere, except for possible differences in the intensities of the *gravitational field*. That follows from the *equivalence principle*. As a clock, we can appeal to perhaps two mirrors at a distance of 1 cm apart, when we count the number of times a light signal goes back and forth between them. We would then operate with a type of local time that Abraham denoted by $l$. It is then related to the universal time by:

$$dl = c\ dt.$$

If we measure the time by $l$ then we will assign a certain velocity dx/dl to a spring of mass m that has been stretched in a certain way by means of the energy of deformation, and independently of how large c is at a location where that process takes place. One has:

$$dx/dl = dx/cdt = a,$$

in which $a$ is independent of c. However, from (8), the *kinetic energy* that corresponds to that motion can be set equal:

$$m/2c\ q^2 = m/2c\ (dx/dt)^2 = m/2c\ a^2c^2 = ma^2/2c\ .\ c.$$

The energy of the spring is then proportional to c, and there is equality between energy and force for any system.

That dependency has a direct physical meaning. I imagine, e.g., a massless wire that stretched between two points $P_1$ and $P_2$ with different gravitational potentials. One of two equally-composed springs is stretched to a point $P_1$ on the wire, while the second one is stretched to $P_2$, in such a way that equilibrium exists. However, the elongations $l_1$ and $l_2$ that the two springs experience in that way will not be equal, since the equilibrium condition will read[2]:

[2] It is generally assumed in that that no forces act on the stretched massless spring in the *gravitational field*. That will be founded in an article that will follow shortly.

$$l_1/c_1 = l_2/c_2.$$

Finally, let it be mentioned that equation (5.b) also agrees with this general result. It will, in fact, follow from that equation and the fact that the gravitational force that acts on a mass, m, equals – m grad c that *the force* $\mathfrak{K}$ *of attraction between two masses that are found at a distance r from each other in a potential c is given by*:

$$\mathfrak{K} = ck\ mm'/4\pi r^2,$$

in the first approximation. That force is also proportional to c then. If we further imagine a "gravitational clock" that consists of a mass m that orbits around a fixed mass m′ at a constant distance R under the action of only the gravitational field then, according to (6.b), that will happen in accord with the equations:

$$m\ddot{x} = c\, \mathfrak{K}_x, \text{ etc.},$$

in the first approximation. It will then follow that:

$$m\omega^2 R = c^2 k\, mm'/4\pi R^2.$$

The rate ω at which the gravitational clock takes is then proportional to c, which should be true for clocks of any type.

***§ 4. – General remarks in regard to space and time.***

*How does the foregoing theory relate to the older theory of relativity (i.e., to the theory of a universal c)*? In Abraham's opinion, the equations of the Lorentz transformation must be true, as before, *in the infinitely small*, i.e., they should give an xt-transformation such that:

$$dx' = (dx - v\, dt)/\sqrt{(1 - v^2/c^2)},$$
$$dt' = (- v/c^2\, dx + dt)/\sqrt{(1 - v^2/c^2)}.$$

dx′ and dt′ must be complete differentials. The following equations must then be true:

$$\partial/\partial t\, \{1/\sqrt{(1 - v^2/c^2)}\} = \partial/\partial x\, \{- v/\sqrt{(1 - v^2/c^2)}\},$$
$$\partial/\partial t\, \{(- v^2/c^2)/\sqrt{(1 - v^2/c^2)}\} = \partial/\partial x\, \{1/\sqrt{(1 - v^2/c^2)}\}.$$

Now let the *gravitational field* in the unprimed system be a static one. c is then an arbitrarily-given function of x, but it is independent of t. Should the primed system be a "uniformly-moving" one, then v would have to be independent of t for a fixed x in any case. The left-hand sides of the equations would then have to vanish, and therefore, the right-hand sides, as well. However, the latter is impossible, since for arbitrarily-given functions c of x, both right-hand sides cannot be made to vanish when one suitably chooses v as a function of x. In that way, *it is then proved that one cannot establish the Lorentz transformation for infinitely-small space-time regions either as soon as one abandons the universal constancy of c.*

It seems to me that the space-time problem consists of the following: If one restricts oneself to a region of constant gravitational potential then the laws of nature will take on a distinctly simpler and invariant form when one refers them to a space-time system of those manifolds that are coupled to each other by the Lorentz transformations with constant c. If one does not restrict oneself to the regions of constant c then the manifold of equivalent systems, as well as the manifold of transformations that leave the laws of nature unchanged, will become larger, but the laws themselves will become more complicated.

## (3) All elementary particles, including electromagnetic waves (photons), are quantized.

**Relativistic Quantum Electrodynamics, Quantum Field Theory; Renormalization.**

There is a lack of convergence in current *relativistic* formulations of *quantum electrodynamics* (*quantum field theory*) due to the interaction of the electromagnetic and matter fields with their own vacuum fluctuations. The question is whether all *divergencies* can be isolated in unobservable *renormalization* factors.

**Underwood, T. G. (2023).** ***Quantum Electrodynamics***, Volume II, Preface, pp. 33-5: "Volume II, covering the period from 1930 up until 1965, when Tomonaga, Feynman, and Schwinger received their Nobel prizes, addresses the

by Dirac, Tomonaga, Schwinger and Feynman. Witness to this is the very comprehensive 2018 *An Introduction to Quantum Field Theory* by attempts to formulate a *relativistic quantum electrodynamics* or *quantum field theory* for the *electron* when the energy of the electron is *relativistic*, and in particular to address, through a process of *renormalization*, the still unresolved *divergencies* arising largely, if not entirely, from the assumption of a *point electron*.

Despite the claims to the contrary in modern textbooks, there have been no significant developments in the quantum electrodynamics or quantum field theory since 1965 to resolve the underlying occurrence of divergencies recognized Michael Peskin and Daniel Schroeder, first published in 1995, which replaced the 1965 two-volume text by James Bjorken and Sidney Drell, *Relativistic Quantum Fields*, which focuses on the application of Feynman diagrams. The former claims that "Quantum Electrodynamics (QED) is perhaps the best fundamental physical theory we have" then devotes Part II (pages 265-470), nearly one third of the book, to *renormalization*.

Born, in his 1954 Nobel prize speech, noted that "Planck, himself, belonged to the sceptics until he died. Einstein, De Broglie, and Schrodinger have unceasingly stressed the unsatisfactory features of quantum mechanics and called for a return to the concepts of classical, Newtonian physics while proposing ways in which this could be done without contradicting experimental facts".

Schwinger, in the Preface of his 1958 book [*Selected Papers on Quantum electrodynamics*.], "questioned whether *renormalization* simply corrected a mathematical error that causes the divergencies, or whether *there is a serious flaw in the structure of field theory*". He concluded that "the observational basis of quantum electrodynamics is self-contradictory" and that "a convergent theory cannot be formulated consistently within the framework of present space-time concepts" … "It can never explain the observed value of the dimensionless coupling constant measuring the electron charge … a full understanding of the electron charge can exist only when the theory of elementary particles has come to a stage of perfection that is presently unimaginable".

Tomonaga, in his 1965 Nobel prize speech, note that "In order to overcome the difficulty of an infinitely large *electromagnetic mass, Lorentz considered the electron not to be point-like but to have a finite size. It is very difficult, however, to incorporate a finite sized electron into the framework of relativistic quantum theory.* Many people tried various means to overcome this problem of infinite quantities, but nobody succeeded".

Feynman, in his 1965 Nobel prize speech, described *renormalization* as "simply a way to sweep the difficulties of the divergences of electrodynamics under the rug".

Dirac's final judgment on *quantum field theory*, in his last paper published in 1987 [The inadequacies of quantum field theory.], was that "These rules of *renormalization* give surprisingly, excessively good agreement with experiments. Most physicists say that these working rules are, therefore, correct. I feel that is not an adequate reason. Just because the results happen to be in agreement with observation does not prove that one's theory is correct.""

## Schwinger, J. (November, 1948). Quantum Electrodynamics. I. A Covariant Formulation.

*Phys. Rev.*, 74, 10, 1439-61; https://doi.org/10.1103/PHYSREV.74.1439; also in Underwood, T. G. (2023). *Quantum Electrodynamics*, Volume II, Preface, pp. 410-36.

Received July 29, 1948.

Harvard University, Cambridge, MA.

*Lack of convergence in current formulations of quantum electrodynamics indicates that revision of electrodynamic concepts at ultra-relativistic energies is necessary*, elementary phenomenon in which divergences occur as a result of virtual transitions involving particles with unlimited energy are *polarization of the vacuum* and *self-energy of the electron* which express *the interaction of the electromagnetic and matter fields with their own vacuum fluctuations*, this alters the constants characterizing the properties of the individual fields and their mutual coupling by infinite factors, the question is whether all divergencies can be isolated in such unobservable *renormalization* factors, *this paper is occupied with the formulation of a completely covariant electrodynamics*, manifest covariance with respect to Lorentz and gauge transformations essential in a divergent theory, customary *canonical commutation relations* fail to exhibit the desired covariance since they refer to field variables at equal times and different points of space, *can be put in covariant form by replacing the four-dimensional surface t = const. by a space-like surface*, offers the advantage over the Schrodinger representation in which all operators refer to the same time providing distinct separation between *kinematical* and *dynamical* aspects, formulation that retains evident covariance of the Heisenberg representation but offers something akin to Schrodinger representation can be based on distinction between the properties of *non-interacting fields*, and the effects of *coupling between fields*, constructs a *canonical transformation* that changes the *field equations* in the *Heisenberg representation* into those of *non-interacting fields*, *supplementary condition* restricting the admissible states of the system and the *commutation relations* must be added to the *equations of motion*, describes the coupling between fields in terms of a varying state vector, then simple matter to evaluate commutators of *field* quantities at arbitrary *space-time* points, one thus obtains an obviously covariant and practical form of quantum electrodynamics expressed in a mixed Heisenberg-Schrodinger representation called the *interaction representation*, discusses *covariant* elimination of longitudinal field in which customary distinction between longitudinal and transverse fields is replaced by a suitable *covariant* definition, describes collision processes in terms of an invariant *collision operator* which is the unitary operator that determines the over-all change in state of a system as the result of interaction, notes that a *second paper* treats the problems of electron and photon *self-energy* together with the *polarization of the vacuum* and a *third paper* is concerned with the determination of the *radiative corrections* to the properties of an electron and the comparison with experiment [this was not addressed, it stated that "*radiative corrections to energy levels* will be treated in the next paper of the series" but *this did not appear, nor are there any references to it*].

[Oppenheimer, J. R. (1948). Electron Theory. *Report to the Solvay Conference for Physics at Brussels*, Belgium, September 27 to October 2, 1948, pages 6-7. (Reprint in Schwinger, J. (ed). (1958). *Selected Papers on Quantum electrodynamics*, Dover, New York, pages 150-1: "Now it is true that the fundamental equations of quantum-electrodynamics are gauge and Lorentz covariant. But they have in a strict sense no solutions expansible in powers of e. If one wishes to explore these solutions, bearing in mind that certain infinite terms will, in a later theory, no longer be infinite, one needs a covariant way of identifying these terms, and for that, not merely the field equations themselves, but *the whole method of approximation and solution must at all stages preserve covariance. This means that the familiar Hamiltonian methods, which imply a fixed Lorentz frame t = constant, must be renounced*; neither Lorentz frame nor gauge can be specified until after, in a given order in e, all terms have been identified, and those bearing on the definition of charge and mass recognized and relegated; then of course, in the actual calculation of transition probabilities and the reactive corrections to them, or in the determination of stationary states in fields which can be treated as static, and in the reactive corrections thereto, the introduction of a definite coordinate system and gauge for these no longer singular and completely well-defined terms can lead to no difficulty.

It is probable that, at least to order $e^2$, more than one covariant formalism can be developed. Thus, *Stueckelberg's four-dimensional perturbation theory*[26] would seem to offer a suitable starting point, as also do the related algorithms of Feynman[27].

[26] Stueckelberg, E. C. G. (September, 1934). Relativistisch invariante Störungstheorie des Diracschen Elektrons I. Teil: Streustrahlung und Bremsstrahlung. (Relativistically invariant perturbation theory of Dirac's electron Part I: scattered radiation and Bremsstrahlung.) *Ann. Phys*., 413, 4, 367-89.

[27] Feynman, R. P. (November, 1948). Relativistic Cut-Off for Quantum Electrodynamics. *Phys. Rev*., 74, 1430-8.

But a method originally suggested by Tomonaga[28], and *independently developed and applied by Schwinger*[22], would seem, apart from its practicality, to have the advantage of very great generality and a complete conceptual consistency.

[28] Tomonaga, S. (1943). On a Relativistically Invariant Formulation of the Quantum Theory of Wave Fields. *Bull. I. P. C. R. Riken-iho*, 22, 545 (in Japanese); translation Tomonaga, S. (August, 1946). *Prog. Theor. Phys*. 1, 2, 27-42; Koba,

Z., Tati, T. & Tomonaga, S. (1947a). On a Relativistically Invariant Formulation of the Quantum Theory of Wave Fields. II. Case of Interacting Electromagnetic and Electron Fields. § 1-4. *Prog. Theor. Phys*., 2, 101–116; Koba, Z., Tati, T. & Tomonaga, S. (1947b). On a Relativistically Invariant Formulation of the Quantum Theory of Wave Fields. III. Case of Interacting Electromagnetic and Electron Fields. § 5-7. *Prog. Theor. Phys*., 2, 198–208.

[22] Schwinger, J. (November, 1948). Quantum Electrodynamics. I. A Covariant Formulation. *Phys. Rev.*, 74, 10, 1439-61, and in press.

It has also been shown by Dyson[29] how Feynman's algorithms can be derived from the Tomonaga equations.

[29] Dyson, F. J., *Phys. Rev*., in press."]

---

***Abstract.***

Attempts to avoid the divergence difficulties of quantum electrodynamics by mutilation of the theory have been uniformly unsuccessful. The lack of convergence does indicate that *a revision of electrodynamic concepts at ultra-relativistic energies is indeed necessary*, but no appreciable alteration of the theory for moderate *relativistic* energies can be tolerated. The elementary phenomena in which divergences occur, in consequence of virtual transitions involving particles with unlimited energy, are the *polarization of the vacuum* and the *self-energy of the electron*, effects *which essentially express the interaction of the electromagnetic and matter fields with their own vacuum fluctuations.* The basic result of these *fluctuation interactions* is to alter the constants characterizing the properties of the individual fields, and their mutual coupling, albeit *by infinite factors. The question is naturally posed whether all divergences can be isolated in such unobservable renormalization factors*; more specifically, *we inquire whether quantum electrodynamics can account unambiguously for the recently observed deviations from the Dirac electron theory*, without the introduction of fundamentally new concepts.

*This paper,* the first in a series devoted to the above question, *is occupied with the formulation of a completely covariant electrodynamics*. Manifest covariance with respect to Lorentz and gauge transformations is essential in a divergent theory since the use of a particular reference system or gauge in the course of calculation can result in a loss of covariance in view of the ambiguities that may be the concomitant of infinities.

[*Covariance* is the property of a function of retaining its form when the variables are linearly transformed. An equation is said to be Lorentz *covariant* if it can be written in terms of Lorentz *covariant quantities*. The key property of such equations is that if they hold in one inertial frame, then they hold in any inertial frame; this follows from the result that if all the components of a tensor vanish in one frame, they vanish in every frame.]

It is remarked, in the first section, that the customary *canonical commutation relations*, which fail to exhibit the desired covariance since they refer to field variables at equal times and different points of space, *can be put in covariant form by replacing the four-dimensional surface t = const. by a space-like surface*. The latter is *such that light signals cannot be propagated between any two points on the surface*. In this manner, a formulation of quantum electrodynamics is constructed in the *Heisenberg representation*, which is obviously *covariant* in all its aspects. It is not entirely suitable, however, as a practical means of treating *electrodynamic* questions, since *commutators of field quantities at points separated by a time-like interval can be constructed only by solving the equations of motion*. This situation is to be contrasted with that of the *Schrodinger representation*, in which all operators refer to the same time, thus providing a distinct separation between kinematical and dynamical aspects. *A formulation that retains the evident covariance of the Heisenberg representation, and yet offers something akin to the advantage of the Schrodinger representation can be based on the distinction between the properties of non-interacting fields, and the effects of coupling between fields.*

In the second section, we construct a *canonical transformation* that changes the *field equations* in the *Heisenberg representation* into those of *non-interacting fields*, and therefore *describes the coupling between fields in terms of a varying state vector*. It is then a simple matter to evaluate commutators of *field* quantities at arbitrary *space-time* points. *One thus obtains an obviously covariant and practical form of quantum electrodynamics, expressed in a mixed Heisenberg-Schrodinger representation, which is called the interaction representation.*

The third section is devoted to a discussion of the *covariant* elimination of the longitudinal field, in which the customary distinction between longitudinal and transverse fields is replaced by a suitable *covariant* definition. *The fourth section is concerned with the description of collision processes in terms of an invariant collision operator, which is the unitary operator that determines the over-all change in state of a system as the result of interaction.* It is shown that the *collision operator* is simply related to the *Hermitian reaction operator*, for which a variational principle is constructed.

***Introduction.***

The predictions of quantum electrodynamics concerning higher order perturbation effects have long been discredited in view of the divergent nature of the results. Several attempts[1] have been made to arbitrarily remove supposedly objectionable features of the theory - the so-called "*subtraction physics*".

[1] Dirac, P. A. M. (March, 1934). Discussion of the infinite distribution of electrons in the theory of the positron. *Proc. Camb. Phil. Soc.*, 30, 2, 150-63 [; attempt by Dirac to address problem with his *relativistic* 'hole' theory which implies an infinite number of negative-energy electrons (per unit volume) with energies extending continuously from $-mc^2$ to $-\infty$, when an electromagnetic field is present positive- and negative-energy states cannot be distinguished in *relativistically* invariant way, need to set up assumptions for production of *electromagnetic field* by the electron distribution such that any finite change in distribution produces a change in the field in agreement with Maxwell's equations and such that the infinite field which would be required by Maxwell's equations from an infinite density of electrons is in some way cut out, *assumes each electron has its own individual wave function in space-time* and each electron moves in an *electromagnetic field* which is the same for all electrons part coming from external causes and part from the electron distribution itself, introduces *relativistic density matrix* $\sum_{oc} \psi_{k'}(x' t') \psi^*_{k''}(x'' t'')$ referring to two points in space and two times, separates density distribution into two parts where one contains the singularities, and the other describes the *electric* and *current densities* physically present.]; Heisenberg, W. (March, 1934). Bemerkungen zur Diracschen Theorie des Positrons. (Remarks on the Dirac theory of positron.) *Zeit. Phys.*, 90, 3-4, 209-31 [; https://doi.org/10.1007/ BF01333516[; Heisenberg's reconstruction of Dirac's *theory of the positron* in the formalism of quantum electrodynamics, demands that symmetry between positive and negative charge should be expressed in the basic equations from the outset, in addition to the well-known difficulties with the divergences no new infinities should appear in the formalism, theory should provide an approximation for the treatment of problems that have been treated by quantum electrodynamics up to now, Dirac [(1934). Discussion of the infinite distribution of electrons in the theory of the positron.] showed that a quantum mechanical system of many electrons that fulfill the Pauli principle and move in a given force field without back-reaction can be characterized by a *density matrix* $(x', t', k' \mid R \mid x'', t'', k'') = \sum n\ \psi^*n\ (x', t', k')\ \psi_n\ (x'', t'', k'')$ when $\psi_n(x', t', k')$ means the normalized eigenfunctions of the states that possess one electron, and $x', t', k'$ ($x'', t'', k''$, resp.) are position, time, and spin variables, all physically-important properties of quantum-mechanical systems like *charge density*, *current density*, etc., can be read off from the *density matrix*, the temporal change in the density matrix is determined by the Dirac differential equation, Dirac made different choice of *density matrix* representing external field resulting in a different energy and impulse density, by restricting oneself to an *intuitive analogue theory of matter fields the negative energy levels in the Dirac theory could be avoided by replacing the homogeneous Dirac differential equation with an inhomogeneous equation* where the inhomogeneity is indicative of pair creation, for most practical applications e.g. pair creation, annihilation, Compton scattering, etc. *the theory described here does not yield anything new compared to the formulation of the Dirac theory*, in the Maxwell theory a continuous charge distribution also leads to a finite self-energy; it is the "quantization" that leads to the infinite self-energy, *if one represents the quantization of the electromagnetic field by point-like light quanta then the infinitude of the self-energy also emerges in the intuitive theory of matter waves.*]; Heitler, W. & Peng, H. W. (1942). *Proc. Camb. Phil. Soc.*, 38, 296.

*All such efforts have been fruitless*; either failing in their avowed purpose, or lacking internal consistency[2].

> [2] Serber, R. (April, 1936). A Note on Positron Theory and Proper Energies. *Phys. Rev.*, 49, 545; https://doi.org/10.1103/PhysRev.49.545; Bethe, H. A. & Oppenheimer, J. R. (October, 1946). Reaction of Radiation on Electron Scattering and Heitler's Theory of Radiation Damping. *Phys. Rev.*, 70, 451; https://doi.org/10.1103/PhysRev.70.451.

The unqualified success of quantum electrodynamics in applications involving the lowest order of perturbation theory indicates its essential validity for moderately *relativistic* particle energies. *The objectionable aspects of quantum electrodynamics are encountered in virtual processes involving particles with ultra-relativistic energies.* The two basic phenomena of this type are the *polarization of the vacuum* and the *self-energy of the electron.*

> [*Vacuum polarization*: The possibility of creating *electron-positron* pairs with unlimited energy, through the virtual creation and annihilation of *electron-positron* pairs by an *electromagnetic field*, results in the generation of unlimited *charge* and *current* in the vacuum.]

The phrase "*polarization of the vacuum*" describes the modification of the properties of an *electromagnetic field* produced by its interaction with the charge fluctuations of the vacuum. In the language of perturbation theory, the phenomenon considered is the generation of *charge* and *current* in the vacuum through the virtual creation and annihilation of *electron-positron* pairs by the *electromagnetic field*. If the *electromagnetic field* is that of a light quantum, the *vacuum polarization* effects are equivalent to ascribing a *proper mass* to the *photon*. Previous calculations have yielded non-vanishing, divergent expressions for the light quantum *proper mass*. However, *the latter quantity must be zero in a proper gauge invariant theory*. The failure to obtain this result from a *gauge invariant* formulation can be ascribed only to a faulty application of the theory, rather than to an essential deficiency thereof. When the *electromagnetic field* is that of a given *current* distribution, one obtains a logarithmically divergent contribution to the *vacuum polarization current* which is everywhere proportional to the given distribution. This divergent result expresses the possibility, according to present theory, of creating *electron-positron* pairs with unlimited energy, a situation that presumably will be corrected in a more satisfactory theory. Thus, the physically significant divergence arising from the *vacuum polarization* phenomenon occurs in a factor that alters the strength of all *charges*, a uniform *renormalization* that has no observable consequences other than the conflict with the empirical finiteness of *charge*.

> [*Self-energy of an electron*: In a Lorentz invariant (*relativistic*) theory, the possibility of an *electron* emitting light quanta with unlimited energy results in the

addition of an infinite *electromagnetic proper mass* to the electron's observable mechanical mass and to the infinite *self-energy* of the electron.]

The *interaction* between the *electromagnetic field vacuum fluctuations* and an *electron*, or more exactly, the *electron-positron matter field,* modifies the properties of the *matter field* and produces the *self-energy* of an *electron.* The mechanism here under discussion is commonly described as the *virtual emission and absorption of a light quantum by an otherwise free electron*, although an equally important effect is *the partial suppression, via the exclusion principle, of the coupled vacuum fluctuations of the electromagnetic and matter fields.*

In a Lorentz invariant theory, self-energy effects for a free *electron* can only result in the addition of an *electromagnetic proper mass* to the *electron*'s mechanical *proper mass.* Calculations performed for a stationary *electron*[3] have yielded a logarithmically divergent *electromagnetic proper mass,* a divergence that results from the possibility of emitting light quanta with unlimited energy.

[3] Weisskopf, V. F. (July, 1939). On the Self-energy and Electromagnetic Field of the Electron. *Phys. R*ev., 56, 1, 72[; the main purpose of this paper is to show the physical significance of the logarithmic divergence of the self-energy of the electron and to demonstrate the reason for its occurrence, the self-energy of the electron is its total energy in free space when isolated from other particles or light quanta,

$$W = T + (1/8\pi) \int (H^2 + E^2)\, dr$$

where T is the *kinetic energy* of the electron and H and E are the *magnetic* and *electric field strengths* at point r, identifies three reasons why quantum theory of the electron results in infinite *self-energy* of the electron, claims that *quantum kinematics shows that the radius of the electron must be assumed to be zero*, resulting in infinite energy of the *electrostatic field*, the contributions of the electric and magnetic fields of the spin to the *self-energy* of the electron cancel one another, *quantum theory of the electromagnetic field postulates the existence of field strength fluctuations in empty space*, gives rise to an additional energy which diverges more strongly that the electrostatic self-energy, induces the electron to perform vibrations with energy that diverges quadratically *for an infinitely small radius*, Dirac's positron theory implies that the charge and magnetic dipole of the electron are extended over a finite region, explains why the *self-energy* is only *logarithmically infinite, divergences are consequence of assumption of point electron*].

It is here, as in the *vacuum polarization* problem, that modifications will be introduced in a more satisfactory theory. However, *the electromagnetic proper mass merely produces a renormalization of the electron mass that has no observable consequences, other than the conflict with the empirical finiteness of mass.*

It is evident that these *two phenomena* are quite analogous and *essentially describe the interaction of each field with the vacuum fluctuations of the other field*. The effect of these fluctuation interactions is simply to alter the fundamental constants e and m, although by logarithmically divergent factors. However, it may be argued that a future modification of the theory, *inhibiting the virtual creation of particles that possess energies many orders of magnitude in excess of* $mc^2$, will ascribe a value to these logarithmic factors not vastly different from unity. The *charge* and *mass renormalization factors* will then differ only slightly from unity, as befits a perturbation theory, in consequence of the small *coupling constant* for the *matter* and *electromagnetic* fields,

$$e^2/4\pi hc = 1/137.$$

We may now ask the fundamental question: *Are all the physically significant divergences of the present theory contained in the charge and mass renormalization factors*? Will the consideration of interactions more complicated than these simple vacuum fluctuation effects introduce new divergences; or will all further phenomena involve only moderate *relativistic* energies, and thus be comparatively insensitive to the high energy modifications that are presumably to be introduced in a more satisfactory theory? This series of papers represents an attempt to supply at least a partial answer to the question, which has acquired an immediate importance in view of recent conclusive evidence that the electromagnetic properties of the *electron* are not fully described by the *Dirac wave equation*. Fine structure measurements on hydrogen, deuterium[4], and ionized helium[5] have revealed energy level displacements that imply the existence of a weak, short range repulsive *interaction* between the *electron* and *proton*.

[4] Lamb, Jr., W. E. & Retherford, R. C. (August, 1947). Fine Structure of the Hydrogen Atom by a Microwave Method. *Phys. Rev.*, 72, 3, 241-3[; the spectrum of hydrogen has a fine structure of the energy levels which according to the *Dirac wave equation* for an electron moving in a Coulomb field is due to the combined effects of *relativistic* variation of mass with velocity and spin-orbit coupling, according to this theory the $2^2S_{1/2}$ state should exactly coincide in energy with the $2^2P_{1/2}$ state which is the lower of the two P states, previous attempts at measurement have alternated between finding confirmation and discrepancies of as much as eight percent, using microwave method depending on a novel property of the $2^2S_{1/2}$ level results indicate that contrary to the *Dirac wave equation* the $2^2S_{1/2}$ state is higher than the $2^2P_{1/2}$ by about 1000 Mc/sec].

[5] Mack, J. E. & Austern, N. (November, 1947). Newly Observed Structure in He II λ 4686. *Phys. Rev.*, 72, 972; https://doi.org/10.1103/PhysRev.72.972.

Experiments on the hyperfine structure of hydrogen and deuterium[6], together with electron g value determinations for several states of gallium and sodium[7], *prove that the electron possesses a small additional spin magnetic moment*.

[6] Nafe, J. E., Nelson, E. B. & Rabi, I. I. (June, 1947). The Hyperfine Structure of Atomic Hydrogen and Deuterium. *Phys. Rev.*, 71, 914; https://doi.org/10.1103/PhysRev.71.914; Nagle, D. E., Julian, R. S. & Zacharias, J. R. (November, 1947). The Hyperfine Structure

of Atomic Hydrogen and Deuterium. *Phys. Rev.*, 72, 971; https://doi.org/10.1103/PhysRev.72.971.

[7] Kusch, P. & Foley, H. M. (December, 1947). Precision Measurement of the Ratio of the Atomic 'g Values' in the $^2P_{3/2}$ and $^2P_{1/2}$ States of Gallium. *Phys. Rev.*, 72, 1256; https://doi.org/10.1103/PhysRev.72.1256.2; Foley, H. M. & Kusch, P. (February, 1948). On the Intrinsic Moment of the Electron. *Phys. Rev.*, 73, 412; https://doi.org/10.1103/PhysRev.73.412, also in Schwinger, J. (ed). (1958). *Selected Papers on Quantum electrodynamics*. Dover, New York, page 135.

Immediately upon completion of the Lamb-Retherford experiment, it was generally recognized[8] that the most probable explanation was to be found in higher order *electrodynamic* effects; the radiative corrections to the properties of a bound *electron* other than *mass* and *charge renormalization*.

[8] Discussion at the Shelter Island Conference on the Foundations of Quantum Mechanics, June 1947.

A provisional *non-relativistic* calculation[9] lent support to this view.

[9] Bethe, H. A. (August, 1947). The Electromagnetic Shift of Energy Levels. *Phys. Rev.*, 72, 339-41[; Lamb and Retherford results show that fine structure of second quantum state of hydrogen does not agree with *Dirac wave equation*, Schwinger, Weisskopf, and Oppenheimer suggest might be due to *shift of energy levels by interaction of electron with the radiation field*, this shift comes out infinite in all existing theories and has therefore always been ignored, possible to identify the most strongly (linearly) divergent term in the level shift with an *electromagnetic mass effect* which must exist for a bound as well as for a free electron, already included in the *observed mass* of the electron so should be subtracted, assumes *relativistic cut-off* in quantum energies (frequencies) of included atomic states, then calculation of Lamb shift for hydrogen atom using *non-relativistic* ordinary radiation theory gives shift of the levels due to *radiation interaction* in close agreement with observed value, removes discrepancy with Dirac theory, did not carried out *relativistic* calculations].

However, it required a completely *relativistic* treatment[10] to demonstrate that radiative corrections could account simultaneously for the two apparently unrelated deviations from the Dirac electron theory.

[10] Schwinger, J. (February, 1948). On quantum-electrodynamics and the magnetic moment of the electron. *Phys. Rev.*, 73, 4, 416-7[; *electrodynamics unquestionably requires revision at ultra-relativistic energies,* desirable to isolate those aspects of the current theory that essentially involve high energies and are subject to modification by a more satisfactory theory, this goal has been achieved by transforming the Hamiltonian of current *hole theory* electrodynamics to exhibit explicitly the logarithmically divergent self-energy of a free electron which arises from the virtual emission and absorption of light quanta, the

electromagnetic *self-energy* of a free electron can be ascribed to an *electromagnetic mass* which must be added to the mechanical mass of the electron, new Hamiltonian involves experimental *electron mass* rather than unobservable *mechanical mass*, electron now interacts with radiation field only in presence of external field such that only an accelerated electron can emit or absorb a light quantum, *interaction energy* of electron with external field subject to *finite* radiative correction, *polarization of the vacuum* still produces logarithmically divergent term proportional to *interaction energy* of electron in an external field, such term equivalent to altering value of *electron charge* by constant factor with only final value being identified with experimental charge, interaction between matter and radiation produces *renormalization* of electron charge and mass, all divergences contained in *renormalization* factors, radiative correction for energy of electron in external magnetic field corresponds to *additional magnetic moment associated with electron spin* of magnitude $\delta\mu/\mu = (\frac{1}{2}\pi)e^2/hc = 0.001162$, experimental measurements on hyperfine splitting of ground states of atomic hydrogen and deuterium larger than expected from directly measured nuclear moments, finds additional *electron spin magnetic moment* to account for measured hydrogen and deuterium *hyperfine structures* to be $\delta\mu/\mu = 0.00126$ and $\delta\mu/\mu = 0.00131$ respectively, these discrepancies accounted for by additional spin magnetic moment to the electron of $\delta\mu/\mu = 0.0018 \pm 0.00003$, *values yielded by relativistic calculation of Lamb shift differ only slightly from those conjectured by Bethe on basis of non-relativistic calculation and are in good accord with experiment*].

It is our major task to enlarge on this development.

*In order to isolate the divergent aspects of quantum electrodynamics in a manner that is Lorentz and gauge invariant, it is necessary to employ a formulation of the theory that preserves these covariant features at all stages.* The use of a particular reference system or gauge in the course of calculation can result in a loss of covariance in view of the ambiguities that may arise in a divergent theory. *The first paper is occupied with the development of a suitable covariant formulation.* In the *second paper* we treat the problems of electron and photon self-energy, together with the polarization of the vacuum.

> [This appeared as Schwinger, J. (February, 1949). Quantum Electrodynamics. II. Vacuum Polarization and Self-Energy. *Phys. Rev.*, 75, 4, 651-79; below.]

The *third paper* is concerned with the major topic, the determination of the radiative corrections to the properties of an electron, and the comparison with experiment.

> [The third paper appeared as Schwinger, J. (September, 1949). Quantum Electrodynamics. III. The Electromagnetic Properties of the Electron—Radiative Corrections to Scattering. *Phys. Rev.*, 76, 6, 790–817; below, but this did not address how *induction of a current in a vacuum by an electron results in an alteration in its electromagnetic properties.* It states that this paper is concerned with the computation of the second-order corrections to the current operator and

the application to electron scattering. *Radiative corrections to energy levels* will be treated in the next paper of the series", but this did not appear.]

Scalar and vector matter fields will be discussed in a *fourth paper*. It is hoped that successive papers of this series will deal with such subjects as the corrections to the Klein-Nishina formula, the scattering of light by light, and by a Coulomb field.

**1. *Covariance in the Heisenberg representation.***

In this section, we employ the following notation: Greek subscripts assume values ranging from 1 to 4, and a repeated index is to be so summed. The coordinate vector of a four-dimensional point x is denoted by $x_\mu = (r, ict)$. The real time coordinate $x_0 = (1/i)x_4 = ct$ is also used. In particular, the four-dimensional element of volume is defined as $d\omega = dx_0dx_1dx_2dx_3$. The *four-vector potential* of the *electromagnetic field* is $A_\mu(x) = \{A(r, t), i\phi(r, t))$, while $\psi_\alpha(x)$ designates the four-component *Dirac spinor*. …

… The notation $\gamma_\mu$ is used for the four Hermitian matrices that obey the *anticommutation relations*

$$\gamma_\mu\gamma_\nu + \gamma_\nu\gamma_\mu = 2\delta_{\mu\nu}. \tag{1.1}$$

The *adjoint spinor* $\psi^-_\alpha(x)$ is defined by

$$\psi^-_\alpha(x) = \psi^+(x)\gamma_4 = \gamma_4^T\psi^+(x), \tag{1.2}$$

where $\psi^-_\alpha(x)$ is the *Hermitian conjugate* of $\psi_\alpha(x)$. The so-called charge *conjugate spinor* $\psi'_\alpha(x)$ and its *adjoint* $\psi^{-\prime}_\alpha(x)$ are represented by

$$\psi'(x) = C\psi^-(x), \qquad \psi^{-\prime}(x) = C^{-1}\psi(x). \tag{1.3}$$

Here C is a matrix such that

$$\gamma_\mu^T = \gamma_\mu^* = -C^{-1}\gamma_\mu C, \tag{1.4}$$

which has the property of being skew-symmetric:

$$C^T = -C \tag{1.5}$$

and unitary:

$$C^+C = 1. \tag{1.6}$$

In the latter equation, $C^+ = C^{T*}$ is the *Hermitian conjugate matrix*. For the particular representation in which all elements of $\gamma_4$ are imaginary, while all elements of the other matrices are real, the conditions on C are satisfied with $C = -\gamma_4$. With this choice, $\psi'(x) = \psi^+(x)$; *charge* and *Hermitian conjugation* are equivalent. Finally,

$$\kappa_0 = m_0c/\hbar \tag{1.7}$$

where $m_0$ is the *mechanical proper mass* of the electron.

The *equations of motion* of the coupled *electromagnetic* and *electron-positron matter fields* can be derived from the *variational principle*:

$$\delta \int L \, d\omega = 0, \tag{1.8}$$

where the *Lagrangian density L* is

$$L = \ldots \tag{1.9}$$

and is so constructed that it is *invariant with respect to Lorentz transformations, gauge transformations and charge conjugation*. The proof of *Lorentz invariance* follows the conventional treatment and need not be repeated. *Gauge invariance*, that is, *invariance under the combined transformations*

$$\begin{aligned} A_\mu(x) &\to A_\mu(x) - \partial\Delta(x)/\partial x_\mu \\ \psi(x) &\to \exp[-ie/hc\ \Delta(x)]/\psi(x) \\ \psi'(x) &\to \exp[ie/hc\ \Delta(x)]/\psi'(x) \end{aligned} \tag{1.10}$$

induced by a *scalar function of position*, $\Delta(x)$, would be generally valid were it not for the term in the *Lagrangian density* that refers to the *electromagnetic field* alone. The addition to $L$ arising therefrom is

…

of which the first term has no effect on the *equations of motion*. Hence *gauge invariance* is restricted to the group of generating functions that obey

$$\partial^2\Delta(x)/\partial x_\mu^2 = \Box^2\Delta(x) = 0 \tag{1.11}$$

*Invariance under charge conjugation expresses the complete symmetry between positive and negative charge*. The interchange of $\psi(x)$ and $\psi'(x)$, together with +e and –e, evidently leaves the *Lagrangian density* unaltered.

In order to obtain the *equations of motion* for the *matter field*, it is necessary to express the *Lagrangian density* entirely in terms of $\psi(x)$ and $\psi^*(x)$, or alternatively, $\psi'(x)$ and $\psi^{*\prime}(x)$. By virtue of Eqs. (1.3), (1.4), and (1.5)

[$\psi'(x) = C\psi^-(x)$, $\quad\psi^{-\prime}(x) = C^{-1}\psi(x)$, (1.3)
$\gamma_\mu{}^T = \gamma_\mu{}^* = -C^{-1}\gamma_\mu C$, (1.4)
$C^T = -C$, (1.5)]

the following relations hold

$$\begin{aligned} \psi^{-\prime}\gamma_\mu\psi' &= C^{-1T}\gamma_\mu C\psi^- = \psi\gamma_\mu{}^T\psi^- \\ \psi^{-\prime}\psi' &= \psi C^{-1T}C\psi^- = -\psi\psi^-, \end{aligned} \tag{1.12}$$

…

We find, as the result of variation, apart from discarded divergencies,

$$\delta L = \dots , \tag{1.13}$$

where

$$j_\mu(x) = iec/2\ [\psi^-(x)\gamma_\mu\psi(x) - \psi^{-\prime}(x)\gamma_\mu\psi'(x)] \tag{1.14}$$

represents the four-vector of *charge* and *current*; $j_\mu = (j_\mu, ic\rho)$. It is consistent with the form of the *commutation relations* imposed on the field quantities to infer that

$$\square^2 A_\mu(x) = -\ 1/c\ j_\mu(x), \tag{1.15}$$

and

$$[\gamma_\mu\{\partial/\partial x_\mu - ie/\hbar c\ A_\mu(x)\} + \kappa_0]\ \psi(x) = 0$$
$$[\gamma_\mu^T\{\partial/\partial x_\mu + ie/\hbar c\ A_\mu(x)\} - \kappa_0]\ \psi^-(x) = 0, \tag{1.16}$$

The *Dirac equations* for the *matter field* can also be cast in the *charge conjugate* form

$$[\gamma_\mu\{\partial/\partial x_\mu + ie/\hbar c\ A_\mu(x)\} + \kappa_0]\ \psi'(x) = 0$$
$$[\gamma_\mu^T\{\partial/\partial x_\mu - ie/\hbar c\ A_\mu(x)\} - \kappa_0]\ \psi^{-\prime}(x) = 0. \tag{1.17}$$

*To the equations of motion must be added a supplementary condition, and the commutation relations.* The *supplementary condition*

$$\partial A_\mu(x)/\partial x_\mu\ \Phi = 0 \tag{1.18}$$

restricts the *admissible states* of the system, as characterized by the constant vector $\Phi$ of our *Heisenberg representation*. The compatibility of (1.18) with the *equations of motion* is a consequence of the *charge conservation equation*

$$\partial j_\mu(x)/\partial x_\mu = 0. \tag{1.19}$$

The customary *Maxwell equations*, involving the *field strengths*

$$F_{\mu\nu} = \partial A_\nu(x)/\partial x_\mu - \partial A_\mu(x)/\partial x_\nu, \tag{1.20}$$

rather than the potentials, appear as derived supplementary conditions:

$$[\partial F_{\mu\nu}(x)/\partial x_\mu + 1/c\ j_\nu(x)]\Phi = 0. \tag{1.21}$$

The *commutation relations*, in their conventional canonical form, read

$$[A_\mu(r,t),\ 1/c\ \partial/\partial t\ A_\nu(r',t)] = ihc\delta_{\mu\nu}\delta(r - r') \tag{1.22a}$$
$$\{\psi_\alpha(r,t),\ (\psi^*(r',t)\gamma_4)_\beta\} = \delta_{\alpha\beta}\delta(r - r') \tag{1.22b}$$

where the bracket symbols signify the *commutator* and *anticommutator*, respectively:

$$[A,B] = AB - BA, \qquad \{A,B\} = AB + BA. \tag{1.23}$$

…

*It should be noted that the particle field commutation relations are invariant with regard to charge conjugation*. Thus,

… (1.24)

A further remark concerns the consistency of the *supplementary condition* and the *commutation relations*. Since (1.18) contains the arbitrary point x, one will obtain additional supplementary conditions by commutation, unless

$$[\partial A_\mu(x)/\partial x_\mu, \partial A_\nu(x')/\partial x_\nu'] = 0 \quad (1.25)$$

for arbitrary x and x'. In actuality, the *canonical commutation relations* are such as to yield (1.25). It must be realized that the *commutator*, considered as a function of x, obeys the *wave equation*, whence the validity of (1.25) is assured provided the *commutator* and its time derivative vanish for t = t'. This is easily verified.

The physical quantities characterizing the distribution of *energy* and *momentum* in the field are combined in the *canonical energy-momentum tensor*

$$T_{\mu\nu} = \ldots \quad (1.26)$$

…

which satisfies the *conservation equation*

$$\partial/\partial x_\mu \, T_{\mu\nu} = 0 \quad (1.27)$$

…

The *canonical tensor* can be replaced by a symmetrical *energy-momentum tensor*

$$\Theta_{\mu\nu} = \ldots \quad (1.29)$$

However, it is only the expectation value of $\Theta_{\mu\nu}$,

$$< \Theta_{\mu\nu}> = < \Phi, \Theta_{\mu\nu}\Phi>, \quad (1.30)$$

that satisfies the conservation equation

$$\partial/\partial x_\mu < \Theta_{\mu\nu}> = 0 \quad (1.31)$$

…

The symmetrical energy-momentum tensor is evidently invariant with respect to *gauge transformations* and *charge conjugation*. …

…

The spatial volume integrals

$$P_\mu = -\, i/c \int T_{4\mu} \, d\upsilon \quad (1.35)$$

form a time independent four-vector that unites the *momentum* and *energy* integrals of the *equations of motion*; $P_\mu = (P, iW/c)$. …

…

The operators $P_\mu$ form the *infinitesimal generators of the coordinate translation group. …*

More generally, if F(x) is an arbitrary function of the field variables at the point x, but does not explicitly involve position coordinates,

$$i/\hbar\ [F(x)_\nu P_\nu] = \partial F(x)/\partial x_\nu, \qquad (1.38)$$

One can exploit this aspect of the operators $P_\mu$ to prove anew that they constitute constants of the motion, and to demonstrate that the *canonical commutation relations* are consistent with the *equations of motion.* In a similar way, one can introduce other operator constants of the motion which compose the *angular momentum* tensor. *These quantities form the infinitesimal generators of the Lorentz group, and with their aid the covariance of the canonical quantization scheme can be demonstrated.* However, it is at this point that we must deviate from the conventional development that here has so briefly been outlined.

*The equations of motion and the supplementary condition are manifestly covariant; the canonical commutation relations lack this essential characteristic since a special Lorentz reference system is employed.* The *commutation relations* involve field variables at two points of a four-dimensional surface characterized by t = const. *We shall achieve the desired covariance by replacing such surfaces with the invariant concept of a space-like surface.* The latter is such that light signals cannot be propagated between any two points on the surface. In terms of the position vectors of two points, $x_\mu$ and $x_\mu'$, it is required that

$$(x_\mu - x_\mu')^2 = (r - r')^2 - c^2(t - t')^2 > 0, \qquad (1.39)$$

which clearly involves no special reference system. Surfaces of the type t = const. form a special non-covariant class of plane *space-like surfaces*. The customary *commutation relations* are essentially an expression of the kinematical independence of field quantities at different points of space for a given time. *It is evident that the proper covariant description of this general property should involve field quantities at two space-time points that cannot be connected by light signals, that is, two points on a space-like surface. Accordingly, we endeavor thus to generalize the commutation relations into a manifestly covariant form.*

The simplest basis for a generalization of (1.22a)

$$[[A_\mu(r,t),\ 1/c\ \partial/\partial t\ A_\nu(r',t)] = i\hbar c\delta_{\mu\nu}\delta(r - r') \qquad (1.22a)]$$

is provided by the two statements that express the properties of $\delta(r - r')$:

$$[A_\mu(r,t),\ 1/c\ \partial/\partial t\ A_\nu(r',t)] = 0, \qquad r \neq r' \qquad (1.40a)$$

$$\int [A_\mu(r,t),\ 1/c\ \partial/\partial t\ A_\nu(r',t)]\ dv' = i\hbar c\delta_{\mu\nu} \qquad (1.40b)$$

in which the spatial volume integration is extended over an arbitrary region that includes the point r. The *proper generalization of (1.40a)*, together with the other vanishing *electromagnetic field* commutators, is simply

$$[A_\mu(x), A_\nu(x')] = 0, \qquad (x_\mu - x_\mu')^2 > 0; \qquad (1.41)$$

that is, *the field quantities associated with two distinct points on a space-like surface commute*. In order to generalize (1.40b), it will prove convenient to define a four-vector differential surface area:

$$d\sigma_\mu = (dx_2dx_3dx_0,\ dx_1dx_3dx_0,\ dx_1dx_2dx_0,\ dx_1dx_2dx_3/i). \qquad (1.42)$$

Considered as defining the direction of the normal to a *space-like surface*, $d\sigma_\mu$ must be a *time-like vector*, that is, $d\sigma_\mu^2 < 0$. It should be noted that our definitions of surface area and volume are such that the volume generated by the displacement $\delta x_\mu$ imparted to the surface area $d\sigma_\mu$ is $\delta\omega = d\sigma_\mu \delta x_\mu$. It is evident from the notation $dv' = id\sigma_4'$, $\partial A_\nu(r',t)/\partial ct = i\partial A_\nu(x')/\partial x_4'$, that the *proper covariant generalization of (1.40b) is*

$$\int_\sigma [A_\mu(x), \partial/\partial x_\lambda'\ A_\nu(x')]\ d\sigma_\lambda' = hc/i\ \delta_{\mu\nu} \qquad (1.43)$$

in which the x' integration is extended over an arbitrary portion of a *space-like surface* σ that includes the point x.

In order to demonstrate the self-consistency of these and further *covariant commutation relations*, we must show that the values attributed to such surface integrals are unaltered as the *space-like surface* σ passing through the point x is varied; and, for a fixed surface relative to the point x, that the *commutation relations* are compatible with an arbitrary displacement of x. The latter requirement involves a detailed consideration of the *equations of motion* and will be discussed at an appropriate place. The verification of the first requirement is facilitated by introducing the notion of the *functional derivative*. The quantity occurring on the left side of (1.43)

$$[\int_\sigma [A_\mu(x), \partial/\partial x_\lambda'\ A_\nu(x')]\ d\sigma_\lambda' = hc/i\ \delta_{\mu\nu} \qquad (1.43)]$$

involves the field variables at all points of the surface σ and is thus a functional of the *space-like surface* σ, say F[σ]. We may compare this with the functional of a neighboring *space-like surface* σ', F[σ'], which surface is such that it deviates from σ only in a neighborhood of the point x. If the volume enclosed between the and surfaces, δω, is allowed to approach zero, we obtain a definition of the functional derivative of F[σ] at the point x:

$$\delta F[\sigma]/\delta\sigma(x) = \mathrm{Lim}_{\delta\omega\to 0}\ \{F[\sigma'] - F[\sigma]\}/\delta\omega, \qquad (1.44)$$

in which the notation emphasizes that we are considering the variation in F produced by a deformation of the surface σ at the point x. …

…

… It can be shown, as before, that the commutation relations are also valid for the charge conjugate matter fields. …

…

An obviously covariant definition of the energy-momentum four-vector, replacing Eq. (1.35)

$$[P_\mu = -\, i/c \int T_{4\mu}\, d\upsilon \qquad (1.35)]$$

is

$$P_\mu = -\, i/c \int_\sigma d\sigma_\lambda T_{\lambda\mu}(x), \qquad (1.55)$$

in which the integration is extended over an entire space-like surface. The conservation laws now have their covariant expression in the statement that $P_\mu$ is independent of the surface σ. Thus

$$\delta/\delta\sigma(x)\, P_\mu c = \delta/\delta x_\lambda\, T_{\lambda\mu}(x) = 0. \qquad (1.56)$$

The conservation law for the total charge

$$Q = 1/c \int_\sigma d\sigma_\mu j_\mu(x)$$

has an analogous expression:

$$\delta/\delta\sigma(x)\, Qc = \delta/\delta x_\mu\, j_\mu(x) = 0. \qquad (1.57)$$

…

It can now be shown that the covariant *commutation relations* are consistent with the *equations of motion*. We examine the change in the *commutator* or *anticommutator* of two field variables associated with two points on a *space-like surface*, produced by a rigid displacement of the surface. In other words, we seek to evaluate

$$\delta/\delta x_\nu\, [F(x),G(x-\xi)] \text{ or } \delta/\delta x_\nu\, \{F(x), G(x-\xi)\},$$

where $\xi_\mu$ is a *space-like vector* and F, G are any two field variables. It is a consequence of elementary identities that if F and G obey the *equations of motion* (1.38)

$$[i/h\, [F(x)_\nu P_\nu] = \partial F(x)/\partial x_\nu, \qquad (1.38)]$$

so also do the brackets

$$[F(x),G(x-\xi)] \text{ and } \{F(x), G(x-\xi)\}.$$

Therefore, the specification of such brackets as ξ-dependent multiples of the unit operator is self-consistent, since both the derivative with respect to $x_\nu$, and the commutator with $P_\nu$, vanish.

*The formulation of quantum mechanics that has now been developed is obviously covariant in all its aspects. However, it is not entirely suitable as a practical means of treating electrodynamic questions*. In the course of application, it is often necessary to evaluate commutators of held quantities at points separated by a time-like interval. Such commutators are to be constructed by solving the *equations of motion* subject to boundary conditions on a *space-like surface*. *This jumbling of the kinematical and dynamical aspects of the situation is a detriment in the systematic discussion of electrodynamic problems*. At the opposite extreme is the Schrodinger picture, in which all operators are time independent, and the time development of the system is represented by a varying *state vector*; a procedure that is non-covariant in its aspect. *We now seek a formulation that enables us to retain the evident covariance of the Heisenberg representation, and yet offers something akin to the advantage of the Schrodinger representation, a distinct separation between kinematical and dynamical aspects*. The desired separation is to be found in that between the *elementary properties of non-interacting fields*, and the *modification of these properties by the coupling between fields*. For *non-interacting fields*, it is a simple matter to carry out the program previously mentioned, and construct commutation relations for field quantities at arbitrary *space-time* points. In order to exploit this advantage, it is necessary to find a canonical transformation that changes the equations of motion for field quantities in the Heisenberg representation into those of non-interacting fields, and therefore describes the coupling between fields in terms of a varying state vector. We shall perform this transformation in the next section, and thus obtain an obviously covariant and practical form of quantum electrodynamics, expressed in a *mixed Heisenberg-Schrodinger representation*, which may be called the *interaction representation*[11].

[11] The *interaction representation* can be regarded as a field generalization of the *many-time formalism*, from which point of view it has already been considered by Tomonaga. [Tomonaga, S. (1943). On a Relativistically Invariant Formulation of the Quantum Theory of Wave Fields. *Bull. I. P. C. R. Riken-iho*, 22, 545 (in Japanese); translation Tomonaga, S. (August, 1946). *Prog. Theor. Phys.* 1, 2, 27-42[; *existing formalism of quantum field theory not perfectly relativistic*, commutation relations refer to points in space at different times, Schrodinger equation for the vector representing the state of the system is a function of time, time variable plays a different role than space variables, *probability amplitude* not *relativistically invariant* in the space-time world, follows Dirac (1932) [Relativistic Quantum Mechanics] in *generalizing the notion of probability amplitude as far as is required by the theory of special relativity*, substitutes four-dimensional form of the commutation relations, *generalizes the Schrodinger equation* following the Dirac (1933) [The Lagrangian in Quantum Mechanics] *many-time* formalism, then introduces his *super many-time theory* $[\{H_{12}(P) + h/i\ \partial/\partial C_P\}\ \Psi[C] = 0$ at point P on surface C with infinitely many time variables representing the local time for each position in the space, results in *relativistic interaction representation*, three-dimensional manifold (space-like "surface") in four-dimensional space-time world, not necessary to also assume time-like surfaces for

the variable surface as was required by Dirac, previous formalism built up in way too analogous to ordinary *non-relativistic* mechanics, theory was divided into one section giving the kinematical relations between various quantities at the same instant of time and another section determining the causal relations between quantities at different instants of time with the *commutation relations* belong to the first section and the *Schrodinger equation* to the second, this way of separating the theory into two sections very *unrelativistic*, "same instant of time" plays a distinct role, *new formalism consists of one section giving laws of behavior of the fields when they are left alone and the other giving the laws determining the deviation from this behavior due to interactions*, can be carried out *relativistically*, although theory in more satisfactory form no new contents added, *divergence difficulties inherited*, fundamental equations admit only catastrophic solutions *due to non-vanishing zero-point amplitudes of the fields which inheres in the operator $H_{12}(P)$, a more profound modification of the theory is required in order to remove this fundamental difficulty*].

*Relativistic* quantum theories have also been discussed recently by Dirac [Dirac, P. A. M. (May, 1948). Quantum Theory of Localizable Dynamical Systems. *Phys. Rev*., 73, 9, 1092-103; a dynamical system is called *localizable* if its *wave functions* can be expressed in terms of variables, each referring to physical conditions at only one point in *space-time*. These variables may be at points on any three-dimensional space-like surface in space-time. A general investigation is made of how the *wave function* varies when the surface is varied in any way. The variation of the *wave function* is given by equations of the Schrödinger type involving certain operators which play the role of Hamiltonians. The *commutation relations* for these operators are obtained. The theory works entirely with *relativistic* concepts and it provides the general pattern which any *relativistic* quantum theory must conform to, provided the dynamical system is *localizable*].

**2. *The interaction representation.***

To alter the *equations of motion* in the above outlined manner, we introduce a *unitary operator* $U[\sigma]$, defined for a *space-like* surface $\sigma$, and construct the *state vector* of the *interaction representation*

$$\Psi[\sigma] = U[\sigma]\Phi, \tag{2.1}$$

which depends upon the surface $\sigma$, in contrast with the constant vector $\Phi$ of the *Heisenberg representation*. The *expectation value* of some *field variable* $\mathbf{F}(x)$ becomes (in this section, the operators of the *Heisenberg representation* will be denoted by bold face letters)

$$(\Phi,\mathbf{F}(x)\Phi) = (\Psi[\sigma],U[\sigma]\mathbf{F}(x)U^{-1}[\sigma]\Psi[\sigma]) = (\Psi[\sigma],F(x)\Psi[\sigma]) \tag{2.2}$$

*which defines the operators of the interaction representation in terms of those of the Heisenberg representation*:

$$F(x) = U[\sigma]\mathbf{F}(x)U^{-1}[\sigma]. \tag{2.3}$$

It is understood that σ is a *space-like* surface passing through the point x. In order, however, that F(x) depend only on the point x and not on the particular surface σ, *the form of U[σ] must be restricted*, as indicated by the following requirement:

$$\delta/\delta\sigma(x')F(x) = \ldots = \ldots = 0$$
$$(x_\mu - x_\mu')^2 > 0. \qquad (2.4)$$

This will be satisfied if

$$\delta U[\sigma]/\delta\sigma(x')\, U^{-1}[\sigma]$$

*is an invariant function of the field operators* at the point x', since the commutation properties on the surface σ are unaffected by the *unitary transformation*. If, further, the unitary character of U[σ] is to be preserved by its *equation of motion*, it is necessary that

$$i\, \delta U[\sigma]/\delta\sigma(x)\, U^{-1}[\sigma]$$

be a Hermitian operator. Therefore, on writing

$$ihc\, \delta U[\sigma]/\delta\sigma(x) = H(x)U[\sigma], \qquad (2.5)$$

we obtain a *covariant equation of motion* for U[σ], in which H(x) is a Hermitian operator, *an invariant function of the field quantities* at the point x, and has the dimensions of an *energy density*. The *equation of motion* for Ψ[σ] is, correspondingly,

$$ihc\, \delta\Psi[\sigma]/\delta\sigma(x) = H(x)\Psi[\sigma], \qquad (2.6)$$

*We have obtained the conditions that must be satisfied by any canonical transformation.*

*It will now be shown that the special transformation desired is attained with H(x) chosen as the negative of the coupling term in the Lagrangian density*, that is,

$$H(x) = -\,(1/c)j_\mu(x)A_\mu(x). \qquad (2.7)$$

*To construct the equations of motion in the interaction representation*, we first note that the gradient of any field quantity can be exhibited as a *functional derivative*, through an obvious generalization of *Gauss' theorem*:

$$\partial F[x]/\partial x_\nu = \delta/\delta\sigma(x) \int_\sigma F(x')\, d\sigma_\nu'$$
$$= \delta/\delta\sigma(x)\, U[\sigma] \int_\sigma \mathbf{F}(x')\, d\sigma_\nu'\, U^{-1}[\sigma]. \qquad (2.8)$$

The *functional derivative* in the latter form affects both the surface of integration and the operator U[σ], whence

$$\partial F[x]/\partial x_\nu = U[\sigma]\, \partial\mathbf{F}[x]/\partial x_\nu\, U^{-1}[\sigma]$$
$$+ \int_\sigma [\delta U[\sigma]/\delta\sigma(x)\, U^{-1}[\sigma], F(x')]\, d\sigma_\nu'$$

$$= U[\sigma]\ \partial \mathbf{F}[x]/\partial x_\nu\ U^{-1}[\sigma] - i/\hbar c \int_\sigma [H(x), F(x')]\ d\sigma_\nu'. \qquad (2.9)$$

If we first place $F(x) = A_\mu(x)$, it is immediately found from the covariant *commutation relations* on the *space-like* surface σ, that

$$\partial A_\mu(x)/\partial x_\nu = U[\sigma]\ \partial A_\mu(x)/\partial x_\nu\ U^{-1}[\sigma], \qquad (2.10)$$

which, indeed, is necessary, in order that the electromagnetic field commutation relations retain their form under this canonical transformation. However, with $F(x) = \delta A_\mu(x)/\delta x_\nu$, one obtains

$$\square^2 A_\mu(x) = \ldots = \ldots = 0; \qquad (2.11)$$

*the equations of motion for the electromagnetic field in the interaction representation are those of an isolated field.* In addition, the *supplementary condition* is unchanged in form:

$$\partial A_\mu(x)/\partial x_\nu\ \Psi[\sigma] = 0 \qquad (2.12)$$

provided the point x lies on the surface σ. Finally, if $F(x) = \gamma_\nu\psi(x)$,

$$(\gamma_\nu\ \partial/\partial x_\nu + \kappa_0)\psi(x) = \ldots . \qquad (2.13)$$

But, according to (1.12) and (1.14)

$$[\psi^{-\prime}\gamma_\mu\psi' = C^{-1T}\gamma_\mu C\psi^- = \psi\gamma_\mu{}^T\psi^-$$

$$\psi^{-\prime}\psi' = \psi C^{-1T}C\psi^- = -\psi\psi^-, \qquad (1.12)$$

$$j_\mu(x) = iec/2\ [\psi^-(x)\gamma_\mu\psi(x) - \psi^{-\prime}(x)\gamma_\mu\psi'(x)]] \qquad (1.14)$$

$$_\mu(x) = iec/2\ [\psi^*(x)\gamma_\mu\psi(x) - \psi(x)\ \psi^*(x)\gamma_\mu], \qquad (2.14)$$

and

$$[j_\mu(x), \gamma_\nu\psi(x')] = -\ iec\ \{\gamma_\nu\psi(x'),\psi^*{}_\alpha(x)\}(\gamma_\mu\psi(x')_\alpha, \qquad (2.15)$$

so that

$$(\gamma_\nu\ \partial/\partial x_\nu + \kappa_0)\psi(x) = \ldots = 0; \qquad (2.16)$$

the *equations of motion* for the *matter field* in the *interaction representation* are those of an isolated field. *This completely proves the correctness of the choice (2.7)*

$$[H(x) = -\ (1/c)j_\mu(x)A_\mu(x). \qquad (2.7)]$$

*for H(x).*

*We may now proceed to construct the general commutation laws for the field quantities in the new representation, by employing their elementary equations of motion.* This process will be facilitated by introducing two *invariant functions of position*, D(x) and Δ(x), which are associated with the *electromagnetic* and *matter fields*, respectively, and have the following covariant definitions:

$$\square^2 D(x) = 0; \qquad D(x) = 0, \qquad x_\mu^2 > 0$$
$$\int \partial D(x)/\partial x_\mu \, d\sigma_\mu = 1, \tag{2.17}$$

$$\{\square^2 D(x) - \kappa_0^2\}\Delta(x) = 0; \qquad \Delta(x) = 0, \qquad x_\mu^2 > 0$$
$$\int \partial \Delta(x)/\partial x_\mu \, d\sigma_\mu = 1, \tag{2.18}$$

In these definitions, the surface integrations are to be extended over a *space-like* surface that includes the origin. It is easily verified that the constant value attributed to the surface integrals for arbitrary $\sigma$ is consistent with the other equations. *The detailed construction of these and related functions will be postponed to the second paper of this series*; the properties contained in the equations of definition will suffice for our present purposes. It is easily deduced for example, that D and $\Delta$ are odd functions of the coordinates:

$$D(-x) = -D(x), \qquad \Delta(-x) = -\Delta(x). \tag{2.19}$$

We note that

$$\ldots, \tag{2.20}$$

which implies that the surface integral is independent of the particular surface $\sigma$. By choosing $\sigma$ to be, successively, a *space-like* surface through the points x' and x", it is inferred that

$$\Delta(x'' - x') = -\Delta(x' - x'') \tag{2.21}$$

which proves the second statement of (2.19). The proof for D(x) is identical.

*The importance of these invariant functions stems from their utility in expressing the solutions of the equations of motion in terms of boundary values prescribed on some space-like surface*. The *electromagnetic potentials* are uniquely determined if $A_\mu(x)$ and its normal derivative are specified on a surface $\sigma$. The explicit realization of this relation is provided by

$$A_\mu(x) = \int_\sigma [D(x - x')\, \partial/\partial x_\nu'\, A_\mu(x') - A_\mu(x')\, \partial/\partial x_\nu'\, D(x - x')]\, d\sigma_\nu'. \tag{2.22}$$

To verify this statement, it is sufficient to observe that, analogously to Eq. (2.20), the right side of (2.22) is independent of $\sigma$, which can be specially chosen as a *space-like* surface through the point x, yielding

$$\int_\sigma A_\mu(x')\, \partial/\partial x_\nu'\, D(x - x')]\, d\sigma_\nu' = A_\mu(x)$$

as the value of the surface integral. The corresponding solution of the boundary value problem for the first order *Dirac equation* is provided by

$$\psi(x) = \int_\sigma S(x - x')\gamma_\mu \psi(x')\, d\sigma_\mu', \tag{2.23}$$

where

$$S(x) = (\gamma_\nu\, \partial/\partial x_\nu - \kappa_0)\Delta(x)$$

Following the general pattern, we remark that the right side of (2.23) is independent of σ:

$$\delta/\delta\sigma(x') \int_\sigma S(x - x')\gamma_\mu\psi(x')\, d\sigma_\mu' = \ldots = \ldots = 0, \qquad (2.25)$$

since

$$(\gamma_\mu\, \partial/\partial x_\mu + \kappa_0)S(x) = \partial/\partial x_\nu\, S(x)\gamma_\mu + \kappa_0 S(x) = \{\Box^2 - \kappa_0^2\}\Delta(x) = 0; \quad (2.26)$$

and that an evaluation with a surface through the point x, gives

$$\int_\sigma \gamma_\nu\gamma_\mu\, \partial/\partial x_\nu\, \Delta(x - x')\psi(x')\, d\sigma_\mu' = \ldots = \psi(x)$$

with the aid of the lemma (1.58). The *adjoint equation*

$$\psi^-(x) = \int_\sigma d\sigma_\mu'\psi^-(x')\gamma_\mu S(x' - x) \qquad (2.27)$$

can be proved directly, or inferred from (2.23).

*The construction of the general commutation relations is now trivial.* To evaluate $[A_\mu(x), A_\nu(x')]$, for example, it is merely necessary to express $A_\mu(x)$in terms of the field variables on a *space-like* surface that includes the point x', and employ the *commutation relations* for such surfaces. Thus

$$[A_\mu(x), A_\nu(x')] = -\int_\sigma D(x - x'') \times [A_\nu(x'), \partial/\partial x_\lambda''\, A_\mu(x'')]\, d\sigma_\lambda'',$$

whence

$$[A_\mu(x), A_\nu(x')] = i\hbar c\delta_{\mu\nu}D(x - x'). \qquad (2.28)$$

In a similar way

$$\{\psi_\alpha(x),\psi^-_\beta(x')\} = \int_{\alpha\gamma} S(x - x'') \times [\{\gamma_\mu\psi(x'')\}_\gamma,\psi^-_\beta(x')]\, d\sigma_\mu''$$

so that

$$\{\psi_\alpha(x),\psi^-_\beta(x')\} = 1/i\, S_{\alpha\beta}(x - x') = 1/i\, (\gamma_\mu\, \partial/\partial x_\mu - \kappa_0)_{\alpha\beta}\Delta(x - x'). \qquad (2.29)$$

All other *matter field* anti-commutators vanish. Of course, the *matter field commutation relations* are invariant with respect to *charge conjugation*.

Finally, we turn to the generalization of the *supplementary condition* (2.12)

$$[\partial A_\mu(x)/\partial x_\nu\, \Psi[\sigma] = 0, \qquad (2.12)]$$

*which consists of removing the restriction that x be situated on the surface σ.* It follows from (2.22)

$$[A_\mu(x) = \int_\sigma [D(x - x')\, \partial/\partial x_\nu'\, A_\mu(x') - A_\mu(x')\, \partial/\partial x_\nu'\, D(x - x')]\, d\sigma_\nu'. \qquad (2.22)]$$

that, for an arbitrary point x,

$$\partial A_\mu(x)/\partial x_\mu\, \Psi(\sigma) = \int_\sigma D(x - x'') \times \partial/\partial x_\nu''\{\partial A_\mu(x'')/\partial x_\mu''\}d\sigma_\nu\, \Psi(\sigma). \qquad (2.30)$$

However, according to (2.9)

$$[\partial F[x]/\partial x_\nu = U[\sigma]\, \partial \mathbf{F}[x]/\partial x_\nu\, U^{-1}[\sigma] + \int_\sigma [\delta U[\sigma]/\delta\sigma(x)\, U^{-1}[\sigma], F(x')]\, d\sigma_\nu' = U[\sigma]\, \partial \mathbf{F}[x]/\partial x_\nu\, U^{-1}[\sigma] - i/\hbar c \int_\sigma [H(x), F(x')]\, d\sigma_\nu', \qquad (2.9)]$$

with $F = \partial A_\mu(x'')/\partial x_\mu''$,

$$\partial/\partial x_\nu''\{\partial A_\mu(x'')/\partial x_\mu''\}\Psi(\sigma) = \ldots = \ldots = \ldots . \qquad (2.31)$$

In the last transformation, we have used the lemma (1.58) and the fact that the *electromagnetic field commutators* contain only the difference in coordinates of the two points involved. On introducing (2.31) into (2.30) and performing the x" integration, we find without further difficulty that

$$[\partial A_\mu(x)/\partial x_\mu - \int_\sigma D(x - x')\, 1/c\, j_\mu(x')\, d\sigma_\mu']\, \Psi(\sigma) = 0, \qquad (2.32)$$

*which is the supplementary condition for the interaction representation.* Although the consistency of the *supplementary condition* is guaranteed by the corresponding property in the *Heisenberg representation*, it is well to verify it directly. However, since the proof involves the commutation properties of the current four-vector, we digress briefly to derive the necessary theorems.

It is easy to deduce from the expression (2.14) for $j_\mu(x)$, and the anti-commutator (2.29), that

$$\ldots . \qquad (2.33)$$

Of course, all components of $j_\mu$ commute at two distinct points on a *space-like* surface. However, the important statement is that a time-like component of $j_\mu$ commutes with all components of the *current* at the same point. We prove this by demonstrating that

$$\int_\sigma [j_\mu(x), j_\nu(x')]\, d\sigma_\nu' = 1/c\, [j_\mu(x), Q] = 0 \qquad (2.34)$$

where σ is any *space-like* surface, which, in particular, can include the point x. The validity of this statement follows immediately from (2.23) and (2.27)

$$[\psi(x) = \int_\sigma S(x - x')\gamma_\mu\psi(x')\, d\sigma_\mu', \qquad (2.23)$$

$$\psi^-(x) = \int_\sigma d\sigma_\mu'\psi^-(x')\gamma_\mu S(x' - x), \qquad (2.27)]$$

since

$$\int_\sigma [j_\mu(x), j_\nu(x')]\, d\sigma_\nu' = \ldots = 0. \qquad (2.35)$$

Indeed, *Eq. (2.34) is an expression of charge conservation*, for, according to (2.9)

$$[\partial F[x]/\partial x_\nu = U[\sigma]\, \partial \mathbf{F}[x]/\partial x_\nu\, U^{-1}[\sigma] + \int_\sigma [\delta U[\sigma]/\delta\sigma(x)\, U^{-1}[\sigma], F(x')]\, d\sigma_\nu' = U[\sigma]\, \partial \mathbf{F}[x]/\partial x_\nu\, U^{-1}[\sigma] - i/\hbar c \int_\sigma [H(x), F(x')]\, d\sigma_\nu', \qquad (2.9)]$$

with $F = j_\nu(x)$:

$$\partial j_\nu(x)/\partial x_\nu = i/\hbar c\ 1/c \int_\sigma [j_\mu(x), j_\nu(x')]\, d\sigma_\nu'\, A_\mu(x)$$
$$= 1/\hbar c\ [j_\mu(x),Q]A_\mu(x). \qquad (2.36)$$

*To prove the suitability of (2.32)*

$$[[\partial A_\mu(x)/\partial x_\mu - \int_\sigma D(x - x')\ 1/c\ j_\mu(x')\, d\sigma_\mu'] \Psi(\sigma) = 0, \qquad (2.32)]$$

*as a supplementary condition, we must show that it is consistent with the field equations of motion, the equation of motion for Ψ(σ), and the commutation relations*. In terms of the operator

$$\Omega[x,\sigma] = \partial A_\mu(x)/\partial x_\mu - \int_\sigma D(x - x')\ 1/c\ j_\mu(x')\, d\sigma_\mu' \qquad (2.37)$$

we must verify that

… , (2.38a)

… , (2.38b)

… , (2.38c)

The first statement is trivial. As to (2.38b), note that

…

while

…

in view of the property of $j_\mu(x)$ just established. Finally, the same property implies that

$$\Omega[x,\sigma],\Omega[x',\sigma] = \ \dots = - i\hbar c\Box^2 D(x - x') = 0.$$

*Gauge invariance has a different aspect in the new representation from that of the Heisenberg representation, since the matter field equations do not involve the electromagnetic field*. On introducing a change in *gauge*

$$A_\mu(x) \rightarrow A_\mu(x) - \partial\Delta(x)/\partial x_\mu,$$

where $\Delta(x)$ is a scalar function of position such that

$$\Box^2\Delta(x) = 0,$$

the *supplementary condition*, *commutation relations* and *field equations of motion* are unaffected, but the *equation of motion* for $\Psi[\sigma]$ becomes

$$i\hbar c\ \delta\Psi[\sigma]/\delta\sigma(x) = [H(x) + \partial/\partial x_\mu\{1/c\ j_\mu(x)\Delta(x)\}]\ \Psi[\sigma], \qquad (2.39)$$

in which the *charge conservation equation* has been used. We shall show that it is possible to restore this equation to its original form, and thus prove *gauge invariance*, by a *canonical transformation* on $\Psi[\sigma]$. Indeed, the proper transformation is

$$\Psi[\sigma] \rightarrow e^{-iG[\sigma]}\Psi[\sigma] \qquad (2.40)$$

where

$$G[\sigma] = 1/\hbar c \int_\sigma 1/c\, j_\mu(x)\Delta(x)\, d\sigma_\mu. \qquad (2.41)$$

The *equation of motion* for the new *state vector* is

$$\ldots, \qquad (2.42)$$

as a consequence of the commutation properties of j,, on a *space-like* surface. We may now employ the simple expansion theorem

$$\ldots \qquad (2\ 43)$$

to deduce that

$$i\hbar c i G\, \delta e^{-iG}/\delta\sigma(x) = \ldots = \partial/\partial x_\mu\{1/c\, j_\mu(x)\Delta(x)\},$$

in which *the commutability of $j_\mu$ with a time-like component of $j_\mu$ on the surface $\sigma$ ensures that only the first term of the series survives*. We have thereby demonstrated the correctness of the transformation (2.40).

The form of the *energy-momentum* quantities, as well as their significance as displacement operators, *is altered by the canonical transformation that generates the interaction representation*. In the *Heisenberg representation*, the *functional derivative* of an operator is of immediate significance in computing the *functional derivative* of the expectation value of that operator:

$$\delta/\delta\sigma(x)(\Phi, F[\sigma]\Phi) = \{\Phi, \delta F[\sigma]/\delta\sigma(x)\, \Phi\}. \qquad (2.44)$$

*In the interaction representation, however, part of the change in the expectation value is accounted for by the variation in $\Psi[\sigma]$*:

$$\delta/\delta\sigma(x)(\Psi[\sigma], F[\sigma]\Psi[\sigma]) = \{\Psi[\sigma], \delta F[\sigma]/\delta\sigma(x)\, \Psi[\sigma]\} + i/\hbar c\, \{\Psi[\sigma], [H(x), F[\sigma]\Psi[\sigma]. \qquad (2.45)$$

Accordingly, it is natural to define the *total functional derivative* of an operator,

$$\Delta F[\sigma]/\Delta\sigma(x) = \ldots = U[\sigma]\, \delta F[\sigma]/\delta\sigma(x)\, U^{-1}[\sigma] \qquad (2.46)$$

which is composed of the partial *functional derivatives, expressing the explicit coordinate variation and the implicit dynamical variation*. With this definition,

$$\delta/\delta\sigma(x)(\Psi[\sigma], F[\sigma]\Psi[\sigma]) = \{\Psi[\sigma], \Delta F[\sigma]/\Delta\sigma(x)\, \Psi[\sigma]\}. \qquad (2.47)$$

If the *functional* is of the form

$$F[\sigma] = \int_\sigma F(x)\, d\sigma_\mu,$$

we are led to write

$$\Delta F[\sigma]/\Delta\sigma(x) = dF(x)/dx_\mu,$$

where

$$dF(x)/dx_\mu = \partial F(x)/\partial x_\mu + i/\hbar c\ [H(x), \int_\sigma F(x')\ d\sigma_\mu']$$

defines the *total coordinate derivative*. It should be clear that *the conservation theorem (1.56)*

$$[\delta/\delta\sigma(x)\ P_\mu c = \delta/\delta x_\lambda\ T_{\lambda\mu}(x) = 0. \qquad (1.56)]$$

*and the equation of motion (1.38)*

$$[i/\hbar\ [F(x)_\nu P_\nu] = \partial F(x)/\partial x_\nu, \qquad (1.38)]$$

*in the interaction representation are to be written*

$$\Delta P_\mu[\sigma]/\Delta\sigma(x) = \delta P_\mu[\sigma]/\delta\sigma(x) + i/\hbar c\ [H(x),P_\mu[\sigma]] = 0, \qquad (2.48)$$

and

$$i/\hbar c\ [F(x),P_\mu[\sigma]] = dF(x)/dx_\mu$$
$$= \partial F(x)/\partial x_\mu + i/\hbar c\ [H(x), \int_\sigma F(x')\ d\sigma_\mu'] \qquad (2.49)$$

Now the partial coordinate derivative $\partial F(x)/\partial x_\mu$ is that to be associated with the behavior of *non-interacting* fields, and can therefore be calculated from the *energy-momentum four-vector* of the isolated fields, $P_\mu^{(0)}$, according to

$$i/\hbar c\ [F(x), P_\mu^{(0)}] = dF(x)/dx_\mu. \qquad (2.50)$$

Therefore,

$$[F(x),P_\mu[\sigma] - P_\mu^{(0)}] = \ldots = -\ 1/c \int_\sigma [F(x'), H(x')]\ d\sigma_\mu' \qquad (2.51)$$

in which we have used the fact that only the point x' = x will contribute to the surface integral. One may infer that

$$P_\mu[\sigma] = P_\mu^{(0)} - 1/c \int_\sigma H(x)]\ d\sigma_\mu, \qquad (2.52)$$

*which, indeed, is compatible with the conservation theorem* (2.48)

$$[\Delta P_\mu[\sigma]/\Delta\sigma(x) = \delta P_\mu[\sigma]/\delta\sigma(x) + i/\hbar c\ [H(x),P_\mu[\sigma]] = 0, \qquad (2.48)]$$

since

$$\Delta P_\mu[\sigma]/\Delta\sigma(x) = \delta P_\mu^{(0)}/\delta\sigma(x)$$
$$-\ 1/c\ \{\partial H(x)/\partial x_\mu - i/\hbar\ [H(x), P_\mu^{(0)}]\} = 0. \qquad (2.53)$$

The statement (2.52) can be confirmed by direct calculation. The appropriate transcription of the *Heisenberg operator* (1.64) involves the introduction of the total derivatives $dA_\lambda/dx_\nu$ and $d\psi/dx_\nu$. Only the latter differs from the explicit coordinate derivative. Now the operator $P_\nu^{(0)}$ is formally identical with (1.64), but expressed in terms of the *interaction representation* operators and their explicit coordinate derivatives. Therefore,

$$P_\nu[\sigma] = P_\nu^{(0)} + 1/2c \int_\sigma d\sigma_\mu'\ [\psi^-(x'),[H(x),\gamma_\mu\psi(x')]]\ d\sigma_\nu'. \qquad (2.54)$$

However,

$$[H(x),\gamma_\mu\psi(x')] = -\, i/c\; [j_\lambda(x),\, \gamma_\mu\psi(x')]A_\lambda(x) = \ldots\,, \qquad (2.55)$$

whence

$$P_\nu[\sigma] - P_\nu^{(0)} = \ldots = \ldots = 1/c^2 \int_\sigma j_\lambda(x)A_\lambda(x)\; d\sigma_\nu. \qquad (2.56)$$

which is the content of Eq. (2.52)

$$[P_\mu[\sigma] = P_\mu^{(0)} - 1/c \int_\sigma H(x)]\; d\sigma_\mu. \qquad (2.52)]$$

**3. *Covariant elimination of the longitudinal field.***

*It is the function of the supplementary condition to ensure that the electromagnetic field contains no spinless light quanta*, which have various unphysical properties. It is possible, indeed to eliminate the *scalar potential* and the longitudinal part of the *vector potential*, leaving only the transverse *vector potential* as the quantity truly descriptive of light waves. *Such conventional procedures suffer from a lack of covariance* which will be remedied in this section.

We shall show that one can replace the *electromagnetic field vector*, $A_\mu(x)$, by two scalar fields, $\Delta(x)$ and $\Delta'(x)$, together with a restricted vector field $G_\mu(x)$, in such a way that the *supplementary condition* involves only the scalar fields, while the *equation of motion* for $\Psi[\sigma]$ contains only $G_\mu(x)$, the sole physically significant part of the field. The decomposition will be conveniently expressed with the aid of an arbitrary time-like unit vector $n_\mu$; $n_\mu^2 = -\,1$. The procedure of the customary theory corresponds to the special choice: $n_\mu = (0, 0, 0, i)$.

We decompose $A_\mu(x)$ into the gradient in the *time-like* direction specified by $n_\mu$ of a scalar operator $\Delta(x)$, the gradient in the *space-like* direction orthogonal to $n_\mu$ of a scalar operator $\Delta'(x)$ and the vector $\mathbf{G}_\mu(x)$ which has no component in the direction $n_\mu$, and is divergence-less. …

…

We have thereby succeeded in constructing an *equation of motion* for $\Psi[\sigma]$ which no longer contains the *electromagnetic field* variables involved in the *supplementary condition. The additional term thus introduced is evidently the covariant generalization of the Coulomb interaction between charges.*

…

**4. *The invariant collision operator.***

While the *interactions* between fields and their vacuum fluctuations are conveniently regarded as modifying the properties of the *non-interacting* fields, other types of *interactions* are often best viewed as producing *transitions* among the *states* of the individual fields. We shall conclude this paper with a brief discussion of a *covariant* manner of describing such *transitions*. The change in *state* of several fields arising from their mutual *interaction* is described by the *equation of motion* (2.6)

[$i\hbar c\, \delta\Psi[\sigma]/\delta\sigma(x) = H(x)\Psi[\sigma]$, (2.6)]

for the *state vector* $\Psi[\sigma]$. The question that must be answered in order to describe collisions between the particles associated with the quantized fields is: given the *state vector* on a surface $\sigma_1$, what is the *state vector* on the surface $\sigma_2$, in the limit as $\sigma_1$ and $\sigma_2$ recede into the remote and past and future, respectively? …

…

As a final remark, we observe that the representation of S as an integral extended over all *space-time* indicates that it is unaffected by a translation of the coordinate system, and therefore commutes with the operator $P_\mu^{(0)}$ (cf. Eq. (2.50)

[$i/\hbar c\, [F(x), P_\mu^{(0)}] = dF(x)/dx_\mu$. (2.50)]

$[S, P_\mu^{(0)}] = 0$. (4.24)

*This is the energy-momentum conservation law for collision processes*, since, according to (4.10), the expectation value of $P_\mu^{(0)}$ is unchanged by the course of *interaction*, for an arbitrary initial *state*.

## (4) All elementary particles have mass, apart from the photon and gluons.

## Masses of elementary particles in the Standard Model.

Only the *photon* and the 8 *gluons* have no *mass*. The large *masses* of the $W^+$, $W^-$, $Z^0$ and *Higgs bosons* and the *top quark* (respectively 85.7, 85.7, 97.2, 133.3 and 184.9 times the *mass* of the *proton*) relative to the *masses* of the *proton* (938.3 MeV/$c^2$) and *neutron* (939.6 MeV/$c^2$) raise questions regarding whether they are really *elementary particles*, in particular in view of how they were created by collisions of high energy *protons* in *proton-antiproton* and *hadron* colliders.

**Masses of elementary particles in the Standard Model.**

| # | *Symbol* | *Description* | *Value* (MeV/$c^2$) | *Proton masses* |
|---|---|---|---|---|
| **Fermions:** | | | | |
| Leptons: | | | | |
| 1 | $m_e$ | Electron mass | 0.511 | 0.00054 |
| 2 | | Electron neutrino mass | 0.511 | 0.00054 |
| 3 | $m_\mu$ | Muon mass | 105.7 | 0.113 |
| 4 | | Muon neutrino mass | 105.7 | 0.113 |
| 5 | $m_\tau$ | Tau mass | 1,776.9 | 1.894 |
| 6 | | Tau neutrino mass | 1,776.9 | 1.894 |
| Quarks*: | | | | |
| 7 | $m_u$ | Up quark mass | 1.9 | 0.0020 |
| 8 | $m_d$ | Down quark mass | 4.4 | 0.0047 |
| 9 | $m_c$ | Charm quark mass | 1,320 | 1.407 |
| 10 | $m_s$ | Strange quark mass | 87 | 0.093 |
| 11 | $m_t$ | Top quark mass | 173,500 | 184.9 |
| 12 | $m_b$ | Bottom quark mass | 4,240 | 4.519 |
| **Bosons:** | | | | |
| Gauge bosons: | | | | |
| 13 | $m_v$ | Photon mass | 0.00 | |
| 14 | $m_{W+}$ | $W^+$ boson mass | 80,400 | 85.69 |
| 15 | $m_{W-}$ | $W^-$ boson mass | 80,400 | 85.69 |
| 16 | $m_{Z0}$ | $Z^0$ boson mass | 91,200 | 97.20 |
| 17-24 | | Gluons mass (8)† | 0.00 | 0.00 |
| 25 | | Graviton (hypothetical) | | |
| Scalar boson: | | | | |
| 26 | $m_{H0}$ | Higgs mass | 125,090 | 133.3 |

## **Higgs, P. W. (October, 1964). Broken Symmetries and the Masses of Gauge Bosons.**

*Phys. Rev. Lett.*, 13, 16, 508–9; https://journals.aps.org/prl/pdf/10.1103/PhysRevLett.13.508; also in Underwood, T. G. (2024). *The Standard Model*, Part I, pp. 338-41.

Tait Institute of Mathematical Physics, University of Edinburgh, Edinburgh, Scotland

Received August 31, 1964.

Brout and Englert showed that *gauge vector fields*, abelian and non-abelian, could acquire *mass* if empty space were endowed with a particular type of structure that one encountered in material systems. [Englert, F. & Brout, R. (August, 1964). Broken Symmetry and the Mass of Gauge Vector Mesons. *Phys. Rev. Lett.*, 13, 9, 321–3; https://journals.aps.org/prl/pdf/10.1103/PhysRevLett.13.321. Other physicists, Peter Higgs and Gerald Guralnik, C. R. Hagen and Tom Kibble had reached similar conclusions at about the same time. The *Brout–Englert–Higgs* (BEH) *mechanism* is believed to give rise to the *masses* of all the elementary particles in the *Standard Model*. This includes the *masses* of the W and Z *bosons*, and the *masses* of the *fermions*, i.e. the *quarks* and *leptons*. In a previous paper, Higgs had shown that the *Goldstone theorem*, that Lorentz-covariant field theories in which spontaneous breakdown of symmetry under an internal Lie group occurs contain zero-mass particles, *failed if and only if the conserved currents associated with the internal group were coupled to gauge fields*. The purpose of the present note was to report that, *as a consequence of this coupling, the spin-one quanta of some of the gauge fields acquired mass*; the longitudinal degrees of freedom of these particles (which would be absent if their *mass* were zero) go over into the *Goldstone bosons when the coupling tends to zero. The model was discussed mainly in classical terms*; nothing was proved about the quantized theory. Higgs noted that it should be understood, therefore, that *the conclusions which were presented concerning the masses of particles were conjectures based on the quantization of linearized classical field equations.*

---

In a recent note[1] it was shown that the *Goldstone theorem*[2], that *Lorentz-covariant field theories* in which *spontaneous breakdown* of *symmetry* under an internal Lie group occurs contain *zero-mass particles*, *fails if and only if the conserved currents associated with the internal group are coupled to gauge fields*.

[1] Higgs, P. W., to be published.
[2] Goldstone, J. (January, 1961). Field theories with "Superconductor" solutions. *Nuovo Cimento*, 19, 154, see above; Goldstone, J., Salam, A. & Weinberg, S. (August, 1962). Broken Symmetries. *Phys. Rev.* 127, 965, see above.

The purpose of the present note is to report that, *as a consequence of this coupling, the spin-one quanta of some of the gauge fields acquire mass*; the longitudinal degrees of freedom of these particles (which would be absent if their *mass* were zero) go over into the Goldstone bosons *when the coupling tends to zero*. This phenomenon is just the *relativistic analog of the plasmon phenomenon* to which Anderson[3] has drawn attention: that the *scalar zero-mass excitations of a superconducting neutral Fermi gas become longitudinal plasmon modes of finite mass when the gas is charged.*

[3] Anderson, P. W. (1963). Plasmons, Gauge Invariance, and Mass. *Phys. Rev.*, 130, 439; https://journals.aps.org/pr/abstract/10.1103/PhysRev.130.439.

The simplest theory which exhibits this behavior is a *gauge-invariant* version of a model used by Goldstone[2] himself: Two real[4] *scalar fields* $\varphi_1$, $\varphi_2$, and a real *vector field* $A_\mu$ *interact* through the *Lagrangian density*

$$L = \tfrac{1}{2}(\nabla\varphi_1)^2 - \tfrac{1}{2}(\nabla\varphi_2)^2 - V(\varphi_1^2 + \varphi_2^2) - \tfrac{1}{4} F_{\mu\nu} F^{\mu\nu}, \qquad (1)$$

where

$$\nabla_\mu\varphi_1 = \partial_\mu\varphi_1 - eA_\mu\varphi_2,$$
$$\nabla_\mu\varphi_2 = \partial_\mu\varphi_2 + eA_\mu\varphi_1,$$
$$F_{\mu\nu} = \partial_\mu A_\nu - \partial_\nu A_\mu,$$

e is a dimensionless *coupling constant*, and the metric is taken as −+++. L is *invariant* under *simultaneous gauge transformations of the first kind* on $\varphi_1 \pm i\,\varphi_2$, and *of the second kind* on $A_\mu$.

[4] *In the present note the model is discussed mainly in classical terms*; nothing is proved about the quantized theory. It should be understood, therefore, that *the conclusions which are presented concerning the masses of particles are conjectures based on the quantization of linearized classical field equations*. However, essentially the same conclusions have been reached independently by F. Englert and R. Brout, *Phys. Rev. Letters*, 13, 321 (1964): These authors discuss the same model *quantum mechanically* in lowest order perturbation theory about the self-consistent vacuum.

Let us suppose that $V'(\varphi_0^2) = 0$, $V''(\varphi_0^2) > 0$; then *spontaneous breakdown of U(1) symmetry* occurs. Consider the equations [derived from (1) by treating $\Delta\varphi_1$, $\Delta\varphi_2$, $A_\mu$ as small quantities] governing the propagation of small oscillations about the "*vacuum*" solution $\varphi_1(x) = 0$, $\varphi_2(x) = \varphi_0$:

$$\partial^\mu \{\partial_\mu (\Delta\varphi_1) - e\varphi_0 A_\mu\} = 0, \qquad (2a)$$
$$\{\partial^2 - 4\,\varphi_0^2\, V''(\varphi_0^2)\}(\Delta\varphi_2) = 0, \qquad (2b)$$

$$\partial F^{\mu\nu} = e\varphi_0 \{\partial^\mu (\Delta\varphi_1) - e\varphi_0 A_\mu\}. \quad (2c)$$

Equation (2b) describes waves whose *quanta* have (bare) *mass* $2\varphi_0\{V''(\varphi_0^2)\}^{1/2}$; Eqs. (2a) and (2c) may be transformed, by the introduction of new variables

$$B_\mu = A_\mu - (e\varphi_0)^{-1}\{\partial_\mu (\Delta\varphi_1), \quad (3)$$
$$G_{\mu\nu} = \partial_\mu B_\nu - \partial_\nu B_\mu = F_{\mu\nu},$$

into the form

$$\partial_\mu B^\mu = 0, \qquad \partial_\nu G^{\mu\nu} + e^2 \varphi_0^2 B^\mu = 0. \quad (4)$$

Equation (4) describes *vector waves* whose *quanta* have (bare) *mass* $e\varphi_0$. In the absence of the *gauge field coupling* ($e = 0$) the situation is quite different: Equations (2a) and (2c) describe *zero-mass scalar* and *vector bosons*, respectively. In passing, we note that the right-hand side of (2c) is just the linear approximation to the *conserved current*: It is linear in the *vector potential*, *gauge invariance* being maintained by the presence of the gradient term[5].

[5] In the *theory of superconductivity* such a term arises from collective excitations of the Fermi gas.

When one considers theoretical models in which *spontaneous breakdown* of *symmetry* under a *semi-simple group* occurs, one encounters a variety of possible situations corresponding to the various distinct irreducible representations to which the *scalar fields* may belong; the *gauge field* always belongs to the *adjoint representation*[6].

[6] See, for example, Glashow, S. L. & Gell-Mann, M. (September, 1961). Gauge Theories of Vector Particles. *Ann. Phys. (N.Y.)*, 15, 437-60. See above.

*The model of the most immediate interest is that in which the scalar fields form an octet under SU(3)*: Here one finds the possibility of two non-vanishing *vacuum expectation values*, which may be chosen to be the two $Y = 0$, $I_3 = 0$ members of the *octet*[7].

[7] These are just the parameters which, if the *scalar octet* interacts with *baryons* and *mesons*, lead to the *Gell-Mann-Okubo* and *electromagnetic mass splittings*: See Coleman, S. & Glashow, S. L. (May, 1964). Departures from the Eightfold Way: Theory of Strong Interaction Symmetry Breakdown. *Phys. Rev.*, 134, B671; https://journals.aps.org/pr/abstract/10.1103/PhysRev.134.B671.

There are two massive *scalar bosons* with just these *quantum numbers*; the remaining six components of the *scalar octet* combine with the corresponding components of the *gauge-*

*field octet* to describe massive *vector bosons*. There are two I = ½ *vector doublets*, degenerate in *mass* between Y = ± 1 but with an *electromagnetic mass splitting* between $I_3$ = ± ½, and the $I_3$ = ± 1 components of a Y = 0, I = 1 *triplet* whose *mass* is entirely *electromagnetic*. The two Y = 0, I = O *gauge fields* remain massless: This is associated with the residual *unbroken symmetry* under the *Abelian group* generated by Y and $I_3$. It may be expected that when a further mechanism (presumably related to the *weak interactions*) is introduced in order to break Y *conservation*, one of these *gauge fields* will acquire *mass*, leaving the *photon* as the only massless *vector particle*. A detailed discussion of these questions will be presented elsewhere.

It is worth noting that an essential feature of the type of theory which has been described in this note is the prediction of incomplete *multiplets* of *scalar* and *vector bosons*[8].

[8] Tentative proposals that *incomplete SU(3) octets* of *scalar particles* exist have been made by a number of people. Such a role, as an isolated Y = ± 1, I = ½ *state*, was proposed for the *K meson* (725 MeV) by Nambu, Y. & Sakurai, J. J. (1963). κ Meson and the Strangeness-Changing Currents of Unitary Symmetry. *Phys. Rev. Letters*, 11, 42. More recently the possibility that the *σ meson* (385 MeV) may be the Y = I = 0 member of an incomplete *octet* has been considered by Brown, L. M. (1964). Apparent Regularity in the Masses of the Mesons. *Phys. Rev. Lett.*, 13, 42; https://journals.aps.org/prl/abstract/10.1103/PhysRevLett.13.42.

It is to be expected that this feature will appear also in theories in which the *symmetry-breaking scalar fields* are not elementary dynamic variables but bilinear combinations of *Fermi fields*[9].

[9] In the *theory of superconductivity* the *scalar fields* are associated with *fermion pairs*; the doubly charged excitation responsible for the quantization of *magnetic flux* is then the surviving member of a *U(1) doublet*.

## Peter W. Higgs – 2013 Nobel Lecture, December 8, 2013. *Evading the Goldstone Theorem.*

Peter Higgs – Nobel Lecture. NobelPrize.org. https://www.nobelprize.org/prizes/physics/2013/higgs/lecture/.

In his Nobel Prize lecture Higgs provided an account of how he came to publish his paper.

---

My story begins in 1960, when I was appointed Lecturer in Mathematical Physics at the University of Edinburgh. Before I took up my appointment, I was invited to serve on the committee of the first Scottish Universities Summer School in Physics. I was asked to act as Steward at the School in July, my principal duty being to purchase and look after supplies of the wine which was to be served at dinner each evening. The students at the school included four who stayed up late into the night in the common room of Newbattle Abbey College (the crypt of a former abbey) discussing theoretical physics, and rarely got up in time for the first lecture of the following day. They were Dr. N. Cabibbo (Rome), Dr. S. L. Glashow (CERN), Mr. D. W. Robinson (Oxford) and Mr. M. J. G. Veltman (Utrecht). Many years later, Cabibbo told me that their discussions had been lubricated by bottles of wine collected after dinner and hidden inside the grandfather clock in the crypt. I did not take part in these discussions, since I had other things to do (such as conserving wine). Consequently, I did not learn about Glashow's paper on *electroweak unification*, which had already been written.

***Broken Symmetries.***

During my first year as a lecturer, I was in search of a worthwhile research program. In the previous four years in London, I had rather lost my way in particle physics and had become interested in quantum gravity. Symmetry had fascinated me since my student days, and I was puzzled by the approximate symmetries (what are now called flavor symmetries) of particle physics.

Then in 1961, I read Nambu's and Goldstone's papers on models of symmetry breaking in particle physics based on an analogy with the theory of super conductivity. (Nambu's models were inspired by the Bardeen, Cooper & Schrieffer theory, based on Bose condensation of Cooper pairs of electrons: Goldstone used scalar fields, with a 'wine bottle' potential to induce Bose condensation, as in the earlier Ginzburg-Landau theory.) What I found very attractive was the concept of a *spontaneously* broken symmetry, one that is exact in the underlying dynamics but appears broken in the observed phenomena as a consequence of an asymmetric *ground state* ("*vacuum*" in *quantum field theory*).

Most particle theorists at the time did not pay much attention to the ideas of Nambu and Goldstone. Quantum field theory was out of fashion, despite its successes in *quantum electrodynamics*; it was failing to describe either the *strong* or the *weak interactions*. Besides, condensed matter physics was commonly viewed as another country. At a Cornell seminar in 1960, Victor Weisskopf remarked (as recalled by Robert Brout):

> "Particle physicists are so desperate these days that they have to borrow from the new things coming up in many body physics—like BCS [Bardeen, Cooper & Schrieffer theory]. Perhaps something will come of it."

***The Goldstone Theorem.***

There was an obstacle to the success of the Nambu-Goldstone program. Nambu had shown how *spontaneous breaking* of a *chiral symmetry* could generate the *masses* of *spin-½ particles*, such as the *proton* and *neutron*, but his model predicted *massless spin-0 particles* (*pions*?), contrary to experimental evidence. (As noted by Weinberg, any such particles would dominate the radiation of energy from stars). Goldstone had argued that such massless particles would always be the result of excitations around the trough of the wine bottle potential.

In 1962 a paper entitled "*Broken Symmetries*" by Goldstone, Salam and Weinberg proved the "*Goldstone Theorem*", that "*In a manifestly Lorentz-invariant quantum field theory*, if there is a continuous symmetry *under which the Lagrangian is invariant*, then either the *vacuum state* is also *invariant* or there must exist *spinless particles of zero mass*". This theorem appeared to put an end to Nambu's program.

***Can one evade the Goldstone Theorem?***

In 1963 the condensed matter theorist Phil Anderson pointed out that in a superconductor the Goldstone mode becomes a massive "*plasmon*" mode due to long-range (Coulomb) forces, and that this mode is just the longitudinal partner of transverse *electromagnetic* modes, which are also massive. Anderson remarked "The Goldstone zero-mass difficulty is not a serious one, because we can probably cancel it off against an equal Yang-Mills zero-mass problem". However, he did not show that there was a flaw in the *Goldstone theorem* and he did not discuss any *relativistic* model, so particle theorists such as myself received his remark with skepticism.

In March 1964 Abe Klein and Ben Lee suggested that, even in *relativistic theories*, a certain equation which was crucial for the proof of the *Goldstone theorem* could be modified by the addition of an extra term, just as in condensed matter theories. But in June, Wally Gilbert (who was in transition from theoretical physics to molecular biology, for which he later won a Nobel Prize for Chemistry) ruled out this term as a violation of *Lorentz invariance*. It was at this point that my intervention took place.

***How to evade the Goldstone Theorem.***

I read Gilbert's paper on July 16, 1964—it had been published a month earlier, but in those days the University of Edinburgh's copies of *Physical Review Letters* came by sea—and I was upset because it implied that there was no way to evade *Goldstone's theorem*.

> [*Goldstone's Theorem* states that "*in a manifestly Lorentz-invariant quantum field theory, if there is a continuous symmetry transformation under which the Lagrangian is invariant, then either the vacuum state is also invariant under the transformation, or there must exist spinless particles of zero mass*".]

But over the following weekend I began to recall that I had seen similar apparent violations of *Lorentz invariance* elsewhere, in no less a theory than *quantum electrodynamics*, as formulated by Julian Schwinger.

*Quantum electrodynamics* is invariant under *gauge transformations* and the *gauge* must be fixed before well-defined quantum formalism can be set up. The fashionable way to do this was to choose a *Lorentz gauge*, which was manifestly compatible with *relativity*. However, such a *gauge* had unsatisfactory features that led Schwinger to prefer a *Coulomb gauge*, which introduces an apparent *conflict with relativity*. Nevertheless, it was well known that this choice did not lead to any conflict between the predicted physics and relativity.

Schwinger had, as recently as 1962, written papers in which he demolished the folklore that it is *gauge invariance* alone that requires *photons* to be massless. He had provided examples of some properties of a *gauge theory* containing massive "*photons*", but without describing explicitly the underlying dynamics.

During the weekend of 18–19 July it occurred to me that Schwinger's way of formulating *gauge theories* undermined the axioms which had been used to prove the *Goldstone theorem*. So, *gauge theories* might save Nambu's program. During the following week I wrote a short paper about this. It was sent to *Physics Letters* on 24 July and was accepted for publication. [Higgs, P. W. (September, 1964). Broken symmetries, massless particles and gauge fields. *Physics Letters*., 12, 2, 132–3; doi:10.1016/0031-9163(64)91136-9.]

By then I had written down the (classical) field equations of the simplest illustrative model that I could imagine, the result of introducing an *electromagnetic interaction* into Goldstone's simplest scalar model. It became obvious that in this model the Goldstone *massless mode* became the longitudinal polarization of a massive spin-1 "*photon*", just as

Anderson had suggested. My second short paper, consisting of a brief account of this model, was sent to *Physics Letters* on 31 July. *It was rejected.* The editor (at CERN) suggested that I develop my ideas further and write a full account for *Il Nuovo Cimento.*

I was indignant; it seemed that the referee had not seen the point of my paper. (Later, a colleague who returned from a month's visit to CERN told me that the theorists there did not think it had any relevance to particle physics.) Besides, it seemed odd that the earlier paper had been accepted but the more physical sequel had not.

I decided to augment the paper by some remarks on possible physical consequences, and to send the revised version across the Atlantic to *Physical Review Letters*. Among the additional material was the remark, "It is worth noting that an essential feature of this type of theory is the prediction of incomplete multiplets of scalar and vector bosons."

The revised paper was received by *Physical Review Letters* on 31 August and was accepted. [Higgs, P. W. (October, 1964). Broken Symmetries and the Masses of Gauge Bosons. See above.] The referee invited me to comment on the relation of my paper to that of Englert and Brout, whose paper (received on 22 June) had been published that day. Until then I had been unaware of their work, but I added a footnote to my paper as soon as I had received a copy of theirs. Twenty years later, at a conference in 1984, I met Nambu, who revealed that he had refereed both papers.

***Postcript.***

It took some time for the work of Englert and Brout and myself (and of Guralnik, Hagen and Kibble, who published a little later) to gain acceptance.

My longer (1966) paper was written in autumn 1965 at Chapel Hill, North Carolina, where I was spending a sabbatical year at the invitation of Bryce DeWitt as a consequence of my interest in *quantum gravity*. [Higgs, P. W.* (May, 1966). Spontaneous Symmetry Breakdown without Massless Bosons. See below.] A preprint sent to Freeman Dyson received a positive response; he invited me to give a talk at I.A.S. Princeton. There, in March 1966, I faced an audience including axiomatic quantum field theorists who still believed that there could be no exceptions to the *Goldstone theorem.*

The next day I gave a talk at Harvard (arranged by Stanley Deser) to another skeptical audience, including Wally Gilbert. I survived this too. After the seminar Shelly Glashow complimented me on having invented 'a nice model', but he did not recognize its relevance to his *electroweak theory*—a missed opportunity!

Like Nambu, the six of us who published in 1964 expected to apply our ideas to the *broken flavor symmetries* of the *strong interactions*, but this did not work. So, it was left to Weinberg and Salam in 1967 to find the right application. [Salam, A. Weak and Electromagnetic Interactions; in (1968). *Elementary Particle Theory* (Ed. N Svartholm), Almqvist and Wiskell, pp. 367–377; Weinberg, S. (1967). A Model of Leptons. *Phys. Rev. Let.*, 19, 1264. See below.]

Four more years passed before Gerard ‘t Hooft, in an extension of Veltman’s program, proved the *renormalizability* of such theories and another two before the discovery of *weak neutral currents* indicated that Glashow’s *electroweak unification* was the correct one. And in 1976 Ellis, Gaillard and Nanopoulos at CERN encouraged experimentalists to look for the massive *spinless boson* that the theory predicted.

## (5) Elementary and composite particles can have electric charge or be neutral.

## Electric charges of elementary particles in the Standard Model.

The *electric charge* of the *electron*, e = –1.602 × $10^{-19}$ coulombs. Particles with the same *mass* but opposite *electric charge* to an existing *particle* are described as *anti-particles*. In the Standard Model, they are a different form of *matter*, known as *antimatter*. The *electric charge* of the *positron* (the *anti-electron*) is – e (i.e. – 1 *electron charge*, positive). *Quarks* have fractional *electric charges*.

**Electric charges of elementary particles in the Standard Model.**

| # | *Description* | *Value (electron Charges, e)* |
|---|---|---|
| ***Fermions:*** | | |
| Leptons: | | |
| 1 | Electron charge | +1 (negative) |
| 2 | Electron neutrino charge | 0 |
| 3 | Muon charge | +1 (negative) |
| 4 | Muon neutrino charge | 0 |
| 5 | Tau charge | +1 (negative) |
| 6 | Tau neutrino charge | 0 |
| Quarks: | | |
| 7 | Up quark charge | + 2/3 (negative) |
| 8 | Down quark charge | − 1/3 (positive) |
| 9 | Charm quark charge | + 2/3 (negative) |
| 10 | Strange quark charge | − 1/3 (positive) |
| 11 | Top quark charge | + 2/3 (negative) |
| 12 | Bottom quark charge | − 1/3 (positive)] |
| **Bosons:** | | |
| Gauge bosons: | | |
| 13 | Photon charge | 0 |
| 14 | $W^+$ boson charge | −1 (positive) |
| 15 | $W^-$ boson charge | +1 (negative) |
| 16 | $Z^0$ boson charge | 0 |
| 17-24 | Gluons charge (8) | 0 |
| 25 | Graviton (hypothetical) | 0 |
| Scalar boson: | | |
| 26 | Higgs boson charge | 0 |

## (6) Elementary particles have a quantum state called spin.

### Electron spin.

The first direct experimental evidence of the *electron spin* was the Stern–Gerlach experiment of 1922 in which silver atoms were observed to possess two possible discrete angular momenta despite having no orbital angular momentum. However, the correct explanation of this experiment was only given in 1927. The original interpretation assumed the two spots observed in the experiment were due to *quantized orbital angular momentum*. However, in 1927 Ronald Fraser showed that Sodium atoms are isotropic with no *orbital angular momentum* and suggested that the observed magnetic properties were due to *electron spin*. In same year, Phipps and Taylor applied the *Stern-Gerlach technique* to hydrogen atoms; the ground state of hydrogen has zero *angular momentum* but the measurements again showed two peaks. Once the quantum theory became established, it became clear that *the original interpretation could not have been correct*: the possible values of *orbital angular momentum* along one axis is always an odd number, unlike the observations. Hydrogen atoms have a single electron with *two spin states* giving the two spots observed; silver atoms have closed shells which do not contribute to the *magnetic moment* and only the unmatched outer electron's *spin* responds to the field.

***The spin of an elementary particles is a quantum state and consequently a non-relativistic concept.***

We could try to determine the behavior of *spin* under general Lorentz transformations, but we would immediately discover a major obstacle. Unlike SO(3), the group of Lorentz transformations SO(3,1) is *non-compact* and therefore does not have any faithful, unitary, finite-dimensional representations.

## Spin of elementary particles in the Standard Model.

The *spin quantum state* is loosely related to the *angular momentum* of the particle. *Fermions* have ½-integer values, and *bosons* have integer values. These values have two directions, + or –.

**Spin of elementary and composite particles in the Standard Model.**

| # | *Description* | *Value* (±) |
|---|---|---|
| **Fermions:** | | |
| Leptons: | | |
| 1 | Electron spin | 1/2 |
| 2 | Electron neutrino spin | 1/2 |
| 3 | Muon spin | 1/2 |
| 4 | Muon neutrino spin | 1/2 |
| 5 | Tau spin | 1/2 |
| 6 | Tau neutrino spin | 1/2 |
| Quarks: | | |
| 7 | Up quark spin | 1/2 |
| 8 | Down quark spin | 1/2 |
| 9 | Charm quark spin | 1/2 |
| 10 | Strange quark spin | 1/2 |
| 11 | Top quark spin | 1/2 |
| 12 | Bottom quark spin | 1/2 |
| **Bosons:** | | |
| Gauge bosons: | | |
| 13 | Photon spin | 1 |
| 14 | $W^+$ boson spin | 1 |
| 15 | $W^-$ boson spin | 1 |
| 16 | $Z^0$ boson spin | 1 |
| 17-24 | Gluons spin (8) | 1 |
| 25 | Graviton spin (hypothetical) | 1 |
| Scalar boson: | | |
| 26 | Higgs boson spin | 1 |

## (7) Elementary and composite particles with mass attract each other through the gravitational interaction or gravitational force, according to Einstein's theory of General Relativity.

**Newton's law of Gravitation.**

*Newton's law of gravitation* states that every point mass in the universe attracts every other point mass with a force that is directly proportional to the product of their masses, and inversely proportional to the square of the distance between them.

**Einstein's theory of General Relativity.**

Einstein's *theory of general relativity* attempted to extend his *theory of special relativity* beyond space and time, to include *matter* and *gravitational fields*. Whilst this allowed Einstein to construct a *relativistic theory* of the effect of a *gravitational field* on *matter*, it also resulted in him rejecting his *postulate on the constancy of light* in the presence of a gravitational field.

*General relativity* is claimed to generalize *special relativity* and refine Newton's *law of universal gravitation*, providing a unified description of *gravity* as a geometric property of *space and time* or four-dimensional *spacetime*. In particular, the *curvature of spacetime* is directly related to the *energy* and *momentum* of whatever *matter* and *radiation* are present. The relation is specified by the *Einstein field equations*, a system of second-order partial differential equations.

In order to make calculations with his theory, Einstein had to import *Newton's law of gravitation*, which itself is an empirical law with no fundamental foundation. Consequently, the only evidence that Einstein could provide for his *theory of general relativity* was effectively Newtonian.

However, reconciliation of *general relativity* with the laws of *quantum physics* remains a problem *as there is a lack of a self-consistent theory of quantum gravity*. It is not yet known how gravity can be unified with the three non-gravitational forces: strong, weak and electromagnetic.

The lack of convergence in current formulations of *relativistic quantum electrodynamics*, or *quantum field theory* for the *electron*, due to the interaction of the electromagnetic and matter fields with their own vacuum fluctuations raised the question of whether the still unresolved *divergencies* arising largely, if not entirely, from the assumption of a *point electron*, could be isolated in unobservable *renormalization* factors.

Schwinger, in the Preface of his 1958 book [*Selected Papers on Quantum electrodynamics*], "questioned whether *renormalization* simply corrected a mathematical error that causes the divergencies, or whether *there is a serious flaw in the structure of field theory*". Feynman, in his 1965 Nobel prize speech, described *renormalization* as "simply a way to sweep the difficulties of the divergences of electrodynamics under the rug".

Despite the claims to the contrary in modern textbooks, there have been no significant developments in the quantum electrodynamics or quantum field theory since 1965 to resolve the underlying occurrence of divergencies.

**Underwood, T. G. (2023). *General Relativity***, Preface, p. 45: "... in Einstein (November 18, 1915), his calculation of the bending of light, was obtained from his approximations for his equation of the geodetic line $\sum_\alpha \partial\Gamma^\alpha_{\mu\nu}/\partial x^\alpha + \sum_{\alpha\beta} \Gamma^\alpha_{\mu\beta} \Gamma^\beta_{\nu\alpha} = 0$, where $\Gamma^\alpha_{\mu\nu} = -\frac{1}{2} \sum_\beta g^{\alpha\beta} (\delta g_{\mu\beta}/\delta x_\nu + \delta g_{\nu\beta}/\delta x_\mu - \delta g_{\mu\nu}/\delta x_\alpha)$, in which the link to the weak attractive force of gravitation was provided by *Newton's law of gravitation.* He noted that according to his *theory of general relativity* $ds^2 = \sum g_{\mu\nu} dx_\mu dx_\nu = 0$, determining the velocity of light, so that light-rays were bent if the $g_{\mu\nu}$ were not constant. ..."

**Underwood, T. G. (2023). *General Relativity***, Conclusion, p.474: "... A detailed examination of Einstein's *theory of general relativity* reveals that it is not a *theory of gravity*; it is a *relativistic* theory about the *effects* of gravitation, or more strictly, of a uniformly accelerated reference frame. There is nothing in any version of this theory that represents or explains or provides any connection to the weak attractive gravitational force between matter. We are no further forward in understanding the origin of this fundamental force. Whilst Einstein's and others' objectives in removing a preferred reference frame and the existence of an ether from physics were admirable intentions, Einstein's subsequent fixation on the constancy of the speed of light, or some form of invariant space-time, in the face of reasonable alternatives, such as Ritz's emission theory on which quantum electrodynamics is founded, was not.

Einstein's *theory of general relativity* attempted to extend his *theory of special relativity* beyond space and time, to include *matter* and *gravitational fields. Gravitation* was introduced through the "*equivalence principle*", the equivalence of the *outcome* of the force of *gravity* and the acceleration of *matter*, first recognized in Newton's *Principia.* This allowed Einstein to construct a *relativistic theory* of the effect of a *gravitational field* on *matter*, but it also resulted in him rejecting his *postulate on the constancy of light* in the presence of a gravitational field, and provided no connection to or explanation of the weak attractive *gravitational force* between *matter*. In order to make calculations with his theory, Einstein had to import *Newton's law of gravitation*, which itself is an empirical law with no fundamental foundation. Consequently, the only evidence that Einstein could provide for his *theory of general relativity* was effectively Newtonian.

In the light of the continued failure of Einstein's efforts to overcome the main objections to his *theory of special relativity* - the Ehrenfest paradox, and its failure to explain the observed Doppler redshift and blueshift of light – or to provide any evidence for it, and in the absence of any supportive evidence for his *theory of general relativity*, both theories must be rejected until such objections are overcome and such evidence is provided."

**Underwood, T. G. (2024).** ***Gravity***, Preface, p.16: "Part I includes an extract from Einstein, A. (February, 1917). *Kosmologische Betrachtungen zur allgemeinen Relativitätstheorie*. (Cosmological Considerations in the General Theory of Relativity.), which describes Einstein's struggles with supplementing the *relativistic differential equations* by *limiting conditions* at *spatial infinity* in order to regard the universe as being of infinite spatial extent. In his treatment of the planetary problem, he chose these limiting conditions on the basis of the assumption that it is possible to select a system of reference so that at spatial infinity all the *gravitational potentials* $g_{\mu\nu}$ become constant, but it was by no means evident that the same limiting conditions could be applied to larger portions of the physical universe. Einstein attempts to resolve this using a method analogous to the extension of Poisson's equation used in the *non-relativistic case*, by adding to the left-hand side of field equation, the fundamental tensor $g_{\mu\nu}$, multiplied by a universal constant, $-\lambda$. As he noted, "*we admittedly had to introduce an extension of the field equations of gravitation which is not justified by our actual knowledge of gravitation*".

Part II addresses "What is Gravity?". It begins by reviewing Einstein's unsuccessful attempts at producing a *classical unified field theory* between 1923 until he died in 1955, during which time Einstein published 31 papers on a *unified theory of electromagnetism and gravity*. …

Part II also includes Heinrich Weyl's attempt in 1929 to incorporate Dirac theory into the scheme of *general relativity* by introducing *gauge invariance* of *theory of coupled electromagnetic potentials* and Dirac *matter waves* [Weyl, H. (May, 1929). *Elektron und Gravitation*. (Electron and gravity.)] Weyl claimed that the barrier which hems progress of quantum theory is *quantization of the field equations*.

It also notes the 1935 paper [Einstein, A., Podolsky, B. & Rosen, N. (May, 1935). Can Quantum-Mechanical Description of Physical Reality Be Considered Complete? *Phys. Rev*., 47, 777-80], which suggests that the description of reality as given by a wave function in quantum mechanics is not complete.

But then it moves on to consider *quantum entanglement*, or some form of *entanglement* between *matter*, as a potential source of *gravity* and to examine the origin of gravity according to the Big Bang theory."

## Einstein, A. (November 25, 1915). Die Feldgleichungen der Gravitation. (The Field Equations of Gravitation.)

*Sitzungsber. d. Preuß. Akad. d. Wiss.*, 844–7; translation in A. Engel (translator), E. Schuckling (Consultant). (1997). *The Collected Papers of Albert Einstein*, Volume 6: The Berlin Years: Writings, 1914-1917, Princeton University Press, Princeton, Doc. 25, 117-20; https://einsteinpapers.press.princeton.edu/vol6-trans/129; translation by T. G. Underwood; also in Underwood, T. G. (2023). *General Relativity*, Part II, pp. 326-31.

Meeting of the physical-mathematical class of November 25, 1915.

Submitted November 25, 1915.

The third of three papers published by Einstein in November 1915 that led to the final *field equations* for *general relativity. This was seen to be the defining paper of general relativity.* At long last, Einstein felt that he had found workable field equations. Einstein noted that in his previous papers the hypothesis had to be introduced that the *scalar of the energy tensor of matter* disappeared. In this paper he described how he could do away with this hypothesis *if the energy tensor of matter was inserted into the field equations in a slightly different way.* From the covariant of the second rank $G_{im} = R_{im} + S_{im}$, where $R_{im} = -\sum_l \partial/\partial x_l \{{}_{l}^{im}\} + \sum_{l\rho} \{{}_{\rho}^{il}\}\{{}_{l}^{m\rho}\}$ and $S_{im} = \sum_l \partial/\partial x_m \{{}_{l}^{il}\} - \sum_{l\rho} \{{}_{\rho}^{im}\}\{{}_{l}^{\rho l}\}$, where $\{{}_{i}^{im}\} = ½\, g^{l\tau} (\partial g_{i\tau}/\partial x_m + \partial g_{m\tau}/\partial x_i - \partial g_{im}/\partial x_\tau)$, the ten generally-covariant equations of the *gravitational field* in spaces *where "matter" was absent* were obtained by setting $G_{im} = 0$. By choosing the frame of reference so that $\sqrt{(-g)} = 1$, $S_{im}$ vanished because $R_{\mu\nu} = \sum_\alpha \partial\Gamma^\alpha_{\mu\nu}/\partial x^\alpha + \sum_{\alpha\beta} \Gamma^\alpha_{\mu\beta} \Gamma^\beta_{\nu\alpha} = -\kappa T_{\mu\nu}$ [Einstein (November 4, 1915), Equ. 16.] and $R_{im} = \sum_l \partial\Gamma^l_{im}/\partial x_l + \sum_{\rho l} \Gamma^l_{i\rho} \Gamma^\rho_{ml} = 0$, where $\Gamma^l_{im} = -\{{}_{l}^{im}\}$ were *the "components" of the gravitational field. If "matter" was present* in the space under consideration, its *energy tensor* occurred on the right side of $G_{im} = 0$ or $R_{im} = \sum_l \partial\Gamma^l_{im}/\partial x_l + \sum_{\rho l} \Gamma^l_{i\rho} \Gamma^\rho_{ml} = 0$. Setting $G_{im} = -\kappa (T_{im} - ½\, g_{im} T)$, where $T = \sum_{\rho\sigma} g^{\rho\sigma} T_{\rho\sigma} = \sum_{\rho\sigma} T_\rho^\sigma$ was the *scalar of the energy tensor of "matter"*, and specializing the coordinate system, Einstein obtained in place of $G_{im} = -\kappa (T_{im} - ½\, g_{im} T)$, $R_{im} = \sum_l \partial\Gamma^l_{im}/\partial x_l + \sum_{\rho l} \Gamma^l_{i\rho} \Gamma^\rho_{ml} = -\kappa (T_{im} - ½\, g_{im} T)$, and $(-g)^{1/2} = 1$. Assuming, that *the divergence of matter vanished*, the *conservation law of matter and the gravitational field combined* became $\sum_\lambda \partial/\delta x_\lambda (T_\sigma^\lambda + t_\sigma^\lambda) = 0$, where $t_\sigma^\lambda$, the *"energy tensor" of the gravitational field*, was given by $\kappa t_\sigma^\lambda = ½\, \delta_\sigma^\lambda \sum_{\mu\nu\alpha\beta} g^{\mu\nu} \Gamma^\alpha_{\mu\beta} \Gamma^\beta_{\nu\alpha} - \sum_{\mu\nu\alpha} g^{\mu\nu} \Gamma^\alpha_{\mu\sigma} \Gamma^\beta_{\nu\alpha}$.

---

In two recent publications[1]

[1] Einstein, A. (November 4, 1915). Zur allgemeinen Relativitätstheorie. (On the General Theory of Relativity). *Sitzungsber. d. Preuß. Akad. d. Wiss.*, 778-86, 799-801.

*I have shown how one can arrive at field equations of gravitation that correspond to the postulate of general relativity*, i.e. that in their general version arbitrary substitutions of space-time variables are covariant to each other.

The course of development was as follows. *First, I found equations that contain Newton's theory as an approximation and were covariant to arbitrary substitutions of determinant 1.* I then found that *these equations generally correspond to covariant ones if the scalar of the energy tensor of "matter" disappears*. The coordinate system was then to be specialized according to the simple rule that $\sqrt{(-g)}$ is made equal to 1, whereby the equations of the theory are eminently simplified. However, as mentioned, *the hypothesis had to be introduced that the scalar of the energy tensor of matter disappears*.

Recently, *I have found that one can do without a hypothesis about the energy tensor of matter if one inserts the energy tensor of matter into the field equations in a slightly different way* than has been done in my two previous communications. The field equations for the vacuum, on which I based the explanation of the perihelion motion of Mercury, remain unaffected by this modification. I will give the whole consideration here again, so that the reader is not compelled to consult the earlier communications without interruption.

From the well-known Riemannian covariant of the fourth rank, the following covariant of the second rank is derived:

$$G_{im} = R_{im} + S_{im} \quad (1)$$
$$R_{im} = -\sum_l \partial/\partial x_l \{_l^{im}\} + \sum_{l\rho} \{_\rho^{il}\}\{_l^{m\rho}\} \quad (1a)$$
$$S_{im} = \sum_l \partial/\partial x_m \{_l^{il}\} - \sum_{l\rho} \{_\rho^{im}\}\{_l^{\rho l}\} \quad (1b)$$

[where $\{_i^{im}\} = \frac{1}{2} g^{l\tau} (\partial g_{i\tau}/\partial x_m + \partial g_{m\tau}/\partial x_i - \partial g_{im}/\partial x_\tau)$]

[Einstein, A. (March, 1916). Die Grundlage der allgemeinen Relativitätstheorie. (The foundation of the general theory of relativity.) *Ann. Phys.*, 49, 7, 769-822: "By the reduction of

$$B^\rho_{\mu\sigma\tau} = -\partial/\partial x_\tau \{_\rho^{\mu\sigma}\} + \partial/\partial x_\sigma \{_\rho^{\mu\tau}\} - \{_\alpha^{\mu\sigma}\}\{_\rho^{\alpha\tau}\} + \{_\alpha^{\mu\tau}\}\{_\rho^{\alpha\sigma}\} \quad (43)$$

with reference to indices to $\tau$ and $\rho$, we get the covariant tensor of the second rank

$$B_{\mu\nu} = R_{\mu\nu} + S_{\mu\nu} \quad (44)$$
$$R_{\mu\nu} = -\partial/\partial x_\alpha \{_\alpha^{\mu\nu}\} + \{_\beta^{\mu\alpha}\}\{_\alpha^{\nu\beta}\}$$
$$S_{\mu\nu} = \partial lg(-g)^{1/2}/\partial x_\mu \partial x_\nu - \{_\alpha^{\mu\nu}\} \partial log(-g)^{1/2}/\partial x_\alpha$$

(where g is the determinant $|g_{\mu\nu}|$ of $g_{\mu\nu}$, and $\{_\alpha^{\mu\nu}\} = \frac{1}{2} g^{\alpha\tau} (\partial g_{\mu\tau}/\partial x_\nu + \partial g_{\nu\tau}/\partial x_\mu - \partial g_{\mu\nu}/\partial x_\tau)$".

...

"... It has already been remarked in § 8, with reference to the equation (18a), that the coordinates can with advantage be so chosen that $(-g)^{1/2} = 1$. A glance at the equations got in the last two paragraphs shows that, through such a choice, the law of formation of the tensors suffers a significant simplification. It is especially true for the tensor $B_{\mu\nu}$, which plays a fundamental role in the theory. By this simplification, $S_{\mu\nu}$ vanishes of itself so that tensor $B_{\mu\nu}$ reduces to $R_{\mu\nu}$."]

The ten generally-covariant equations of the *gravitational field* in spaces where "matter" is absent are obtained by setting

$$G_{im} = 0. \qquad (2)$$

These equations can be made simpler if one chooses the frame of reference in such a way that $\sqrt{(-g)} = 1$. $S_{im}$ then vanishes because of (16) [in Einstein, A. (November 4, 1915). Zur allgemeinen Relativitätstheorie. (On the General Theory of Relativity)]

$$[R_{\mu\nu} = -\kappa\, T_{\mu\nu} \qquad (16)$$

$$\sum_\alpha \partial\Gamma^\alpha{}_{\mu\nu}/\partial x^\alpha + \sum_{\alpha\beta} \Gamma^\alpha{}_{\mu\beta}\, \Gamma^\beta{}_{\nu\alpha} = -\kappa\, T_{\mu\nu} \qquad (16a)]$$

and one gets instead of (2)

$$R_{im} = \sum_l \partial\Gamma^l{}_{im}/\partial x_l + \sum_{\rho l} \Gamma^l{}_{i\rho}\, \Gamma^\rho{}_{ml} = 0 \qquad (3)$$

We have set here

$$\Gamma^l{}_{im} = -\left\{{}_l{}^{im}\right\} \qquad (4)$$

which quantities we call *the "components" of the gravitational field.*

[Einstein, A. (March, 1916). Die Grundlage der allgemeinen Relativitätstheorie. (The foundation of the general theory of relativity.) *Ann. Phys.*, 49, 7, 769-822: "Therefore, it is clear that, for a gravitational field free from matter, it is desirable that the symmetrical tensors $B^\rho{}_{\mu\sigma\tau}$ deduced from the tensors $B_{\mu\nu}$ should vanish. We thus get 10 equations for 10 quantities $g_{\mu\nu}$, which are fulfilled in the special case when $B^\rho{}_{\mu\sigma\tau}$ all vanish.

Remembering

$$B_{\mu\nu} = R_{\mu\nu} + S_{\mu\nu} \qquad (44)$$

$$R_{\mu\nu} = -\partial/\partial x_\alpha \left\{{}_\alpha{}^{\mu\nu}\right\} + \left\{{}_\beta{}^{\mu\alpha}\right\}\left\{{}_\alpha{}^{\nu\beta}\right\}$$

$$S_{\mu\nu} = \partial \lg(-g)^{1/2}/\partial x_\mu\, \partial x_\nu - \left\{{}_\alpha{}^{\mu\nu}\right\} \partial \log(-g)^{1/2}/\partial x_\alpha$$

[and substituting

$$\Gamma^\tau{}_{\mu\nu} = -\left\{{}_\tau{}^{\mu\nu}\right\} \qquad (45)$$

so $\left\{{}_\alpha{}^{\mu\nu}\right\} = -\Gamma^\alpha{}_{\mu\nu}$

$\left\{{}_\beta{}^{\mu\alpha}\right\} = -\Gamma^\beta{}_{\mu\alpha}$

$\left\{{}_\alpha{}^{\nu\beta}\right\} = -\Gamma^\alpha{}_{\nu\beta}$

in $R_{\mu\nu} = \partial/\partial x_\alpha\, \Gamma^\alpha{}_{\mu\nu} + \Gamma^\beta{}_{\mu\alpha}\, \Gamma^\alpha{}_{\nu\beta}$,

where $R_{\mu\nu} = 0$],

we see that in the absence of matter the *field equations* [for the motion of the point *in a frame moving with uniform acceleration* relative to the reference frame] come out as follows; (when referred to the special coordinate-system chosen)

$\partial\Gamma^{\alpha}_{\mu\nu}/\partial x_\alpha + \Gamma^{\alpha}_{\mu\beta}\,\Gamma^{\beta}_{\nu\alpha} = 0$ (47)

and $(-g)^{1/2} = 1$

[where $\Gamma^{\tau}_{\mu\nu} = -\tfrac{1}{2}\, g^{\tau\alpha}\,(\partial g_{\mu\alpha}/\partial x_\nu + \partial g_{\nu\alpha}/\partial x_\mu - \partial g_{\mu\nu}/\partial x_\alpha)$, and $d^2x_\tau/ds^2 = \Gamma^{\tau}_{\mu\nu}\, dx_\mu/ds\, dx_\nu/ds$ is the equation of the *geodetic line* in pseudo-Riemannian space which is assumed to be the *equation of motion* of a freely moving body in a frame moving with uniform acceleration relative to the reference frame; or substituting for $\Gamma^{\tau}_{\mu\nu}$,

$2\,\partial g^{\alpha\tau}\,(\partial g_{\mu\tau}/\partial x_\nu + \partial g_{\nu\tau}/\partial x_\mu - \partial g_{\mu\nu}/\partial x_\tau)/\partial x_\alpha$
$- g^{\alpha\tau}\,(\partial g_{\mu\tau}/\partial x_\beta + \partial g_{\beta\tau}/\partial x_\mu - \partial g_{\mu\beta}/\partial x_\tau)\,(\partial g_{\nu\tau}/\partial x_\alpha + \partial g_{\alpha\tau}/\partial x_\nu - \partial g_{\nu\alpha}/\partial x_\tau) = 0$"]

*If "matter" is present* in the space under consideration, its *energy tensor* occurs on the right side of (2) or (3)

[ $G_{im} = 0.$ (2)

where $G_{im} = R_{im} + S_{im}$ (1)

$R_{im} = \sum_l \partial\Gamma^{l}_{im}/\partial x_l + \sum_{\rho l} \Gamma^{l}_{i\rho}\,\Gamma^{\rho}_{ml} = 0$ (3)]

respectively. We set

$G_{im} = -\kappa\,(T_{im} - \tfrac{1}{2}\, g_{im} T)$ (2a)

where

$\sum_{\rho\sigma} g^{\rho\sigma} T_{\rho\sigma} = \sum_{\rho\sigma} T_\rho{}^\sigma = T.$ (5)

T is the *scalar of the energy tensor of "matter"*, the right side of (2a) is a tensor. If we again specialize the coordinate system in the usual way, we get the equivalent equations

$R_{im} = \sum_l \partial\Gamma^{l}_{im}/\partial x_l + \sum_{\rho l} \Gamma^{l}_{i\rho}\,\Gamma^{\rho}_{ml} = -\kappa\,(T_{im} - \tfrac{1}{2}\, g_{im} T)$ (6)

$(-g)^{1/2} = 1$ (3a)

As always, *we assume that the divergence of the energy tensor of matter disappears* in the sense of the general differential calculus (energy- momentum theorem). Specializing the choice of coordinates according to (3a), this comes down to the fact that the $T_{im}$ should meet the conditions

$\sum_\lambda \partial T_\sigma{}^\lambda/\delta x_\lambda = -\tfrac{1}{2} \sum_{\mu\nu} \partial g^{\mu\nu}/\partial x_\sigma\, T_{\mu\nu}$ (7)

or

$\sum_\lambda \partial T_\sigma{}^\lambda/\delta x_\lambda = -\sum_{\mu\nu} \Gamma^{\mu}_{\sigma\nu}\, T_\mu{}^\nu.$ (7a)

If one multiplies (6) by $dg_{im}/dx_\sigma$ and sums over i and m, one obtains[1]

[1] About the derivative see Einstein, A. (November 4, 1915). Zur allgemeinen Relativitätstheorie. (On the General Theory of Relativity). *Sitzungsber. d. Preuß. Akad. d. Wiss.*, 44, 778-86, pp. 784-5. I ask the reader to refer to the developments given there on p.785 for comparison.

because of (7) and the relation

$$\tfrac{1}{2} \sum_{im} g_{im} \partial g_{im}/\partial x_\sigma = - d\log(-g)^{1/2}/dx_\sigma = 0$$

that follows from (3a)

$$[(-g)^{1/2} = 1 \qquad (3a)]$$

and the *law of conservation for matter and gravitational field together* in the form

$$\sum_\lambda \partial/\delta x_\lambda (T_\sigma{}^\lambda + t_\sigma{}^\lambda) = 0, \qquad (8)$$

where $t_\sigma{}^\lambda$, (the "energy tensor" of the gravitational field) is given by

$$\kappa t_\sigma{}^\lambda = \tfrac{1}{2} \delta_\sigma{}^\lambda \sum_{\mu\nu\alpha\beta} g^{\mu\nu} \Gamma^\alpha{}_{\mu\beta} \Gamma^\beta{}_{\nu\alpha} - \sum_{\mu\nu\alpha} g^{\mu\nu} \Gamma^\alpha{}_{\mu\sigma} \Gamma^\beta{}_{\nu\alpha} \qquad (8a)$$

The reasons, *which led me to introduce the second link* on the right side of (2a) and (6),

$$[G_{im} = - \kappa (T_{im} - \tfrac{1}{2} g_{im}T) \qquad (2a)$$

$$R_{im} = \sum_l \partial\Gamma^l{}_{im}/\partial x_l + \sum_{\rho l} \Gamma^l{}_{i\rho} \Gamma^\rho{}_{ml} = - \kappa (T_{im} - \tfrac{1}{2} g_{im}T) \qquad (6)]$$

are only clear from the following considerations, which are completely analogous to those given in the passage just quoted (p. 785).

If we multiply (6) by $g^{im}$ and sum over the indices i and m, we get after simple calculation

$$\Sigma_{\alpha\beta} \partial^2 g^{\alpha\beta}/\partial x_\alpha \partial x_\beta - \kappa(T + t) = 0 \qquad (9)$$

where corresponding to (5) we used the abbreviation

$$\sum_{\rho\sigma} g^{\rho\sigma} t_{\rho\sigma} = \sum_{\rho\sigma} t_\rho{}^\sigma = t. \qquad (8b)$$

Note that our additional element entails that in (9)

$$[\Sigma_{\alpha\beta} \partial^2 g^{\alpha\beta}/\partial x_\alpha \partial x_\beta - \kappa(T + t) = 0 \qquad (9)]$$

the energy tensor of the *gravitational field* occurs in the same way as that of matter, which is not the case in equation (21) *loc. cit*.

$$[\sum_{\alpha\beta} \partial^2 g^{\alpha\beta}/\partial x_\alpha \partial x_\beta - \Sigma_{\sigma\tau\alpha\beta}\, g^{\sigma\tau}\Gamma^\alpha{}_{\alpha\beta}\Gamma^\beta{}_{\tau\alpha} + \sum_{\alpha\beta} \partial/\partial x_\alpha \{g^{\alpha\beta}\, \partial lg\sqrt{(-g)}/\partial x_\beta\} = - \kappa \sum_\sigma T_\sigma{}^\sigma. \quad (21)]$$

Furthermore, instead of equation (22), *loc. cit.*,

$$[\Sigma_\beta\, \partial g^{\alpha\beta}/\partial x_\beta = 0. \qquad (22)]$$

the energy equation is used to derive the following relations:

$$\partial/\partial x_\mu \{\Sigma_{\alpha\beta} \partial^2 g^{\alpha\beta}/\partial x_\alpha \partial x_\beta - \kappa(T + t)\} = 0. \qquad (10)$$

Our additional element ensures that these equations do not contain any new conditions compared to (9),

$$[\Sigma_{\alpha\beta} \partial^2 g^{\alpha\beta}/\partial x_\alpha \partial x_\beta - \kappa(T + t) = 0 \qquad (9)$$

where $t = - \tfrac{1}{2} g_{im}T$, (8b)

T is the *scalar of the energy tensor of "matter"*, and

t is the *scalar of the energy tensor of the gravitational field*.]

so that we do not need to make other hypotheses about the energy tensor of matter other than that it complies with the energy momentum theorem.

Thus, the *general theory of relativity* is finally completed as a logical edifice. The *postulate of relativity* in its most general version, *which makes space-time coordinates physically meaningless parameters*, leads with imperative necessity to a very specific *theory of gravitation*, which explains the perihelion motion of Mercury. On the other hand, the general *postulate of relativity* cannot reveal anything about the nature of the other natural processes that the *special theory of relativity* has not already taught. My opinion in this regard the other day was erroneous. *Any physical theory corresponding to the special theory of relativity can be placed in the system of general relativity by means of absolute differential calculus, without the latter providing any criterion for the admissibility of that theory.*

**Einstein, A. (February, 1917). Kosmologische Betrachtungen zur allgemeinen Relativitätstheorie. (Cosmological Considerations in the General Theory of Relativity.)**

*Koniglich Preußische Akademie der Wissenschaften, Sitzungsberichte* (Berlin), 142–52; translation by W. Perrett & G. B. Jeffery in A. Engel (translator), E. Schuckling (consultant). (1997). *The Collected Papers of Albert Einstein*, Volume 6: The Berlin Years: Writings, 1914-1917, Princeton University Press, Princeton, Doc. 43, 421-32; https://einsteinpapers.press.princeton.edu/vol6-trans/433; also in Underwood, T. G. (2024). *Gravity*, Part I, pp. 173-83.

Describes Einstein's struggles with supplementing the *relativistic differential equations* by *limiting conditions* at *spatial infinity* in order to regard the universe as being of infinite spatial extent. In his treatment of the planetary problem, he chose these limiting conditions on the basis of the assumption that it is possible to select a system of reference so that at spatial infinity all the gravitational potentials $g_{\mu\nu}$ become constant, but it was by no means evident that the same limiting conditions could be applied to larger portions of the physical universe. Einstein attempts to resolve this using a method analogous to the extension of Poisson's equation used in the *non-relativistic case*, by adding to the left-hand side of field equation, the fundamental tensor $g_{\mu\nu}$, multiplied by a universal constant, $-\lambda$. As he noted, "*we admittedly had to introduce an extension of the field equations of gravitation which is not justified by our actual knowledge of gravitation*".

---

It is well known that Poisson's equation

$$\nabla^2\varphi = 4\pi K\rho \qquad (1)$$

in combination with the *equations of motion of a material point* is not as yet a perfect substitute for *Newton's theory of action at a distance*. There is still to be taken into account the condition that at spatial infinity the potential $\varphi$ tends toward a fixed limiting value. There is an analogous state of things in the *theory of gravitation* in *general relativity*. Here, too, we must supplement the differential equations by limiting conditions at spatial infinity, if we really have to regard the universe as being of infinite spatial extent.

In my treatment of the planetary problem, I chose these limiting conditions in the form of the following assumption: it is possible to select a system of reference so that at spatial infinity all the gravitational potentials $g_{\mu\nu}$ become constant. But *it is by no means evident a priori that we may lay down the same limiting conditions when we wish to take larger portions of the physical universe into consideration*. In the following pages the reflections will be given which, up to the present, I have made on this fundamentally important question.

***§ 1. The Newtonian Theory.***

It is well known that Newton's limiting condition of the constant limit for φ at spatial infinity leads to the view that *the density of matter becomes zero at infinity*. For we imagine that there may be a place in universal space round about which the *gravitational field of matter*, viewed on a large scale, possesses spherical symmetry. It then follows from Poisson's equation that, in order that φ may tend to a limit at infinity, the mean density ρ must decrease toward zero more rapidly than $1/r^2$ as the distance r from the center increases*.

> * ρ is the mean density of matter, calculated for a region which is large as compared with the distance between neighboring fixed stars, but small in comparison with the dimensions of the whole stellar system.

In this sense, therefore, *the universe according to Newton is finite, although it may possess an infinitely great total mass*.

From this it follows in the first place that *the radiation emitted by the heavenly bodies* will, in part, *leave the Newtonian system of the universe*, passing radially outwards, to become ineffective and lost in the infinite. *May not entire heavenly bodies fare likewise*? It is hardly possible to give a negative answer to this question. For it follows from the assumption of a finite limit for φ at spatial infinity that a heavenly body with finite kinetic energy is able to reach spatial infinity by overcoming the Newtonian forces of attraction. By statistical mechanics this case must occur from time to time, as long as the total energy of the stellar system-transferred to one single star-is great enough to send that star on its journey to infinity, whence it never can return.

We might try to avoid this peculiar difficulty by assuming a very high value for the limiting potential at infinity. That would be a possible way, if the value of the gravitational potential were not itself necessarily conditioned by the heavenly bodies. The truth is that we are compelled to regard the occurrence of any great differences of potential of the gravitational field as contradicting the facts. These differences must really be of so low an order of magnitude that the stellar velocities generated by them do not exceed the velocities actually observed.

If we apply Boltzmann's law of distribution for gas molecules to the stars, by comparing the stellar system with a gas in thermal equilibrium, we find that the Newtonian stellar system cannot exist at all. For there is a finite ratio of densities corresponding to the finite difference of potential between the center and spatial infinity. A vanishing of the density at infinity thus implies a vanishing of the density at the center.

It seems hardly possible to surmount these difficulties on the basis of the Newtonian theory. We may ask ourselves the question whether they can be removed by a modification of the Newtonian theory. First of all, we will indicate a method which does not in itself claim to be taken seriously; it merely serves as a foil for what is to follow. In place of Poisson's equation, we write

$$\nabla^2\varphi - \lambda\varphi = 4\pi\kappa\rho \qquad (2)$$

where $\lambda$ denotes a universal constant. If $\rho_0$ be the *uniform density of a distribution of mass*, then

$$\varphi = -4\pi\kappa/\lambda\ \rho_0 \qquad (3)$$

is a solution of equation (2). This solution would correspond to the case *in which the matter of the fixed stars was distributed uniformly through space*, if the density $\rho_0$ is equal to the actual mean density of the matter in the universe. The solution then corresponds to an infinite extension of the central space, filled uniformly with matter. If, without making any change in the mean density, we imagine matter to be non-uniformly distributed locally, there will be, over and above the $\varphi$ with the constant value of equation (3), an additional $\varphi$, which in the neighborhood of denser masses will so much the more resemble the Newtonian field as $\lambda\varphi$ is smaller in comparison with $4\pi\kappa\rho$.

*A universe so constituted would have, with respect to its gravitational field, no center*. A decrease of density in spatial infinity would not have to be assumed, but both the mean potential and mean density would remain constant to infinity. The conflict with statistical mechanics which we found in the case of the Newtonian theory is not repeated. With a definite but extremely small density, matter is in equilibrium, without any internal material forces (pressures) being required to maintain equilibrium.

### *§ 2. The Boundary Conditions According to the General Theory of Relativity.*

In the present paragraph I shall conduct the reader over the road that I have myself travelled, rather a rough and winding road, because otherwise I cannot hope that he will take much interest in the result at the end of the journey. The conclusion I shall arrive at is that *the field equations of gravitation which I have championed hitherto still need a slight modification, so that on the basis of the general theory of relativity those fundamental difficulties may be avoided which have been set forth in § 1 as confronting the Newtonian theory*. This modification corresponds perfectly to the transition from Poisson's equation

(1) to equation (2) of § 1. *We finally infer that boundary conditions in spatial infinity fall away altogether, because the universal continuum in respect of its spatial dimensions is to be viewed as a self-contained continuum of finite spatial (three-dimensional) volume.*

The opinion which I entertained until recently, as to the limiting conditions to be laid down in spatial infinity, took its stand on the following considerations. In a consistent theory of relativity there can be no inertia *relatively to "space,"* but only an inertia of masses *relatively to one another*. If, therefore, I have a mass at a sufficient distance from all other masses in the universe, its inertia must fall to zero. We will try to formulate this condition mathematically.

According to the *general theory of relativity* the negative momentum is given by the first three components, the energy by the last component of the covariant tensor multiplied by $\sqrt{(-g)}$

$$m \sqrt{(-g)}\, g_{\mu\alpha}\, dx_\alpha/ds \qquad (4)$$

where, as always, we set

$$ds^2 = g_{\mu\nu}\, dx_\mu dx_\nu \qquad (5)$$

In the particularly perspicuous case of the possibility of choosing the system of coordinates so that the gravitational field at every point is spatially isotropic, we have more simply

$$ds^2 = -A(dx_1^2 + dx_2^2 + dx_3^2) + Bdx_4^2.$$

If, moreover, at the same time

$$\sqrt{(-g)} = 1 = \sqrt{(A^3B)}$$

we obtain from (4), to a first approximation for small velocities,

$$mA/\sqrt{(B)}\, dx_1/dx_4, \qquad mA/\sqrt{(B)}\, dx_2/dx_4, \qquad mA/\sqrt{(B)}\, dx_3/dx_4,$$

for the components of *momentum*, and for the *energy* (in the static case)

$$m\sqrt{(B)}.$$

From the expressions for the momentum, it follows that $mA/\sqrt{(B)}$ plays the part of the *rest mass*. As m is a constant peculiar to the point of mass, independently of its position, this expression, *if we retain the condition $\sqrt{(-g)} = 1$ at spatial infinity*, can vanish only when A diminishes to zero, while B increases to infinity. *It seems, therefore, that such a degeneration of the coefficients $g_{\mu\nu}$ is required by the postulate of relativity of all inertia.*

This requirement implies that *the potential energy m√(B) becomes infinitely great at infinity*. Thus, a point of mass can never leave the system; and a more detailed investigation shows that the same thing applies to light-rays. A system of the universe with such behavior of the *gravitational potentials* at infinity would not therefore run the risk of wasting away which was mooted just now in connection with the Newtonian theory.

I wish to point out that the simplifying assumptions as to the *gravitational potentials* on which this reasoning is based, have been introduced merely for the sake of lucidity. It is possible to find general formulations for the behavior of the $g_{\mu\nu}$ at infinity which express the essentials of the question without further restrictive assumptions.

At this stage, with the kind assistance of the mathematician J. Grommer, I investigated *centrally symmetrical, static gravitational fields, degenerating at infinity* in the way mentioned. The gravitational potentials $g_{\mu\nu}$ were applied, and from them the *energy-tensor* $T_{\mu\nu}$ *of matter was calculated on the basis of the field equations of gravitation*. But here it proved that for the system of the fixed stars no boundary conditions of the kind can come into question at all, as was also rightly emphasized by the astronomer de Sitter recently.

For the contravariant energy-tensor $T_{\mu\nu}$ of ponderable matter is given by

$$T^{\mu\nu} = \rho\, dx_\mu/ds\; dx_\nu/ds,$$

where ρ is the *density of matter* in natural measure. With an appropriate choice of the system of co-ordinates the stellar velocities are very small in comparison with that of light. We may, therefore, substitute $\sqrt{(g_{44})}\, dx4$ for ds. This shows us that all components of $T^{\mu\nu}$ must be very small in comparison with the last component $T^{44}$. *But it was quite impossible to reconcile this condition with the chosen boundary conditions*. In the retrospect this result does not appear astonishing. The fact of the small velocities of the stars allows the conclusion that wherever there are fixed stars, the gravitational potential (in our case √B) can never be much greater than here on earth. *This follows from statistical reasoning, exactly as in the case of the Newtonian theory*. At any rate, *our calculations have convinced me that such conditions of degeneration for the $g_{\mu\nu}$ in spatial infinity may not be postulated.*

After the failure of this attempt, two possibilities next present themselves.
(a) We may require, as in the problem of the planets, that, with a suitable choice of the system of reference, the $g_{\mu\nu}$ in spatial infinity approximate to the values

$$\begin{matrix} -1 & 0 & 0 & 0 \\ 0 & -1 & 0 & 0 \\ 0 & 0 & -1 & 0 \\ 0 & 0 & 0 & 1 \end{matrix}$$

(b) We may refrain entirely from laying down boundary conditions for spatial infinity claiming general validity; but at the spatial limit of the domain under consideration we have to give the $g_{\mu\nu}$ separately in each individual case, as hitherto we were accustomed to give the initial conditions for time separately.

*The possibility (b) holds out no hope of solving the problem, but amounts to giving it up.* This is an incontestable position, which is taken up at the present time by de Sitter.*

*de Sitter, *Akad. van Wetensch. te Amsterdam*, Nov. 8, 1916.

*But I must confess that such a complete resignation in this fundamental question is for me a difficult thing.* I should not make up my mind to it until every effort to make headway toward a satisfactory view had proved to be vain.

*Possibility (a) is unsatisfactory in more respects than one.* In the first place those boundary conditions presuppose a definite choice of the system of reference, *which is contrary to the spirit of the relativity principle.* Secondly, if we adopt this view, *we fail to comply with the requirement of the relativity of inertia.* For the inertia of a material point of mass m (in natural measure) depends upon the $g_{\mu\nu}$; but these differ but little from their postulated values, as given above, for spatial infinity. Thus, inertia would indeed be influenced, but would not be conditioned by matter (present in finite space). If only one single point of mass were present, according to this view, it would possess inertia, and in fact an inertia almost as great as when it is surrounded by the other masses of the actual universe. Finally, those statistical objections must be raised against this view which were mentioned in respect of the Newtonian theory.

From what has now been said it will be seen that I have not succeeded in formulating boundary conditions for spatial infinity. Nevertheless, there is still a possible way out, without resigning as suggested under (b). *For if it were possible to regard the universe as a continuum which is finite (closed) with respect to its spatial dimensions, we should have no need at all of any such boundary conditions.* We shall proceed to show that both the *general postulate of relativity* and the fact of the small stellar velocities are compatible with the hypothesis of a spatially finite universe; though certainly, in order to carry through this idea, *we need a generalizing modification of the field equations of gravitation.*

### *§ 3. The Spatially Finite Universe with a Uniform Distribution of Matter.*

*According to the general theory of relativity the metrical character (curvature) of the four-dimensional space-time continuum is defined at every point by the matter at that point and the state of that matter.* Therefore, on account of the lack of uniformity in the distribution of matter, the metrical structure of this continuum must necessarily be extremely complicated.

But if we are concerned with the structure only on a large scale, we may represent matter to ourselves as being uniformly distributed over enormous spaces, so that its density of distribution is a variable function which varies extremely slowly. Thus, our procedure will somewhat resemble that of the geodesists who, by means of an ellipsoid, approximate to the shape of the earth's surface, which on a small scale is extremely complicated.

The most important fact that we draw from experience as to the distribution of matter is that the relative velocities of the stars are very small as compared with the velocity of light. So, I think that for the present we may base our reasoning upon the following approximative assumption. *There is a system of reference relatively to which matter may be looked upon as being permanently at rest.* With respect to this system, therefore, the contravariant energy-tensor $T^{\mu\nu}$ of matter is, by reason of (5), of the simple form

$$\begin{matrix} 0 & 0 & 0 & 0 \\ 0 & 0 & 0 & 0 \\ 0 & 0 & 0 & 0 \\ 0 & 0 & 0 & \rho \end{matrix} \qquad (6)$$

The scalar $\rho$ of the (mean) *density of distribution* may be a priori a function of the space co-ordinates. But if we assume the universe to be spatially finite, we are prompted to the hypothesis that $\rho$ is to be independent of locality. On this hypothesis we base the following considerations.

As concerns the *gravitational field*, it follows from the equation of motion of the material point

$$d^2x_\nu/ds^2 + \{\alpha\beta, \nu\}\ dx_\alpha/ds\ dx_\beta/ds = 0$$

that a material point in a static gravitational field can remain at rest only when $g_{44}$ is independent of locality. Since, further, we presuppose independence of the time coordinate $x_4$ for all magnitudes, we may demand for the required solution that, for all $x_\nu$,

$$g_{44} = 1. \qquad (7)$$

Further, as always with static problems, we shall have to set

$$g_{14} = g_{24} = g_{34} = 0 \qquad (8)$$

It remains now to determine those components of the gravitational potential which define the purely spatial-geometrical relations of our continuum ($g_{11}$, $g_{12}$, … $g_{33}$). *From our assumption as to the uniformity of distribution of the masses generating the field, it follows that the curvature of the required space must be constant.* With this distribution of mass, therefore, the required finite continuum of the $x_1$, $x_2$, $x_3$, with constant $x_4$, will be a spherical space.

We arrive at such a space, for example, in the following way. We start from a Euclidean space of four dimensions, $\xi_1$, $\xi_2$, $\xi_3$, $\xi_4$, with a linear element $d\sigma$; let, therefore,

$$d\sigma^2 = d\xi_1^2 + d\xi_2^2 + d\xi_3^2 + d\xi_4^2 \qquad (9)$$

In this space we consider the hyper-surface

$$R^2 = \xi_1^2 + \xi_2^2 + \xi_3^2 + \xi_4^2 \qquad (10)$$

where R denotes a constant. The points of this hyper-surface form a three-dimensional continuum, a spherical space of radius of curvature R.

The four-dimensional Euclidean space with which we started serves only for a convenient definition of our hypersurface. Only those points of the hypersurface are of interest to us which have metrical properties in agreement with those of physical space with a uniform distribution of matter. For the description of this three-dimensional continuum, we may employ the co-ordinates $\xi_1$, $\xi_2$, $\xi_3$ (the projection upon the hyperplane $\xi_4 = 0$) since, by reason of (10), $\xi_4$ can be expressed in terms of $\xi_1$, $\xi_2$, $\xi_3$. Eliminating $\xi_4$ from (9), we obtain for the linear element of the spherical space the expression

$$d\sigma^2 = \gamma_{\mu\nu}\, d\xi_\mu d\xi_\nu \qquad (11)$$
$$\gamma_{\mu\nu} = \delta_{\mu\nu} + \xi_\mu\xi_\nu/(R^2 - \rho^2)$$

where $\delta_{\mu\nu} = 1$, if $\mu = \nu$; $\delta_{\mu\nu} = 0$, if $\mu$ # $\nu$, and $\rho^2 = \xi_1^2 + \xi_2^2 + \xi_3^2$.

The co-ordinates chosen are convenient when it is a question of examining the environment of one of the two points $\xi_1 = \xi_2 = \xi_3 = 0$.

Now the linear element of the required four-dimensional space-time universe is also given us. For the potential $g_{\mu\nu}$, both indices of which differ from 4, we have to set

$$g_{\mu\nu} = -\,[\delta_{\mu\nu} + x_\mu x_\nu/\{R2 - (x_1^2 + x_2^2 + x_3^2)\}] \qquad (12)$$

which equation, in combination with (7) and (8), perfectly defines the behavior of measuring-rods, clocks, and light-rays.

### *§ 4. On an Additional Term for the Field Equations of Gravitation.*

My proposed *field equations of gravitation* for any chosen system of co-ordinates run as follows: -

$$G_{\mu\nu} = -\kappa(T_{\mu\nu} - \tfrac{1}{2} g_{\mu\nu} T), \qquad (13)$$

$$G_{\mu\nu} = -\partial/\partial x_\alpha \{\mu\nu, \alpha\} + \{\mu\alpha, \beta\} \{\nu\beta, \alpha\} \\ + \partial^2 \log \surd(-g)/\partial x_\mu \partial x_\nu - \{\mu\nu, \alpha\} \partial \log \surd(-g)/\partial x_\alpha$$

The system of equations (13) *is by no means satisfied when we insert for the $g_{\mu\nu}$ the values given in (7), (8), and (12), and for the (contravariant) energy-tensor of matter the values indicated in (6).* It will be shown in the next paragraph how this calculation may conveniently be made. So that, if it were certain that the field equations (13) which I have hitherto employed were the only ones compatible with the postulate of general relativity, *we should probably have to conclude that the theory of relativity does not admit the hypothesis of a spatially finite universe.*

However, the system of equations (14) allows a readily suggested extension which is compatible with the relativity postulate, and is perfectly analogous to the extension of Poisson's equation given by equation (2). For on the left-hand side of field equation (13) *we may add the fundamental tensor $g_{\mu\nu}$, multiplied by a universal constant, $-\lambda$, at present unknown, without destroying the general covariance.* In place of field equation (13)

$$[G_{\mu\nu} = -\kappa(T_{\mu\nu} - \tfrac{1}{2} g_{\mu\nu} T), \qquad (13)]$$

we write

$$G_{\mu\nu} - \lambda g_{\mu} = -\kappa(T_{\mu\nu} - \tfrac{1}{2} g_{\mu\nu} T). \qquad (13a)$$

This field equation, with $\lambda$ sufficiently small, is in any case also compatible with the facts of experience derived from the solar system. It also satisfies laws of conservation of momentum and energy, because we arrive at (13a) in place of (13) by introducing into Hamilton's principle, instead of the scalar of Riemann's tensor, this scalar increased by a universal constant; and Hamilton's principle, of course, guarantees the validity of laws of conservation. It will be shown in § 5 that field equation (13a) is compatible with our conjectures on field and matter.

### *§ 5. Calculation and Result.*

Since all points of our continuum are on an equal footing, it is sufficient to carry through the calculation for one point, e.g. for one of the two points with the co-ordinates

$$x_1 = x_2 = x_3 = x_4 = 0.$$

Then for the $g_{\mu\nu}$ (13a) we have to insert the values

$$\begin{matrix} -1 & 0 & 0 & 0 \\ 0 & -1 & 0 & 0 \\ 0 & 0 & -1 & 0 \\ 0 & 0 & 0 & 1 \end{matrix}$$

wherever they appear differentiated only once or not at all. We thus obtain in the first place

$$G_{\mu\nu} = \partial/\partial x_1[\mu\nu,1] + \partial/\partial x_2[\mu\nu,2] + \partial/\partial x_3[\mu\nu,3] + \partial^2 \log \sqrt{(-g)}/\partial x_\mu \partial x_\nu.$$

From this we readily discover, taking (7), (8), and (13)

$$[g_{44} = 1 \qquad (7)$$

$$g_{14} = g_{24} = g_{34} = 0 \qquad (8)$$

$$G_{\mu\nu} = -\kappa(T_{\mu\nu} - \tfrac{1}{2} g_{\mu\nu} T), \qquad (13)]$$

into account, that all equations (13a) are satisfied if the two relations

$$-2/R^2 + \lambda = -\kappa\rho/2, \quad -\lambda = -\kappa\rho/2,$$

or

$$\lambda = \kappa\rho/2 = 1/R^2 \qquad (14)$$

are fulfilled.

Thus, the newly introduced universal constant λ defines both the *mean density of distribution* ρ which can remain in equilibrium and also the *radius* R and the v*olume* $2\pi^2 R^3$ *of spherical space*. The total mass M of the universe, according to our view, is finite, and is in fact

$$M = \rho \,.\, 2\pi^2 R^3 = 4\pi^2 R/\kappa = \pi^2 \sqrt{(32/\kappa^3 \rho)}. \qquad (15)$$

Thus, the theoretical view of the actual universe, if it is in correspondence with our reasoning, is the following. *The curvature of space is variable in time and place, according to the distribution of matter, but we may roughly approximate to it by means of a spherical space.* At any rate, this view is logically consistent, and from the standpoint of the *general theory of relativity* lies nearest at hand; whether, from the standpoint of present astronomical knowledge, it is tenable, will not here be discussed. In order to arrive at this consistent view, *we admittedly had to introduce an extension of the field equations of gravitation which is not justified by our actual knowledge of gravitation*. It is to be emphasized, however, that a positive curvature of space is given by our results, even if the supplementary term is not introduced. *That term is necessary only for the purpose of making possible a quasi-static distribution of matter, as required by the fact of the small velocities of the stars.*

[Peebles, P. J. E. & Ratra, B. (April 2003). The Cosmological Constant and Dark Energy. *Reviews of Modern Physics*, 75, 2, 559–606: "The record shows Einstein never liked the Λ term. His view of how general relativity might fit Mach's principle was disturbed by de Sitter's (1917) solution to Eq. (34) for empty space ($T_{\mu\nu} = 0$) with $\Lambda > 0$. Pais, A., 1982, *Subtle is the Lord ...* (Oxford University, New York, p. 288) points out that Einstein in a letter to Weyl in 1923 comments on the effect of Λ in Eq. (24): "According to De Sitter two material points that are sufficiently far apart, continue to be accelerated and move apart. If there is no quasistatic world, then away with the cosmological term." We do not know whether at this time Einstein was influenced by Slipher's redshifts or Friedmann's expanding world model. … Further to this point, in the appendix of the second edition of his book, *The Meaning of Relativity*, Einstein (1945, p. 127) states that the "introduction of the 'cosmologic member' " — Einstein's terminology for Λ — "into the equations of gravity, though possible from the point of view of relativity, *is to be rejected from the point of view of logical economy*", and that if "Hubble's expansion had been discovered at the time of the creation of the general theory of relativity, the cosmologic member would never have been introduced. It seems now so much less justified to introduce such a member into the field equations, since its introduction loses its sole original justification, — that of leading to a natural solution of the cosmologic problem." Einstein knew that without the cosmological constant the expansion time derived from Hubble's estimate of $H_0$ is uncomfortably short compared to estimates of the ages of the stars, and opined that that might be a problem with the star ages. The big error, the value of $H_0$, was corrected by 1960 [Sandage, A. (1958). *Astrophys. J.* 127, 513; (1962). *Problems of Extragalactic Research*, edited by G. C. McVittie (McMillan, New York), p. 359.

Gamow, G. (1970). [*My World Line*, Viking, New York, p. 44] recalls that "when I was discussing cosmological problems with Einstein, he remarked that the introduction of the cosmological term was the biggest blunder he ever made in his life." This certainly is consistent with all of Einstein's written comments we have seen on the cosmological constant per se; we do not know whether Einstein was also referring to the missed chance to predict the evolution of the universe."]

## Weyl, H. (April, 1929). The Electron and Gravitation.

*PNAS*, 15, 4, 323–34, https://doi.org/10.1073/pnas.15.4.323; also in Weyl, H. (May, 1929). Elektron und Gravitation. (The Electron and Gravitation.) *Zeit. Phys.*, 56, 330–352; https://doi.org/10.1007/BF01339504; extract in Underwood, T. G. (2023). *Gravity*, pp. 193-7.

Communicated March 7, 1929. Translated by H. P. Robertson.

Palmer Physical Laboratory, Princeton University.

Attempt to incorporate *Dirac theory** into the scheme of *general relativity*, introduces *gauge invariance* of *theory of coupled electromagnetic potentials* and Dirac *matter waves*, explains why "anti-symmetric" Pauli-Fermi statistics for electrons lead to "symmetric" Bose-Einstein statistics for photons, *barrier which hems progress of quantum theory is quantization of field equations.*

[* Dirac, P. A. M. (February, 1928). The Quantum Theory of the Electron. *Roy. Soc. Proc., A*, 117, 778, 610–24; https://doi.org/10.1098/rspa.1928.0023. Also in Underwood, T. G. *Quantum Electrodynamics – annotated sources.* Volume I, pp. 534-46: "The new quantum mechanics applied to the problem of the *structure of the atom with point-charge electrons* results in discrepancies consisting of "duplexity" phenomena, observed number of stationary states for an electron in an atom twice the number given by the theory, Goudsmit and Uhlenbeck introduced the idea of an electron with a *spin*, previous *relativity* treatments by Gordon and Klein obtain the operator of the wave equation by the same procedure as in the *non-relativity* theory, substitution of classical *quantum differential operators* for the *momentum vector* in the amended *relativistic Hamiltonian equation* and application of resulting differential operator to the *wave function* to obtain the *Klein-Gordon equation*, gives rise to two difficulties, the *first difficulty* is in the physical interpretation of solutions of ψ as the *charge* and the *current*, satisfactory for emission and absorption of radiation, provides probability of any dynamical variable at any specific time having a value between specified limits if they refer to the position of the electron, but, unlike the *non-relativity* theory, *not if they refer to its momentum or any other dynamical variable*, the *second difficulty* is that the conjugate imaginary of the wave equation is the same as that for an electron with charge – e and negative energy, *this paper is concerned only with the removal of the first of difficulties*, the resulting theory is only an approximation but appears sufficient to address duplexity problems without further assumptions, applies the method of *q-numbers* and using non-commutative algebra exhibits the properties of a free electron and of an electron in a central field of electric force, shows that simplest Hamiltonian for a *point charge electron satisfying requirements of both* *relativity and the general transformation theory* of quantum mechanics leads to explanation of all duplexity phenomena of number of stationary states being twice the observed value without further assumption about spin, in contrast to the Schrödinger equation which described wave functions of only one complex value Dirac introduces *vectors of four complex numbers* (known as bispinors), results in a *relativistic equation of motion* for the *wave function of the electron* $\{p_0 + \rho_1 (\boldsymbol{\sigma}, \mathbf{p}) + \rho_3 mc\} \psi = 0$, referred to as the *Dirac equation*, where $\mathbf{p}$ is the *momentum* vector, and $\boldsymbol{\sigma}$ denotes the vector $(\sigma_1, \sigma_2, \sigma_3)$, includes term equal to spin correction given by Darwin and Pauli, describes all spin-½ particles with mass, does not address second class of solutions of the wave equation in which *charge of the electron is positive* and *energy of a free electron is negative*."

> [This work led Dirac to predict the existence of the *positron*, the *electron*'s antiparticle, which he interpreted in terms of what came to be called the *Dirac sea*. The *positron* was observed by Carl Anderson in 1932.]

Dirac, P. A. M. (March, 1928). The quantum theory of the Electron. Part II. *Roy. Soc. Proc., A*, 118, 779, 351-61; https://doi.org/10.1098/rspa.1928.0056. See Underwood, T. G. *Quantum Electrodynamics – annotated sources.* Volume I, pp. 547-57: "Application of the *Dirac equation* to the conservation theorem, the selection principle, the relative intensities of the lines of a *multiplet*, and to the Zeeman effect."]

---

***The Problem***. - The translation of Dirac's theory of the electron into *general relativity* is not only of formal significance, for, as we know, the Dirac equations applied to an electron in a spherically symmetric *electrostatic field* yield *in addition to the correct energy levels those - or rather the negative of those - of an "electron" with opposite charge but the same mass*. In order to do away with these superfluous terms the *wave function* $\psi$ must be robbed of one of its pairs $\psi_1^+, \psi_2^+; \psi_1^-, \psi_2^-$ of components. These two pairs occur unmixed in the *action principle* except for the term

$$m\,(\psi_1^+\psi^*{}_1^- + \psi_2^+\psi^*{}_2^- + \psi_1^-\psi^*{}_2^+ + \psi_2^-\psi^*{}_2^+) \qquad (1)$$

which contains the *mass* m of the electron as a factor. *But mass is a gravitational effect: it is the flux of the gravitational field through a surface enclosing the particle in the same sense that charge is the flux of the electric field.* In a satisfactory theory it must therefore be as impossible to introduce a non-vanishing *mass* without the *gravitational field* as it is to introduce *charge* without *electromagnetic field. It is therefore certain that the term (1) can at most be right in the large scale, but must really be replaced by one which includes gravitation*; this may at the same time remove the defects of the present theory.

The direction in which such a modification is to be sought is clear: the *field equations* arising from an *action principle* - which shall give the true laws of *interaction* between *electrons, protons* and *photons* only after quantization - contain at present only the Schrodinger-Dirac quantity $\psi$, *which describes the wave field of the electron,* in addition to the four *potentials* $\varphi_p$ of the *electromagnetic field. It is unconditionally necessary to introduce the wave field of the proton before quantizing.* But since the $\psi$ of the electron can only involve two components, $\psi_1^+$, $\psi_2^+$should be ascribed to the electron and $\psi_1^-$, $\psi_2^-$ to the proton. *Obviously, the present expression,* $-e\,\psi^\wedge\,\psi$ *for charge-density*[#],

[#] The circumflex indicates transition to the conjugate of the transposed matrix (Hermitean conjugate). The four components of $\psi$ are considered as the elements of a matrix with four rows and one column.

*being necessarily negative, runs counter to this, and something must consequently be changed in this respect. Instead of one law for the conservation of charge we must have two, expressing the conservation of the number of electrons and protons separately.*

If one introduces the quantities $e\varphi_p/ch$ instead of $\varphi_p$ (and calls them $\varphi_p$), the *field equations* contain only the following combinations of atomistic constants: the pure number $\alpha = e^2/ch$ and h/mc, the "wave-length" of the electron[#].

[#] $h/2\pi$ is Planck's constant.

*Hence the equations certainly do not alone suffice to explain the atomistic behavior of matter with the definite values of e, m and h.* But *the subsequent quantization introduces the quantum of action h, and this together with the wave-length h/mc will be* sufficient, *since the velocity of light c is determined as an absolute measure of velocity by the theory of relativity.*

The introduction of the atomic constants by the quantum theory - or at least that of the *wavelength* - into the *field equations* has removed the support from under my *principle of gauge-invariance*, by means of which I had hoped to unify *electricity* and *gravitation.* But as I have remarked, *it possesses an equivalent in the field equations of quantum theory* which is its perfect counterpart in formal respects: the laws are *invariant* under the simultaneous substitution of $e^{i\lambda}\psi$ for $\psi$ and $\varphi_p - \partial\lambda/\partial x_p$ for $\varphi_p$, where $\lambda$ is an arbitrary function of position in space and time. The connection of this invariance with the *conservation law of electricity* remains exactly as before: the fact that the *action integral* is unaltered by the infinitesimal variation

$$\delta\psi = i\,\lambda\psi, \qquad \delta\varphi_p = -\,\partial\lambda/\partial x_p$$

($\lambda$ an arbitrary infinitesimal function) *signifies the identical fulfilment of a dependence between the material and the electromagnetic laws* which arise from the *action integral* by variations of the $\psi$ and $\varphi$, respectively; it means that the *conservation of electricity* is a double consequence of them, that *it follows from the laws of matter as well as electricity*. This new *principle of gauge invariance*, which may go by the same name, has the character of *general relativity* since it contains an arbitrary function $\lambda$, and can certainly only be understood with reference to it.

*It was such considerations as these, and not the desire for formal generalizations, which led me to attempt the incorporation of the Dirac theory into the scheme of general relativity*. We establish the metric in a world point P by a "Cartesian" system of axes (instead of the $g_{pq}$) consisting of four vectors e($\alpha$) {$\alpha$ = 0, 1, 2, 3} of which e(l), e(2), e(3) are real space-like vectors while e(0)/i is a real time-like vector of which we expressly demand that it be directed toward the future. A rotation of these axes is an orthogonal or *Lorentz transformation* which leaves these conditions of reality and sign unaltered. The laws shall remain invariant when the axes in the various points P are subjected to arbitrary and independent rotations. In addition to these we need four (real) coordinates $x_p$ (p = 0, 1, 2, 3) for the purpose of analytic expression. The components of e($\alpha$) in this coordinate system are designated by $e^p(\alpha)$. We need such local cartesian axes e($\alpha$) in each point P in order to be able to describe the quantity $\psi$ by means of its components $\psi_1^+$, $\psi_2^+$; $\psi_1^-$, $\psi_2^-$ for the law of transformation of the components $\psi$ can only be given for orthogonal transformations as it corresponds to a representation of the orthogonal group which cannot be extended to the group of all linear transformations. *The tensor calculus is consequently an unusable instrument for considerations involving the $\psi$*[7].

[7] Attempts to employ only the tensor calculus have been made by Tetrode [Tetrode, H. (May, 1928). Allgemein-relativistische Quantentheorie des Elektrons. (General-relativistic quantum theory of the electron.) *Zeit. Phys.*, 50, 336-46; https://doi.org/10.1007/BF01347512; translation by D. H. Delphenich; https://neo-classical-physics.info/uploads/3/4/3/6/34363841/tetrode_-_impulse-energy_theorem.pdf)]; Whittaker, J. M. (1928). *Proc. Camb. Phil. Soc.*, 25, 501, and others; I consider them misleading.

In formal aspects our theory resembles the more recent attempts of Einstein to unify *electricity* and *gravitation*[8].

[8] Einstein, A. (1928). Riemanngeometrie mit Aufrechterhaltung des Begriffes des Fern-Parallelismus. (Riemannian Geometry with Preservation of the Concept of Distant Parallelism.) *Sitzungsber. Berl. Akad.*, 217-21; (1929). Neue Möglichkeit für eine einheitliche Feldtheorie von Gravitation und Elektrizität. (New Possibility for a Unified Field Theory of Gravity and Electricity.) *Ibidem*, 224-7.

But here there is no talk of "distant parallelism"; there is no indication that Nature has availed herself of such an artificial geometry. *I am convinced that if there is a physical content in Einstein's latest formal developments it must come to light in the present connection.* It seems to me that it is now hopeless to seek a unification of *gravitation* and *electricity* without taking *material waves* into account.

***Use of the Indices.*** - If $t(\alpha)$ be the components of an arbitrary vector at point P with respect to the axes $\mathbf{e}(\alpha)$, then

$$t^p = \Sigma_\alpha e^p(\alpha)\, t(\alpha) \qquad (2)$$

are its contravariant components in the coordinate system $x_p$. Conversely, from the covariant components $t_p$ referred to the coordinates one obtains the components $t(\alpha)$ along the axes by the equations

$$t(\alpha) = \Sigma_p e^p(\alpha)\, t_p. \qquad (3)$$

Equations (2), (3) regulate the transition from one kind of indices to the other (Greek indices referring to the axes, Latin sub- or super-scripts to coordinates.) In the inverse transitions the quantities $e_p(\alpha)$, which are defined by

$$e_p(\alpha)\, e^q(\alpha) = \delta_p{}^q$$

and which also satisfy

$$e_p(\alpha)\, ep(\beta) = \delta(\alpha, \beta)$$

occur as coefficients. The Kronecker $\delta$ is 1 or 0 according to whether its indices agree or not.

***Symmetry and Conservation of the Energy Density.*** - The invariant *action*

$$\int \mathbf{H}\, dx \qquad (dx = dx_0\, dx_1\, dx_2\, dx_3)$$

contains *matter* (in the extended sense) and *gravitation*, the first being represented by the $\psi$ and possibly such additional quantities as the electromagnetic potentials $\varphi_p$, the latter by the components $e^p(\alpha)$ of the $\mathbf{e}(\alpha)$. Variation of the first kind of quantities gives rise to the equations of *matter*, variation of the $e^p(\alpha)$ to the *gravitational equations*. We disregard for the present that part of the *action* which depends only on the $e^p(\alpha)$, as introduced by Einstein in his classical theory of gravitation (1916), and consider only that part **H** which occurs even in the *special theory of relativity*. By an arbitrary infinitesimal variation of the $e^p(\alpha)$ which shall vanish outside of a finite portion of the world, an equation

$$\delta \int \mathbf{H}\, dx = \int \mathbf{t}_p(\alpha)\, \delta e^p(\alpha)\, .\, dx \qquad (4)$$

is obtained which defines the components $\mathbf{t}_p(\alpha)$ of the "*energy density*." In consequence of the *equations of matter*, which are assumed to hold, it is immaterial if or how the quantities describing *matter* are varied. Because of the invariance of the *action* (4) must vanish for variations $\delta\, e^p(\alpha)$ obtained by 1) subjecting the axes $\mathbf{e}(\alpha)$ to an infinitesimal rotation which may depend arbitrarily on position and 2) subjecting the coordinates $x_p$ to an arbitrary infinitesimal transformation, the axes $\mathbf{e}(\alpha)$ being unaltered. The first process is described by

$$\delta e^p(\alpha) = o(\alpha\beta)\, e^p(\beta)$$

where $o(\alpha\beta)$ constitutes an anti-symmetric matrix, whose elements are arbitrary (infinitesimal) functions of position. This requirement yields the *symmetry law*:

$$\mathbf{t}_p(\beta) e^p(\alpha) = \mathbf{t}(\alpha, \beta)$$

depends symmetrically on the two indices, $\alpha$ and $\beta$. But it must be observed that this law is not identically fulfilled, as in the old field theory, but only in consequence of the *equations of matter*, for if the wave field $\psi$ be held unchanged the components $\psi_p$ must undergo a transformation which is induced by the rotation of the axes. If $\delta x_p = \xi^p$ be the change which the coordinates of point P undergo in the second process, then the components of the unaltered vector $\mathbf{e}(\alpha)$ in P will undergo the change

$$\delta' e^p(\alpha) = \partial\xi^p/\partial x_q \,.\, e^q(\alpha).$$

This must, on the other hand, be given by

$$\delta e^p(\alpha) + \partial e^p(\alpha)/\partial x_q\, \xi^q$$

where $\delta$ means the difference at two points P which have the same values $x_p$ of coordinates before and after the deformation. From this there arises in the usual way - again assuming the validity of the *equations of matter* - the differential quasi-*conservation law for energy* (and *linear momentum*) whose four components are

$$\partial \mathbf{t}_p{}^q/\partial x_q + \mathbf{t}_q(\alpha)\, \partial e^q(\alpha)/\partial x_p = 0. \qquad (5)$$

(Only in the *special theory of relativity*, where the second member is lacking, is it a true conservation law.)

It is not necessary that the integral of **H** be invariant, but only that its variation be. This is the case when **H** differs from a *scalar density* by a *divergence*; we then say that the integral is "*practically invariant*". Similarly, it is only necessary that it be "*practically real*", i.e., that the difference between **H** and its complex conjugate be a *divergence*.

***Gradient of $\psi$.*** - Let the *wave field* $\psi$ be given. The invariant change $\delta\psi$ of $\psi$ on going from the point P to a neighboring point P' is to be determined as follows. The axes **e**($\alpha$) in P are taken to P' by parallel displacement: **e**'($\alpha$). $\psi_p = \psi_p(P)$ being the components of $\psi$ with respect to the axes **e**($\alpha$) at P, let $\psi'_p$ be the components of $\psi$ in P' relative to this displaced system: $\delta\psi_p = \psi'_p - \psi_p$. These $\delta\psi_p$ depend only on the choice of axes in P and transform in the same way as the $\psi_p$ on rotation of these axes. The axes **e**'($\alpha$) are obtained from the **e**($\alpha$) in P' by an infinitesimal orthogonal transformation do($\alpha\beta$); consequently the $\psi'_p$ are obtained from the components $\psi_p(P')$ by the corresponding linear transformation[10] dE and we have, $d\psi_p$ being the differential $\psi_p(P') - \psi_p(P)$:

$$\delta\psi = d\psi + dE.\psi.$$

[10] Capital Latin letters (except P for point) denote linear transformations of the four components of $\psi$.

If **e**($\alpha$) be taken as the vector $\overrightarrow{PP'}$ (multiplied by an infinitesimal factor) we write (on ignoring this factor) o($\alpha$,$\beta\gamma$), E($\alpha$), $\psi(\alpha)$ in place of do($\beta\gamma$), dE, $\delta\psi$:

$$\psi(\alpha) = \{e^p(\alpha)\, \partial/\partial x_p + E(\alpha)\}\psi \text{ or } \psi_p = \{\partial/\partial x_p + E_p\}\psi.$$

The calculation of o is accomplished by means of the formula

$$e^p(\gamma)\ \{o(\alpha,\beta\gamma) + o(\beta,\gamma\alpha)\} = \partial e^p(\alpha)/\partial x_q\, e_q(\beta) - \partial e^p(\beta)/\partial x_q\, e_q(\alpha).$$

The right-hand side of this expression is the "*commutator product*" of the two vector fields **e**($\alpha$) and **e**($\beta$), an invariant (under transformations of the coordinates $x_p$), known from the Lie theory.

***Introduction of Dirac's Action.*** - Let S($\alpha$) denote linear transformations which transform $\psi_1^+$, $\psi_2^+$, and $\psi_1^-$, $\psi_2^-$ among themselves. They are described by the matrices

$$S(0) = \begin{Vmatrix} i & 0 \\ 0 & i \end{Vmatrix},\qquad S(1) = \begin{Vmatrix} 0 & 1 \\ 1 & 0 \end{Vmatrix},\qquad S(2) = \begin{Vmatrix} 0 & -i \\ i & 0 \end{Vmatrix},\qquad S(3) = \begin{Vmatrix} 1 & 0 \\ 0 & -1 \end{Vmatrix}$$

for $\psi_1^+$, $\psi_2^+$; for $\psi_1^-$, $\psi_2^-$ the expression for S(0) is unchanged, S(1), S(2) and S(3) assume the opposite sign. The essential fact is that the quantities

$$\psi^{\sim\prime} S(\alpha)\psi$$

transform like the four components t($\alpha$) of a vector on rotation of the axes **e**($\alpha$), $\psi'$ being a quantity of the same kind as $\psi$. (In particular, $\psi^{\sim\prime}S(\alpha)\psi$ is the four-vector flux of probability.) Therefore

$$\psi^{\sim} S(\alpha)\psi(\alpha) = \psi^{\sim} e^p(\alpha)S(\alpha)\, \partial\psi/\partial x_p + \psi^{\sim} S(\alpha)E(\alpha)\psi$$

is a scalar, and after dividing by the absolute value ε of the determinant

$$| e^p(\alpha) |$$

we obtain a scalar density **iH,** whose integral can be employed as *action*. Division by ε will be indicated by changing the ordinary letter into the corresponding gothic. The calculation yields

$$1/\varepsilon\ S(\alpha)E(\alpha) = \tfrac{1}{2}\, \partial \mathbf{e}^p(\alpha)/\partial x_p\ S(a) + \tfrac{1}{2}\, I(\alpha)S'(\alpha),$$

where S'(α) is a transformation analogous to S(α): it agrees with S(α) for $\psi_1^+$, $\psi_2^+$, but is –S(α) for $\psi_1^-$, $\psi_2^-$.

$$\begin{aligned} \varepsilon\, \mathbf{I}(\alpha) &= o(\beta,\gamma\delta) + o(\gamma,\delta\beta) + o(\delta,\beta\gamma) \\ &= \Sigma = \partial \mathbf{e}^p(\beta)/\partial x_q\ e_q(\gamma)\ e^p(\delta) \end{aligned}$$

where the summation (in addition to that over p and q) is alternating and extends over the six permutations of βγδ while αβγδ is an even per mutation of the indices 0, 1, 2, 3.

We have yet to investigate whether **H** is practically real. Since

$$S^p = e^p(\alpha)S(\alpha) \text{ and } S(\alpha)E(\alpha)$$

are Hermitean matrices

$$-\,\mathbf{iH}^{-} = 1/\varepsilon\ \{\partial\psi^{\sim}/\partial x_p\ e^p(\alpha)S(\alpha)\psi + \psi^{\sim} S(\alpha)E(\alpha)\psi\}.$$

The first part is, on neglecting a complete *divergence*,

$$-\,\psi^{\sim}\ \partial \mathbf{e}^p(\alpha)S(\alpha)/\partial x_p = -\,\psi^{\sim}\ \mathbf{e}^p(\alpha)S(\alpha)\ \partial\psi/\partial x_p - \partial \mathbf{e}^p(\alpha)/\partial x_p\ .\ \psi^{\sim}\ S(\alpha)\psi.$$

On adding and subtracting **H** and **H**$^{-}$ we obtain the two *action* quantities **m**, **m'** (**m** = *matter*):

$$\mathbf{im} = \psi^{\sim}\ S^p\ \partial\psi/\partial x_p + \tfrac{1}{2}\, \partial \mathbf{e}^p(\alpha)/\partial x_p\ .\ \psi^{\sim}\ S(\alpha)\psi \qquad (6)$$

and

$$\mathbf{m}' = \mathbf{I}(\alpha).\ \psi^{\sim}\ S'(\alpha)\psi \qquad (7)$$

Both are practically, not actually, invariant, the first is practically and the second actually real.

The first is the essential content of the *Dirac theory*, written in general invariantive form.

> [See reference above; Dirac, P. A. M. (February, 1928). The Quantum Theory of the Electron.]

The corresponding *tensor density* of *energy*

$$\mathbf{t}_p(\alpha) = \mathbf{s}_p(\alpha) - e_p(\alpha)\mathbf{s}$$

where

$$\varepsilon\, \mathbf{s}_p(\alpha) = 1/2i\ \{\psi^{\sim}S(\alpha)\ \partial\psi/\partial x_p - \partial\psi^{\sim}/\partial x_p\ S(\alpha)\psi\}$$

and **s** the contracted

$$\mathbf{s}_p(\alpha)\ e^p(\alpha).$$

It has already been given in the literature for the Dirac theory (*special relativity*). For the *electron* of *hydrogen* in the normal *state* we find that the integral

$$\int t_0{}^0\ dx_1\ dx_2\ dx_3$$

extended over a section $x_0 = t =$ const., which should yield the *mass*, has the value $m/\sqrt{(1-\alpha^2)}$ ($\alpha$ the *fine structure constant*); this is a reasonable result, since m is to be taken as *proper mass* and in the Bohr theory $\alpha c$ is the *velocity* of the *electron* in the *normal state*.

It is worthy of note that there occurs in addition the *action* **m**', which is unknown to *special relativity* since it vanishes for constant $e^p(\alpha)$.

***The Electromagnetic Field.*** - In the Dirac theory the influence of an *electromagnetic field* is taken into account by replacing the operator $\partial/\partial x_p$, affecting the $\psi$ by $\partial/\partial x_p + i\varphi_p$. This yields an additional term of the form $i\varphi(\alpha).\psi^{\sim}S(\alpha)\psi$ in the *action*. On comparing this with (6) and (7) one might think that $\varepsilon.\varphi(\alpha)$ is to be identified with $\partial\mathbf{e}^p(\alpha)/\partial x_p$ or $\mathbf{I}(\alpha)$. Disregarding the *material waves* $\psi$ one would then have a *theory of electricity* of the same kind as the latest development of Einstein; the $\varphi$ are expressed in terms of the $\mathbf{e}^p(\alpha)$ in a way which is *invariant* under transformations of the coordinates, but only permit the cogredient rotations of the axes in all points P (distant parallelism). However, I believe that one must proceed otherwise in order to bring in the *electromagnetic field*. We have previously not mentioned the fact that the linear transformation of the $\psi$ corresponding to a rotation of axes is not uniquely determined. It is indeed possible to *normalize* this transformation, and we have tacitly based our calculations on such a *normalization*; but *the normalizing is itself a double-valued process and this reveals its artificial character*. The mathematical connection is this: we are to find a linear transformation of the four components of $\psi$ such that the quantities

$$\psi^{\sim}S(\alpha)\psi \quad (8)$$

suffer the given rotation[12].

[12] It is to be borne in mind that under the influence of a proper Lorentz transformation the $\psi^+$ components - as well as the $\psi^-$ - are transformed among themselves. Only when the improper operations of the Lorentz group, the reflection

$$\mathbf{e}(0) \rightarrow \mathbf{e}(0), \quad \mathbf{e}(\alpha) \rightarrow -\mathbf{e}(\alpha)\ [\alpha = 1, 2, 3],$$

is taken into account is it necessary to use both pairs of components together.

Obviously, it remains unaltered when $\psi_1^+$, $\psi_2^+$ are multiplied by $e^{i\lambda+}$; $\psi_1^-$, $\psi_2^-$ by $e^{i\lambda-}$ (transformation L) where $\lambda^+$ and $\lambda^-$ are arbitrary real numbers. It is readily seen that the transformations L are the only ones which induce the identical transformation of the quantities (8); i.e., which satisfy the four equations

$$L^{\sim}S(\alpha)L = S(\alpha).$$

A transformation of the four components of $\psi$ which multiplies $\psi_1^+$, $\psi_2^+$ by a number $a$+ and $\psi_1^-$, $\psi_2^-$ by a number $a^-$ will be called *spinless quantity* A = [$a^+$, $a^-$].

In consequence of this the dE employed above is only determined to within the addition of an arbitrary *spinless imaginary quantity* idF. Hence, we obtain.an additive term

$$iF(\alpha)\,.\,\psi^{\sim}\mathbf{S}(\alpha)\psi$$

in the *action* **m**, in which the F(α) are real *spinless quantities* [$\varphi^+(\alpha)$, $\varphi^-(\alpha)$] constituting the components of a vector with respect to the axes. $\partial/\partial x_p$ is to be replaced by $\partial/\partial x_p + iF_p$. We now employ the letter **m** to denote this completed *action*. The introduction of F obviously brings with it *invariance* under the simultaneous replacement of

$$\psi \text{ by } e^{iL}\psi \text{ and } F_p \text{ by } F_p - \partial L/\partial x_p$$

where L is a real *spinless* quantity which depends arbitrarily on *position*. This "*gauge invariance*" shows why it is impossible to employ the scalar (1) as an *action* function.
We must naturally interpret $F_p$ as the components of *electromagnetic potential*. The two fields $\varphi_p^+$, $\varphi_p^-$ are independent of the choice of the axes **e**(α) except that they are interchanged on transition from right- to a left-handed system of axes.

$$\partial\varphi_q^+/\partial x_p - \partial\varphi_p^+/\partial x_q = \varphi_{pq}^+ \text{ and the corresponding } \varphi_{pq}^-$$

are *gauge-invariant anti-symmetric tensors.*

In analogy to the Maxwellian *action* quantity, and in accordance with the above properties, the two functions **f**, **f'** (**f** = *electromagnetic field*) defined by

$$\varepsilon \mathbf{f} = \varphi_{pq}^{+}\varphi^{pq}_{-} \text{ and } \mathbf{ef'} = \varphi_{pq}^{+}\varphi^{pq}_{+} + \varphi_{pq}^{-}\varphi^{pq}_{-} \qquad (9)$$

are to be considered in choosing the *scalar density* of *electromagnetic action*. The identification of F with the *electromagnetic potential* is then justified by the fact that the entity which is represented by F influences and is influenced by *matter* in exactly the same way as the *electromagnetic potentials. If our view is correct, then the electromagnetic field is a necessary accompaniment of the matter-wave field and not of gravitation.*

If the action, insofar as its dependence on the $\psi$ and $\varphi$ is concerned, is an additive combination of **m**, **m'**, **f**, **f'** we obtain the two conservation theorems

$$\partial\rho^{p}_{+}/\partial x_p = 0 \text{ and } \partial\rho^{p}_{-}/\partial x_p = 0 \qquad (10)$$

where

$$\rho^{p}_{+} = \psi^{\sim +}\mathbf{S}^{p}\psi^{+}$$

contains only $\psi_1^{+}$ and $\psi_2^{+}$ and similarly for $\rho^{p}_{-}$. They are a double consequence of the *field laws*, that is, of the *equations of matter* in the narrow sense as well as of the *electromagnetic equations*. These two identities obtaining thus between the *field equations* are an immediate consequence of the *gauge invariance*. In consequence of (10) the two integrals $n^{+}$ and $n^{-}$ of $\rho = \rho^{0}$:

$$n = \int \rho \, dx_1 dx_2 dx_3$$

which are to be extended over a section $x_0$ = const. are *invariants* which are independent of the "time" $x_0$. Since they are interchanged on transition from right- to left-handed axes their values, which are *absolute constants of nature*, must be equal if both kinds of axes are to be equally permissible. We *normalize* them, in accordance with the interpretation of $\psi$ as probability, by

$$n^{+} = 1, \qquad n^{-} = 1. \qquad (11)$$

We have already mentioned how the *quasi-conservation laws for energy, momentum* and *moment of momentum* are related to *invariance* under transformation of coordinates and rotation of axes $\mathbf{e}(\alpha)$.

A stationary solution is characterized by the fact that $e^{p}(\alpha)$ and $F(\alpha)$ are independent of time $x_0 = t$, while the $\psi^{+}$ contain the time in an exponential factor $e^{i\nu t}$, the $\psi^{-}$ in a factor $e^{i\nu' t}$; $\nu$ and $\nu'$ need not be equal.

***Gravitation.*** - We consider as the *gravitational part of the action* the practically (not actually) invariant *density* **g** (**g** = *gravitation*) which underlies Einstein's "*classical*" theory of *gravitation* and which depends only on the $e^p(\alpha)$ and their first derivatives. It is most appropriate to carry through anew the entire calculation, which leads through the *Riemann curvature tensor*, in terms of the $e^p(\alpha)$; we find

$$\varepsilon\mathbf{g} = o(\alpha, \alpha\gamma)\, o(\beta, \beta\gamma) + o(\alpha, \beta\gamma)\, o(\beta, \alpha\gamma).$$

The "*cosmological term*" **g'** is given by $\varepsilon\mathbf{g}' = 1$.

If *gravitation* be represented by **g**, one can, as is well known, add to the *material* + *electromagnetic energy* $\mathbf{t}_p{}^q$ a *gravitational energy* $\mathbf{v}_p{}^q$ in such a way that the sum satisfies a true conservation law. Designating the total-differential of **g** considered as a function of $e^p(\alpha)$ and $e_q{}^p(\alpha) = \partial e^p(\alpha)/\partial x_p$ by

$$\delta\mathbf{g} = \mathbf{g}_p(\alpha)\, \delta e^p(\alpha) + \mathbf{g}_p{}^q(\alpha)\, \delta e_q{}^p(\alpha),$$
$$-\mathbf{v}_p{}^q = \mathbf{g}_r{}^q(\alpha)\, \partial e^{r(\alpha)}/\partial x_p - \delta_p{}^q\, \mathbf{g}.$$

*We obtain thus an invariant constant mass m which must be one of the characteristic universal constants of Nature.*

***Doubts, Prospects.*** - (11) is to be interpreted as the *law of conservation of the number (or charge) of electrons and protons*. Therefore, we ascribe $\psi^+$ and $\psi^-$ to the *electron* and to the *proton*, respectively. Taking **f** as the *electromagnetic* part of the *action*, which seems plausible to me, we then obtain *Maxwell's equations* in the sense that the *proton* generates the field $\varphi^+$ and the *electron* $\varphi^-$; whereas in accordance with the *equations of matter* $\varphi^+$ will effect only the *electron* and $\varphi^-$ the *proton*. This is not as obstruse as it may sound; on the contrary, the previous theory leads to entirely false results if the *potential* due to the *electron*, which at large distances neutralizes that due to the *nucleus*, reacts on the *electron* itself, as Schrodinger has pointed out with emphasis[15].

[15] Schrodinger, E. (1927). Der Energieimpulssatz der Materiewellen. (The Energy Momentum Law of Matter Waves.) *Ann. Phys*., 82, 2, 265-72; https://doi.org/10.1002/andp.19273870211; in particular p. 270.

It may indeed seem queer that $\psi^+$ and $\psi^-$ are here equally permissible, since we know that *positive* and *negative* electricity are fundamentally different - that *protons* and *electrons* have different *mass*. But if we neglect the *gravitational* and *electrical energy* in comparison with the *material* the mass $m$ falls into two parts $m^+$ and $m^-$ which are, however, not strictly constant. It is possible that $m^+$ and $m^-$ are different if our equations admit two classes of solutions which are interchanged on transition from right- to left-handed axes - as in the

*Dirac theory*, the spherically symmetric hydrogen problem admits several solutions for the normal *state* which are not themselves spherically symmetric but which are transformed among themselves by rotation.

As far as we know **m**, g and one of the two quantities **f** or, **f'** are in dispensable for the explanation of the phenomena. I am inclined to believe that the *action* is composed additively of **m**, **f** and **g**.

It should be noted that our *field equations* contain neither the theory of a single *electron* nor that of a single *proton. One might rather consider them as the laws governing a hydrogen atom consisting of an electron and a proton*; but here again, the problem of *interaction* between the two may first require *quantization. What we have obtained is solely a field scheme which can only be applied to and compared-with experience after the quantization has been accomplished.* We know from the Pauli *exclusion principle*[16] what commutation rules are to be applied in the *quantization* of $\psi^+$; those for $\psi^-$ must be the same in our theory.

[16] Jordan, P. & Wigner, E. (September, 1928). über das Paulische Äquivalenzverbot. (On the Paulian prohibition of equivalence.) *Zeit. Phys.*, 47, 631; https://doi.org/10.1007/BF01331938.

*The commutation relations between* $\psi^+$ *and* $\psi^-$ *are as yet entirely unknown.* Those of the *electromagnetic field* (photons) are almost completely known. In this respect we know nothing concerning the *gravitational field.* The commutation rules for F are here almost completely fixed by those for $\psi$, by the condition that these latter be unaltered when $\psi$ is given the increment $\delta\psi = iF(\alpha)\psi$. *That the rules thus obtained are in agreement with experience is indeed a support for our theory; i.e., it tells us why the "anti-symmetric," Pauli-Fermi statistics for electrons leads to the "symmetric" Bose-Einstein statistics for photons.* A definite decision can, however, first be reached *when the barrier which hems the progress of quantum theory is overcome: the quantization of the field equations.*

## Einstein, A. (March, 1921). Eine naheliegende Ergänzung des Fundaments der allgemeinen Relativitätstheorie. (On a natural addition to the foundation of the general theory of relativity.)

*Sitzungsberichte*, 261-4; translation in A. Engel (translator), E. Schuckling (consultant). (2002). *The Collected Papers of Albert Einstein*, Volume 7: The Berlin Years: Writings, 1918-1921, Princeton University Press, Princeton, Doc. 54, 224-8; https://einsteinpapers.press.princeton.edu/vol7-trans/240; also in Underwood, T. G. (2023). *General Relativity*, Part II, pp. 428-32.

Submitted March 3, 1921.

Einstein's comments on Hermann Weyl's attempt to supplement the *general theory of relativity* by adding a further condition of invariance. Weyl's theory was based on two ideas: (1) the *ratios* of components $g_{\mu\nu}$ of the *gravitational potential* have a far more fundamental physical meaning than the components themselves, to which Einstein raised the question "Can the theory of relativity be modified by the assumption that not the quantity ds itself, but only the equation $ds^2 = 0$ has an invariant meaning? (2) Weyl's second idea was related to the method of generalization of the Riemannian metric and to the physical interpretation of the newly arising quantities $\varphi_\nu$ in it. Riemannian geometry contains two assumptions: I. *The existence of transferable measuring rods*. II. *The independence of their length from the path of transfer*. Weyl's generalization of Riemann's metric retained (I) but dropped (II). He allowed the measured length of a measuring rod to depend upon its path of transfer by means of an integral extended over the path of transfer; in general, the integral $\int\varphi_\nu\, dx_\nu$ depended on this path where the $\varphi_\nu$ were *space functions* which, consequently, codetermined the metric. In the physical interpretation of the theory, these were identified with the *electromagnetic potentials*. Einstein raised a second question "Under these circumstances, one can ask if a distinct theory can be obtained by dropping from the beginning not only Weyl's assumption (II), but also assumption (I) about the existence of transferable measuring rods (and clocks, resp.)". In his effort to formulate such a theory, Einstein asked his colleague Wirtinger in Vienna if there was a *generalization of the equation of a geodesic line* such that only the ratios of the $g_{\mu\nu}$ played a role. Wirtinger showed how such a theory could be obtained starting out from only the invariant meaning of the equation $ds^2 = g_{\mu\nu}\, dx_\mu\, dx_\nu = 0$ without using the concept of distance ds, i.e. *without using measuring rods or measuring clocks*. Weyl had shown that the tensor $H_{iklm} = R_{iklm} - 1/(d-2)\ g_{il}R_{km} + g_{km}R_{il} - g_{im}R_{kl} - g_{kl}R_{im} + 1/(d-1)(d-2)\ (g_{il}g_{km} - g_{im}g_{kl})R$ was a Weyl tensor of weight 1, where $R_{iklm}$ was the *Riemann curvature tensor*, and $R_{km}$ the tensor of rank 2 that resulted from the previous one by means of one contraction; R was the scalar resulting from one further contraction, and d was the number of dimensions; a Weyl tensor (of weight n) is a Riemann tensor in which the value of a tensor component is multiplied by $\lambda^n$ if $g_{\mu\nu}$ is replaced by $\lambda g_{\mu\nu}$, where $\lambda$ is an arbitrary function of the coordinates. The desired generalization of the *geodesic line* was then given by the equation $\delta \{ \int d\sigma \} = 0$, where $d\sigma^2 = Jg_{\mu\nu}\, dx_\mu\, dx_\nu$ if J was a Weyl invariant of weight –1.

---

It is well known that H. Weyl tried to supplement the *general theory of relativity* by adding a further condition of invariance. He arrived at a theory which deserves high regard due to its consequential and daring mathematical structure. The theory is essentially based on two ideas.

a. The *ratios* of components $g_{\mu\nu}$ of the *gravitational potential* have a far more fundamental physical meaning than the components themselves. The totality of the world directions issuing from a world point in which light signals can be emitted by it, i.e., the light cone, seems to be given directly with the *space-time continuum*. This light cone, however, is determined by the equation

$$ds^2 = g_{\mu\nu}\, dx_\mu\, dx_\nu = 0$$

into which only the ratios of the $g_{\mu\nu}$ enter. Into the electromagnetic equations of the vacuum too, only the ratios of the $g_{\mu\nu}$ enter. In contrast, the quantity ds, which is determined by the $g_{\mu\nu}$ themselves, does not represent a property of the *space- time continuum* because its quantitative measurement requires a material object (clock). This suggests the question: *Can the theory of relativity be modified by the assumption that not the quantity ds itself, but only the equation $ds^2 = 0$ has an invariant meaning*?

b. Weyl's second idea is related to the method of generalization of the Riemannian metric and to the physical interpretation of the newly arising quantities $\varphi_\nu$ in it. The idea can be sketched as follows: *metric requires the transfer of lengths (measuring rods)*. Furthermore, Riemannian geometry requires that the state (length) of a measuring rod in one place is independent of the path used to get to this place, i.e., it contains two assumptions:

I. *The existence of transferable measuring rods.*

II. *The independence of their length from the path of transfer.*

Weyl's generalization of Riemann's metric retains (I) but drops (II). He allows the measured length of a measuring rod to depend upon its path of transfer by means of an integral extended over this path of transfer; in general, the integral

$$\int \varphi_\nu\, dx_\nu$$

depends on this path where the $\varphi_\nu$ are *space functions* which, consequently, codetermine the metric. *In the physical interpretation of the theory, these are then identified with the electromagnetic potentials.*

Notwithstanding the admirable consistency and beauty of Weyl's framework of ideas, it does not—in my opinion—measure up to physical reality. We do not know things in nature that can be utilized in measuring and whose relative extension depends upon their past history. It also does not appear that the straightest line, introduced by Weyl, and the *electric potentials* explicitly occurring in its equation and in the other equations of Weyl's theory, have direct physical meaning.

On the other hand, the idea elaborated by Weyl under (a) seems to me to be a lucky and natural one, even though one cannot a priori know whether or not it can lead to a useful physical theory. *Under these circumstances, one can ask if a distinct theory can be obtained by dropping from the beginning not only Weyl's assumption (II), but also assumption (I) about the existence of transferable measuring rods (and clocks, resp.).* In what follows, it shall be shown that one arrives freely and easily at a theory by starting out only from the invariant meaning of the equation:

$$ds^2 = g_{\mu\nu}\, dx_\mu\, dx_\nu = 0$$

without using the concept of distance ds, or—to put it in terms of physics—*without using the concepts of measuring rods or measuring clocks.*

In my effort to formulate such a theory, my colleague Wirtinger in Vienna gave me efficient support.

> [Wilhelm Wirtinger (July 19, 1865 – January 16, 1945) was an Austrian mathematician, working in complex analysis, geometry, algebra, number theory, Lie groups and knot theory.]

*I asked him if there is a generalization of the equation of a geodesic line such that only the ratios of the* $g_{\mu\nu}$ *play a role.* He answered me as follows:

By "Riemann tensor" or "Riemann invariant" we understand a tensor or invariant (*relative to an arbitrary point transformation*), resp., such that their invariant character is assured under the postulated invariance of $ds^2 = g_{\mu\nu}\, dx_\mu\, dx_\nu$. Furthermore, we understand as "Weyl tensor" or "Weyl invariant" (of weight n), resp., a Riemann tensor or a Riemann invariant, resp., with the following additional property: the value of a tensor component or invariant, resp., is multiplied by $\lambda^n$ if $g_{\mu\nu}$ is replaced by $\lambda g_{\mu\nu}$, where $\lambda$ is an arbitrary function of the coordinates. This condition can be expressed symbolically by the equation

$$T(\lambda g) = \lambda^n\, T(g).$$

Now, if J is a Weyl invariant of weight −1, depending only upon the $g_{\mu\nu}$ and their derivatives, then

$$d\sigma^2 = J g_{\mu\nu}\, dx_\mu\, dx_\nu \qquad (1)$$

is an invariant of weight 0, i.e., an invariant that depends only upon the ratios of the $g_{\mu\nu}$. *The desired generalization of the geodesic line is then given by the equation*

$$\delta \{ \int d\sigma \} = 0. \qquad (2)$$

This solution, of course, presupposes the existence of a Weyl invariant of the kind defined above. Weyl's investigations show the way to *one* such invariant. He has shown that the tensor

$$H_{iklm} = R_{iklm} - 1/(d-2)\, g_{il}R_{km} + g_{km}R_{il} - g_{im}R_{kl} - g_{kl}R_{im} + 1/(d-1)(d-2)\,(g_{il}g_{km} - g_{im}g_{kl})R \qquad (3)$$

is a Weyl tensor of weight 1. $R_{iklm}$ is here the *Riemann curvature tensor*, and $R_{km}$ the tensor of rank 2 that results from the previous one by means of one contraction; R is then the scalar resulting from one further contraction, and d is the number of dimensions. From this, one immediately gets that

$$H = H_{iklm}\, H^{iklm} \qquad (4)$$

is a Weyl scalar of weight – 2. Therefore,

$$J = \sqrt{H} \qquad (5)$$

is a Weyl invariant of weight – 1. In combination with (1) and (2), this result provides a generalization of the geodesic line according to the method outlined by Wirtinger. Of course, in order to judge the significance of this and the following results, it is a question of great importance whether or not J is the only Weyl invariant of weight – 1 that does not contain higher than second derivatives of the $g_{\mu\nu}$.

Based upon what has been developed so far, it is now easy to assign a Weyl tensor to every Riemann tensor and with this to establish laws of nature in the form of differential equations that depend only upon the ratios of the $g_{\mu\nu}$. If we put

$$g'_{\mu\nu} = Jg_{\mu\nu},$$

Then

$$d\sigma^2 = g'_{\mu\nu}\, dx_\mu\, dx_\nu$$

is an invariant that depends only upon the ratios of the $g_{\mu\nu}$. All Riemann tensors formed as fundamental invariants from in the customary manner are—when seen as functions of the $g_{\mu\nu}$ and their derivatives—Weyl tensors of weight 0. This can be symbolically expressed as follows. If T(g) is a Riemann tensor which depends not only upon the $g_{\mu\nu}$ and their derivatives but also upon other quantities, e.g., the components $\varphi_{\mu\nu}$ of an *electromagnetic field*, then —seen as a function of the $g_{\mu\nu}$ and their derivatives—is also a Weyl tensor of weight 0. Therefore, to every law of nature T(g) = 0 of the *general theory of relativity*, there corresponds a law which contains only the ratios of the $g_{\mu\nu}$.

This result becomes even more distinct by the following consideration. Since there is an arbitrary factor in the $g_{\mu\nu}$, it will be possible to select this factor such that everywhere

$$J = J_0, \tag{6}$$

where $J_0$ is a constant. The $g'_{\mu\nu}$ are then equal to the $g_{\mu\nu}$ up to a constant factor; and the laws of nature in the new theory again take the form

$$T(g) = 0.$$

Compared to the original form of the *general theory of relativity*, the whole novelty consists then only in the addition of the differential equation (6), which the $g_{\mu\nu}$ must obey.

Our only intention was to point out a logical possibility that is worthy of publication*; it may be useful for physics or not*. Only further investigations can show whether one or the other is the case, and if there is more to be considered than one Weyl invariant $J = \sqrt{H}$.

**(8) Elementary and composite particles with the same electric charge attract each other, and elementary and composite particles with opposite electric charge are repulsed, through the electromagnetic interaction or electromagnetic force, according to Coulomb's law.**

The Standard Model adds nothing to the classical *non-relativistic* theory.

**Electromagnetism.**

*Electromagnetism* is an *interaction* that occurs between particles with *electric charge* via *electromagnetic fields*. The *electromagnetic force* is one of the four fundamental forces of nature. It is the dominant force in the interactions of atoms and molecules. *Electromagnetism* can be thought of as a combination of *electrostatics* and *magnetism*, which are distinct but closely intertwined phenomena. *Electromagnetic forces* occur between any two *charged particles*. *Electric forces* cause an attraction between particles with opposite charges and repulsion between particles with the same charge, while *magnetism* is an *interaction* that occurs between *charged particles in relative motion*. These two forces are described in terms of *electromagnetic fields*. Macroscopic charged objects are described in terms of *Coulomb's law for electricity* and *Ampère's force law for magnetism*; the *Lorentz force* describes microscopic charged particles.

**Coulomb's Law.**

*Coulomb's Law* states that the magnitude, or absolute value, of the attractive or repulsive electrostatic force between two point-charges is directly proportional to the product of the magnitudes of their charges and inversely proportional to the squared distance between them. [See Underwood, T. G. (2024). *Electricity & Magnetism*, pp. 52-4.]

**Ampère's Force Law for Magnetism.**

*Ampere's Force Law* is a relationship between the *magnetic field* of a closed path and the *current* around this path. It states that there is an attractive or repulsive force between two parallel wires carrying an *electric current* which is proportional to their *lengths* and to the *intensities* of their *currents*. [See Underwood, T. G. (2024). *Electricity & Magnetism*, pp. 124-5.]

## (9) The spin of elementary and composite particles creates an attractive force between two particles – the weak interaction or weak force - through exchange interaction.

**Exchange interaction.**

According to the *quark formulation* in the *Standard Model*, a *weak interaction occurs when two particles (typically, but not necessarily, half-integer spin fermions) exchange integer-spin, force-carrying bosons*. In the *weak interaction, fermions* can *exchange* three types of force carriers, namely W+, W−, and Z *bosons*. The *masses* of these *bosons* are far greater than the *mass* of a *proton* or *neutron*. The *weak interaction* is the only fundamental *interaction* that breaks *parity symmetry*, and similarly, but far more rarely, the only *interaction* to break *charge–parity symmetry*. The *weak interaction* is considered unique in that it allows *quarks* to *swap* their flavor for another. *Quarks*, which make up composite particles like *neutrons* and *protons*, come in six "*flavors*" – *up*, *down*, *charm*, *strange*, *top* and *bottom* – which give those composite particles their properties. The *swapping* of those properties is mediated by the force carrier *bosons*. For example, during *beta-minus decay*, a *down quark* within a *neutron* is changed into an *up quark*, thus converting the *neutron* to a *proton* and resulting in the emission of an *electron* and an *electron antineutrino*.

**Weak isospin.**

In the 1960s, Sheldon Glashow, Abdus Salam and Steven Weinberg unified the *electromagnetic force* and the *weak interaction* by showing them to be two aspects of a single force, now termed the *electroweak force*. In Glashow, S. L. (February, 1961). Partial-symmetries of weak interactions. *Nuclear Physics*, 22, 4, 579–88, proposed that a relation similar to the Gell-Mann–Nishijima formula for *charge* to *isospin* would also apply to the *weak interaction*[1]:

[1] Glashow, S. L. (February, 1961). Partial-symmetries of weak interactions. *Nuclear Physics*, 22, 4, 579–88. See below.

Here the *charge* is related to the projection of *weak isospin* and the *weak hypercharge*. Glashow combined the *electromagnetic* and *weak interactions* and extended *electroweak unification models* due to Schwinger by including a short-range *neutral current*, the $Z_0$. The resulting *symmetry structure* that Glashow proposed, SU(2) × U(1), forms the basis of the accepted *theory of the electroweak interactions*.

[The *hypercharge* of a particle is a quantum number conserved under the *strong interaction*. The concept of *hypercharge* provides a single *charge operator* that accounts for properties of *isospin*, *electric charge*, and *flavor*. The *hypercharge* is useful to classify *hadrons*; the similarly named *weak hypercharge* has an analogous role in the *electroweak interaction*.

*Hypercharge* is one of two *quantum numbers* of the SU(3) model of *hadrons*, alongside *isospin I3*. The *isospin* alone was sufficient for two *quark flavors* — namely u and d — whereas presently 6 *flavors* of *quarks* are known.]

*Isospin* and *weak isospin* are related to the same symmetry but for different forces. *Weak isospin* is the *gauge symmetry* of the *weak interaction* which connects *quark* and *lepton doublets* of left-handed particles in all generations; for example, *up* and *down quarks*, *top* and *bottom quarks*, *electrons* and *electron neutrinos*. By contrast (strong) *isospin* connects only *up* and *down quarks*, acts on both *chiralities* (left and right) and is a *global* (not a *gauge*) *symmetry*.

## Glashow, S. L. (February, 1961). Partial-symmetries of weak interactions†.

*Nuclear Physics*, 22, 4, 579–88; https://doi.org/10.1016/0029-5582(61)90469-2.

† The article is hidden behind pay walls but can be purchased from the publisher.

Glashow combined the *electromagnetic* and *weak interactions* and extended *electroweak unification models* due to Schwinger by including a short-range *neutral current*, the $Z_0$. The resulting *symmetry structure* that Glashow proposed, SU(2) × U(1), forms the basis of the accepted *theory of the electroweak interactions*. The *W and Z bosons* were predicted in detail by Sheldon Glashow, Mohammad Abdus Salam, and Steven Weinberg. For this discovery, Glashow along with Steven Weinberg and Abdus Salam, was awarded the 1979 Nobel Prize in Physics.

---

***Abstract.***

*Weak and electromagnetic interactions* of the *leptons* were examined under the hypothesis that the *weak interactions* are mediated by *vector bosons*. With only an *isotopic* triplet of *leptons* coupled to a *triplet* of *vector bosons* (two charged decay intermediaries and the photon), the theory possesses no *partial symmetries*. Such symmetries may be established if additional vector *bosons* or additional *leptons* are introduced. Since the latter possibility yields a theory disagreeing with experiment, *the simplest partially symmetric model reproducing the observed electromagnetic and weak interactions of leptons requires the existence of at least four vector-boson fields* (including the *photon*). Corresponding partially conserved quantities *suggest leptonic analogs to the conserved quantities associated with strong interactions: strangeness and isobaric spin.*

[A *lepton* is an elementary particle of half-integer *spin* (*spin* ½) that does not undergo *strong interactions*. Two main classes of *leptons* exist: *charged leptons* (also known as the *electron-like leptons* or *muons*), including the *electron*, *muon*, and *tauon*, and *neutral leptons*, better known as *neutrinos*. *Charged leptons* can combine with other particles to form various composite particles such as atoms and *positronium*, while *neutrinos* rarely interact with anything, and are consequently rarely observed. The best known of all *leptons* is the *electron*.

There are six types of *leptons*, known as flavors, grouped in three generations. The *first-generation leptons*, also called *electronic leptons*, comprise the *electron* ($e^-$) and the *electron neutrino* ($\nu_e$); the second are the *muonic leptons*, comprising the *muon* ($\mu^-$) and the *muon neutrino* ($\nu\mu$); and the third are the *tauonic leptons*, comprising the *tau* ($\tau^-$) and the *tau neutrino* ($\nu\tau$). *Electrons* have the least mass of all the *charged leptons*. The heavier *muons* and *taus* will rapidly change into *electrons* and *neutrinos* through a process of particle decay: the transformation from a higher mass state to a lower mass state. Thus, *electrons* are stable and the most common charged *lepton* in the universe, whereas *muons* and *taus* can only be produced in high-energy collisions (such as those involving cosmic rays and those carried out in particle accelerators).

*Leptons* have various intrinsic properties, including *electric charge*, *spin*, and *mass*. Unlike *quarks*, however, *leptons* are not subject to the *strong interaction*, but they are subject to the other three fundamental interactions: *gravitation*, the *weak interaction*, and to *electromagnetism*, of which the latter is proportional to charge, and is thus zero for the electrically neutral *neutrinos*.

For every *lepton flavor*, there is a corresponding type of *antiparticle*, known as an *antilepton*, that differs from the *lepton* only in that some of its properties have equal magnitude but opposite sign. According to certain theories, *neutrinos* may be their own antiparticle. It is not currently known whether this is the case.

The first *charged lepton*, the *electron*, was theorized in the mid-19th century by several scientists and was discovered in 1897 by J. J. Thomson. The next *lepton* to be observed was the *muon*, discovered by Carl D. Anderson in 1936, which was classified as a meson at the time. After investigation, it was realized that the *muon* did not have the expected properties of a meson, but rather behaved like an *electron*, only with higher mass. It took until 1947 for the concept of "*leptons*" as a family of particles to be proposed. The term *lepton* was first used by physicist Léon Rosenfeld in 1948.

The first *neutrino*, the *electron neutrino*, was proposed by Wolfgang Pauli in 1930 to explain certain characteristics of *beta decay*. It was first observed in the Cowan–Reines *neutrino* experiment conducted by Clyde Cowan and Frederick Reines in 1956. The *muon neutrino* was discovered in 1962 by Leon M. Lederman, Melvin Schwartz, and Jack Steinberger, and the *tau* discovered between 1974 and 1977 by Martin Lewis Perl and his colleagues from the Stanford Linear Accelerator Center and Lawrence Berkeley National Laboratory. The *tau neutrino* remained elusive until July 2000, when the DONUT collaboration from Fermilab announced its discovery.

*Leptons* are an important part of the *Standard Model*. *Electrons* are one of the components of atoms, alongside *protons* and *neutrons*. Exotic atoms with *muons* and *taus* instead of *electrons* can also be synthesized, as well as *lepton–antilepton* particles such as *positronium*.

A *boson* is a subatomic particle whose *spin* quantum number has an integer value (0, 1, 2, ...). *Bosons* form one of the two fundamental classes of subatomic particle, the other being *fermions*, which have odd half-integer spin (1⁄2, 3⁄2, 5⁄2, ...). Every observed subatomic particle is either a *boson* or a *fermion*.

Some *bosons* are elementary particles occupying a special role in particle physics, distinct from the role of *fermions* (which are sometimes described as the constituents of "ordinary matter"). Certain elementary *bosons* (e.g. *gluons*) act as force carriers, which give rise to forces between other particles, while one (the *Higgs boson*) contributes to the phenomenon of *mass*. Other *bosons*, such as *mesons*, are composite particles made up of smaller constituents.

Paul Dirac coined the name *boson* to commemorate the contribution of Satyendra Nath Bose, an Indian physicist.]

**(10) Composite particles, such as protons and neutrons, can exist as different quantum states, referred to as isospin states, which create an attractive force – the strong interaction or strong force - through exchange interaction between two quantum isospin states.**

**Isospin.**

Particles, such as protons and neutrons, can exist as different *quantum states*, referred to as *isospin states*. The name of the concept contains the term *spin* because its quantum mechanical description is mathematically similar to that of *angular momentum* (in particular, in the way it *couples*; for example, a *proton–neutron pair* can be *coupled* either in a *state* of *total isospin* 1 or in one of 0. But unlike angular momentum, it is a dimensionless quantity and is not actually any type of spin.

Before the concept of *quarks* was introduced, particles that are affected equally by the *strong force* but had different *electric charges* (e.g. *protons* and *neutrons*) were considered different states of the same particle, but having *isospin* values related to the number of *charge states*. A close examination of *isospin symmetry* ultimately led directly to the discovery and understanding of *quarks* and to the development of *Yang–Mills theory*.

***The particle zoo.***

These considerations would also prove useful in the analysis of *meson-nucleon interactions* after the discovery of the *pions* in 1947. The three *pions* ($\pi^+$, $\pi^0$, $\pi^-$) could be assigned to an *isospin triplet* with $I = 1$ and $I_3 = +1$, 0 or −1. By assuming that *isospin* was conserved by *nuclear interactions*, the new *mesons* were more easily accommodated by nuclear theory.

> [*Mesons* are made of an even number of *quarks* (usually two *quarks*: one *quark* and one *anti-quark*); and *baryons* are made of an odd number of *quarks* (usually three *quarks*). *Mesons* and *baryons* are *hadrons*.
>
> *Pions* are an example of a *meson. The lightest baryons are the nucleons: the proton and neutron* (which make the majority of the *mass* of an atom).
>
> A *hadron* (from Ancient Greek ἁδρός (hadrós) 'stout, thick') is a *composite subatomic particle made of two or more quarks held together by the strong interaction*. They are analogous to molecules, which are held together by the electric force. Most of the *mass* of ordinary matter comes from two *hadrons*: the *proton* and the *neutron*, while most of the *mass* of the *protons* and *neutrons* is in turn due to the *binding energy* of their constituent *quarks*, due to the *strong force*.

Almost all "free" *hadrons* and *anti-hadrons* (meaning, in isolation and not bound within an atomic nucleus) are believed to be unstable and eventually decay into other particles. The only known possible exception is free *protons*, which appear to be stable, or at least, take immense amounts of time to decay (order of $10^{34}$+ years). By way of comparison, free-standing *neutrons* are the longest-lived unstable particle, and decay with a *half-life* of about 611 seconds, and have a *mean lifetime* of 879 seconds.

A *meson* is a *composite particle* composed of *an equal number of quarks and antiquarks, usually one of each, bound together by the strong interaction*. Because *quarks* have a *spin* ½ , the difference in quark number between *mesons* and *baryons* results in *mesons* being *boson*s, whereas *baryons,* the other members of the *hadron* family, composed of *odd numbers of valence quarks* (at least three), are *fermions*.

Some experiments show evidence of exotic *mesons*, which do not have the conventional *valence quark* content of two *quark*s (one *quark* and one *antiquark*), but four or more.

Because *mesons* are composed of *quark* sub-particles, they have a meaningful physical size, a diameter of roughly one femtometre ($10^{-15}$ m), which is about 0.6 times the size of a *proton* or *neutron*. *All mesons are unstable*, with the longest-lived lasting for only a few tenths of a nanosecond. Heavier *mesons* decay to lighter *mesons* and ultimately to stable *electrons*, *neutrinos* and *photons*.

Each type of *meson* has a corresponding antiparticle (*antimeson*) in which *quarks* are replaced by their corresponding *antiquarks* and vice versa.

Because *mesons* are composed of *quarks*, they participate in both the *weak interaction* and *strong interaction*. *Mesons* with net *electric charge* also participate in the *electromagnetic interaction*. *Mesons* are classified according to their *quark content*, *total angular momentum*, *parity* and various other properties, such as *C-parity* and *G-parity*. Although no *meson* is stable, those of lower mass are nonetheless more stable than the more massive, and hence are easier to observe and study in particle accelerators or in cosmic ray experiments. The lightest group of *mesons* is less massive than the lightest group of *baryons*, meaning that they are more easily produced in experiments, and thus exhibit certain higher-energy phenomena more readily than do *baryons*. But *mesons* can be quite massive: for example, the J/Psi meson (J/ψ) containing the charm quark, first seen 1974, is about three times as massive as a *proton*, and the upsilon meson (ϒ) containing the bottom quark, first seen in 1977, is about ten times as massive as a *proton*.

The existence of *mesons* was predicted by Hideki Yukawa's 1935 *theory of mesons* that postulated the particle as mediating the nuclear force. [Yukawa, H. (1935). On the Interaction of Elementary Particles. *Proceedings of the Physico-Mathematical Society of Japan*, 17, 48; also in (January, 1955). *Progr. Theoret. Phys. Suppl.* 1, 1–10, https://doi.org/10.1143/PTPS.1.1.

A *pion* or *pi meson*, is any of three subatomic composite particles: $\pi^0$, $\pi^+$, and $\pi^-$. In the *quark model*, the neutral *pion* $\pi^0$ is a combination of an *up quark* (0.002 times the *mass* of a *proton*) with an *anti-up quark* (0.002 times the *mass* of a *proton*), or a *down quark* (0.005 times the *mass* of a *proton*) with an *anti-down quark* (0.005 times the *mass* of a *proton*). The two combinations have identical *quantum numbers*, and hence they are only found in superpositions. The lowest-energy superposition of these is the $\pi^0$, which is its own *antiparticle*. An *up quark* (0.002 times the *mass* of a *proton*) and an *anti-down quark* (0.005 times the *mass* of a *proton*) make up a $\pi+$, whereas a *down quark* and an *anti-up quark* make up the $\pi-$, and these are the *antiparticles* of one another. Together, the *pions* form a *triplet* of *isospin*. Each *pion* has overall *isospin* (I = 1) and third-component isospin equal to its charge ($I_z = +1, 0$, or $-1$). Each *pion* consists of a *quark* and an *antiquark* and is therefore a *meson*. *Pions* are the lightest *mesons* with the $\pi\pm$ *mesons* having a *mass* of 139.6 MeV/$c^2$ (0.15 times the *mass* of a *proton*) and a *mean lifetime* of $2.6033\times10^{-8}$ s, and the $\pi^0$ *meson* having a *mass* of 135.0 MeV/$c^2$ (0.15 times the *mass* of a *proton*) and a *mean lifetime* of $8.5\times10^{-17}$ s. The $\pi^0$ *meson* decays via the *electromagnetic force*, which explains why its *mean lifetime* is much smaller than that of the *charged pion* (which can only decay via the *weak force*). *Charged pions* most often decay into *muons* and *muon neutrinos*, while *neutral pions* generally decay into *gamma rays*.

The exchange of virtual *pions*, along with *vector, rho* and *omega mesons*, provides an explanation for the residual *strong force* between *nucleons*. *Pions* are not produced in radioactive decay, but commonly are in high-energy collisions between *hadrons*. *Pions* also result from some *matter–antimatter annihilation* events. All types of *pions* are also produced in natural processes when high-energy cosmic-ray *protons* and other hadronic cosmic-ray components interact with matter in Earth's atmosphere. In 2013, the detection of characteristic *gamma rays* originating from the decay of *neutral pions* in two supernova remnants has shown that *pions* are produced copiously after supernovas, most probably in conjunction with production of high-energy protons that are detected on Earth as cosmic rays

In 1947, the *pion* (or *pi meson*) was discovered by members of C. F. Powell's group at the University of Bristol, in England, including César Lattes, Giuseppe Occhialini and Hugh Muirhead. In the same year, Lattes, together with Powell and Occhialini, determined the new particle's mass. [Lattes, C. M. G., Muirhead, H.,

Occhialini, G. P. S. & Powell, C. F. (1947). Processes involving charged mesons. *Nature*. 159, 4047, 694–7; https://doi.org/ 10.1038/159694a0.]

In 1948, Lattes, Eugene Gardner, and their team first artificially produced *pions* at the University of California's cyclotron in Berkeley, California, by bombarding carbon atoms with high-speed alpha particles.]

As further particles were discovered, they were assigned into *isospin multiplet*s according to the number of different *electric charge states* seen: 2 *doublets* $I = 1/2$ of K *mesons* ($K^-$, $K^0$), ($K^+$,$K^0$), a *triplet* $I = 1$ of *Sigma baryons* ($\Sigma^+$, $\Sigma^0$, $\Sigma^-$), a *singlet* $I = 0$ *Lambda baryon* ($\Lambda^0$), a *quartet* $I = 3/2$ *Delta baryons* ($\Delta^{++}$, $\Delta^+$, $\Delta^0$, $\Delta^-$), and so on.

[A *multiplet* is the *state space* for 'internal' degrees of freedom of a particle, that is, degrees of freedom associated to a particle itself, as opposed to 'external' degrees of freedom such as the particle's position in space. Examples of such degrees of freedom are the *spin state* of a particle in *quantum mechanics*, or the *color*, *isospin* and *hypercharge state* of particles in the *Standard Model* of particle physics. Formally, this state space is described by a *vector space* which carries the *action* of a group of *continuous symmetries*.]

The power of *isospin symmetry* and related methods comes from the observation that families of particles with similar *masses* tend to correspond to the *invariant subspaces* associated with the irreducible representations of the Lie algebra SU(2). In this context, an *invariant subspace* is spanned by basis vectors which correspond to particles in a family. Under the action of the *Lie algebra* SU(2), which generates rotations in *isospin space*, elements corresponding to definite particle *states* or superpositions of *states* can be rotated into each other, but can never leave the space (since the *subspace* is in fact *invariant*). This is reflective of the *symmetry* present. The fact that *unitary matrices* will commute with the Hamiltonian means that the physical quantities calculated do not change even under *unitary transformation*. In the case of *isospin*, this machinery is used to reflect the fact that the mathematics of the *strong force* behaves the same if a *proton* and *neutron* are swapped around (in the modern formulation, the *up* and *down quark are switched* [or the quantum states are *entangled*?]).

***An example: Delta baryons.***

For example, the *composite particles* known as the *Delta baryons* – *baryons* of *spin* 3/2 – were grouped together because they all have nearly the same *mass* (approximately 1,232 MeV/$c^2$, or 1.31 times the *mass* of a *proton*) and interact in nearly the same way. Four closely related Δ *baryons* exist: $\Delta^{++}$ (constituent *quarks*: uuu), $\Delta^+$ (uud), $\Delta^0$ (udd), and

$\Delta^-$ (ddd), which respectively carry an *electric charge* of +2e, +1e, 0e, and −1e. They are all very unstable, with the same *mean lifetime* 5.63 x 10−24 s.

They could be treated as the same particle, with the difference in *electric charge* being due to the particle being in different *states*. *Isospin* was introduced in order to be the variable that defined this difference of *state*. In an analogue to *spin*, an *isospin projection* (denoted $I_3$) is associated to each *charged state*; since there were four Deltas, four projections were needed. Like *spin*, *isospin* projections were made to vary in increments of 1. Hence, in order to have four increments of 1, an *isospin* value of 3/2 is required (giving the projections $I_3$ = +3/2, +1/2, −1/2, −3/2). Thus, all the Deltas were said to have *isospin* $I$ = 3/2, and each individual *charge* had different $I_3$ (e.g. the $\Delta^{++}$ was associated with $I_3$ = +3/2).

In the *isospin* picture, the four *Delta*s and the two *nucleons* were thought to simply be the different *states* of two particles. In the *quark formulation*, the *Delta baryons* are understood to be made of a mix of three *up* and *down quarks* (respectively, 0.002 and 0.005 times the *mass* of a *proton*) – uuu ($\Delta^{++}$), uud ($\Delta^+$), udd ($\Delta^0$), and ddd ($\Delta^-$); the difference in *electric charge* being difference in the *charges* of *up* and *down quarks* (+2/3 *e* and −1/3 *e* respectively); yet, they can also be thought of as the excited *states* of the *nucleons*.

***Relation to hypercharge.***

The *charge operator* can be expressed in terms of the projection of *isospin* and *hypercharge*, Y:

$$Q = \tfrac{1}{2} Y + T_3, \qquad T_3 = T, T-1, \ldots, -T.$$

This is known as the *Gell-Mann–Nishijima formula*.

> [The *Gell-Mann–Nishijima formula* (sometimes known as the *NNG formula*) relates the *baryon number* B, the *strangeness* S, the *isospin* $I_3$ of *quarks* and *hadrons* to the *electric charge* Q. It was originally given by Kazuhiko Nishijima and Tadao Nakano in 1953, and led to the proposal of *strangeness* as a concept, which Nishijima originally called "eta-charge" after the eta meson. Murray Gell-Mann proposed the formula independently in 1956. The modern version of the formula relates all *flavor* quantum numbers (*isospin up* and *down*, *strangeness*, *charm*, *bottomness*, and *topness*) with the *baryon number* and the *electric charge*.]

The *hypercharge* is the center of splitting for the *isospin multiplet*:

$$\tfrac{1}{2} Y = \tfrac{1}{2} (Q_{min} + Q_{max}).$$

This relation has an analog in the *weak interaction* where T is the *weak isospin.*

***Quark content and isospin.***

In the *quark formulation, isospin* (*I*) is defined as a vector quantity in which *up and down quarks* have a value of $I = 1/2$, with the 3rd-component ($I_3$) being +1/2 for *up quarks*, and −1/2 for *down quarks*, while all other *quarks* have $I = 0$. Therefore, for *hadrons* in general, where $n_u$ and $n_d$ are the numbers of *up* and *down quarks* respectively,

$$I_3 = ½ (n_u – n_d).$$

In any combination of *quarks*, the 3rd component of the *isospin vector* ($I_3$) could either be aligned between a pair of *quarks*, or face the opposite direction, giving different possible values for *total isospin* for any combination of *quark flavors*. *Hadrons* with the same *quark* content but different *total isospin* can be distinguished experimentally, verifying that *flavor* is actually a *vector* quantity, not a scalar (*up* vs *down* simply being a projection in the quantum mechanical *z* axis of *flavor* space).

For example, a *strange quark* can be combined with an *up* and a *down quark* to form a *baryon*, but there are two different ways the *isospin* values can combine – either adding (due to being flavor-aligned) or cancelling out (due to being in opposite flavor directions). The *isospin*-1 state (the $\Sigma^0$) and the *isospin*-0 state (the $\Lambda^0$) have different experimentally detected *masses* and *half-lives*.

***Isospin and symmetry.***

In the *quark formulation, isospin* is regarded as a *symmetry* of the *strong interaction* under the action of the *Lie group* SU(2), the two *states* being the *up flavor* and *down flavor*. In quantum mechanics, when a Hamiltonian has a *symmetry*, that symmetry manifests itself through a set of *states* that have the same *energy* (the states are described as being *degenerate*). In simple terms, the *energy operator* for the *strong interaction* gives the same result when an *up quark* and an otherwise identical *down quark* are swapped around.

Like the case for regular *spin*, the *isospin operator* **I** is *vector-valued*: it has three components $\mathbf{I}_x$, $\mathbf{I}_y$, $\mathbf{I}_z$, which are coordinates in the same 3-dimensional *vector space* where the **3** *representation* acts. Note that this *vector space* has nothing to do with the physical space, except similar mathematical formalism. *Isospin* is described by two *quantum numbers*: *I* – the *total isospin*, and $I_3$ – an *eigenvalue* of the $\mathbf{I}_z$ projection for which *flavor states* are *eigenstates*. In other words, each $I_3$ *state* specifies a certain *flavor state* of a *multiplet*. The third coordinate (*z*), to which the "3" subscript refers, is chosen due to notational conventions that relate bases in **2** and **3** *representation spaces*. Namely, for the *spin-1/2* case, components of **I** are equal to *Pauli matrices* divided by 2, and so $\mathbf{I}_z = 1/2\ \tau_3$, where

$$\tau_3 = \begin{pmatrix} 1 & 0 \\ 0 & -1 \end{pmatrix}.$$

While the forms of these matrices are isomorphic to those of *spin*, *these* Pauli matrices only act within the *Hilbert space* of *isospin*, not that of *spin*, and therefore it is common to denote them with $\boldsymbol{\tau}$ rather than $\boldsymbol{\sigma}$ to avoid confusion.

Although *isospin symmetry* is actually very slightly broken, SU(3) *symmetry* is more badly broken, *due to the much higher mass of the strange quark compared to the up and down*. The discovery of *charm*, *bottomness* and *topness* could lead to further expansions up to SU(6) *flavor symmetry*, which would hold if all six *quarks* were identical. However, the very much larger *masses* of the *charm*, *bottom*, and *top quarks* means that SU(6) *flavor symmetry* is very badly broken in nature (at least at low energies), and *assuming this symmetry leads to qualitatively and quantitatively incorrect predictions*. In modern applications, such as lattice QCD, *isospin symmetry* is often treated as exact for the three light *quarks* (uds), while the three heavy *quarks* (cbt) must be treated separately.

*Quark*s carry three types of *color charge*; *antiquarks* carry three types of *anti-color*. *Quarks*, which have electric charges +2/3 e or – 1/3 e, are never found in isolation; they can be found only within *hadrons*, which include *baryons* (such as *protons* and *neutrons*) and *mesons*, or in *quark–gluon plasmas*. For this reason, much of what is known about *quarks* has been drawn from observations of *hadrons*. The key evidence for the existence of *quarks* came from a series of inelastic electron-nucleon scattering experiments conducted between 1967 and 1973 at the Stanford Linear Accelerator Center.

*Gluons* are also subject to the *color charge* phenomena. *Gluons* carry both *color* and *anti-color*. This gives nine *possible* combinations of *color* and *anti-color* in *gluons*, of which 8 are independent. These *possible* combinations are only *effective* states, not the *actual* observed *color states* of *gluons*. To understand how they are combined, it is necessary to consider the mathematics of *color charge* in more detail. Neither can *gluons* be observed. The most direct available evidence for the existence of *gluons* are three-jet events in which is an event with many particles in a final state that appear to be clustered in three jets. A single jet consists of particles that fly off in roughly the same direction. One can draw three cones from the interaction point, corresponding to the jets, and most particles created in the reaction will appear to belong to one of these cones. These were first observed by the TASSO experiment at the PETRA accelerator at the DESY laboratory.

**Exchange Interaction of isospin.**

Under the *six-quark* formulation of the *Standard Model*, *isospin* (*I*) is a *quantum number*, referred to as the *baryon number*, related to the *up- and down quark* content of the particle. *Isospin* is also known as *isobaric spin* or *isotopic spin*. In the terms of the *six-quark* formulation, *isospin states* are quantum states in which the *up* and *down quarks* are switched around. *Isospin* symmetry is a subset of the *flavor* symmetry seen more broadly in the *interactions* of *baryons* and *mesons*.

> [*Protons* and *neutrons* are *baryons*, a composite subatomic particle *that contains an odd number of valence quarks and antiquarks*, conventionally three. *Baryons* belong to the *hadron* family of particles; *hadrons* are composed of *quarks*. Because *quarks* have a *spin* ½, the difference in *quark* number results in being *fermions* because they have half-integer *spin*.
>
> Each *baryon* has a corresponding antiparticle (*antibaryon*) where their corresponding *antiquarks* replace *quarks*. *Baryons* participate in the residual *strong force*, which is mediated by *mesons*.
>
> The most familiar *baryons* are *protons* and *neutrons*, both of which contain three *quarks*, and for this reason they are sometimes called *triquarks*. A *proton* is made of two *up quarks* and one *down quark*; and its corresponding antiparticle, the *antiproton*, is made of two *up antiquarks* and one *down antiquark*. A *neutron* consists of one *up quark* and two *down quark*s.
>
> *Protons* and *neutrons* make up most of the mass of the visible matter in the universe and compose the nucleus of every atom (*electrons*, the other major component of the atom, are members of a different family of particles called *leptons*; leptons do not interact via the *strong force*). Exotic *baryons* containing five quarks, called *pentaquarks*, have also been discovered and studied.
>
> *Baryons* are strongly interacting *fermions*; that is, they are acted on by the *strong nuclear force* and are described by *Fermi–Dirac statistics*, which apply to all particles obeying the Pauli exclusion principle. This is in contrast to the *bosons*, which do not obey the exclusion principle.
>
> The name "*baryon*", introduced by Abraham Pais, comes from the Greek word for "heavy" (βαρύς, barýs), because, at the time of their naming, most known elementary particles had lower masses than the *baryons*.]

***Gauged isospin symmetry.***

Attempts have been made to promote *isospin* from a *global* to a *local symmetry*. In 1954, Chen Ning Yang and Robert Mills suggested that the notion of *protons* and *neutrons*, which are continuously rotated into each other by *isospin*, should be allowed to vary from point to point. To describe this, the *proton* and *neutron* direction in *isospin space* must be defined at every point, giving *local* basis for *isospin*. A *gauge* connection would then describe how to transform *isospin* along a path between two points.

> [Yang, C. N. & Mills, R. (October, 1954). Conservation of Isotopic Spin and Isotopic Gauge Invariance. *Phys. Rev*., 96, 1, 191–5; https://journals.aps.org/pr/pdf/10.1103/PhysRev.96.191: Chen-Ning Yang and Robert Mills extended the concept of *gauge theory* for *abelian* groups, e.g. *quantum electrodynamics*, to *nonabelian* groups to provide an explanation for *strong interactions*. The *Yang–Mills theory* is a *quantum field theory* for *nuclear binding*. It is a *gauge theory* based on a *special unitary group* SU(n), or more generally any compact Lie group. It seeks to describe the behavior of *elementary particles* using these non-abelian Lie groups and *is at the core of the unification of the electromagnetic force and weak forces* (i.e. U(1) × SU(2)) as well as *quantum chromodynamics*, the theory of the *strong force* (based on SU(3)). Thus, it forms the basis of the understanding of the *Standard Model* of particle physics. See Underwood, T. G. (2024). *The Standard Model*.]

The Yang–Mills theory describes interacting *vector bosons*, like the *photon* of *electromagnetism*.

> [A *vector boson* is a *boson* whose *spin* equals one. *Vector bosons* that are also elementary particles are *gauge bosons*, the force carriers of fundamental interactions. Some composite particles are *vector bosons*, for instance any *vector meson* (*quark* and *antiquark*). During the 1970s and 1980s, intermediate *vector bosons* (the *W* and *Z bosons*, which mediate the *weak interaction*) drew much attention in particle physics.]

Unlike the *photon*, the SU(2) gauge theory would contain self-interacting *gauge bosons*. The condition of *gauge invariance* suggests that they have *zero mass*, just as in *electromagnetism*.

Ignoring the *massless* problem, as Yang and Mills did, the theory makes a firm prediction: the vector particle should *couple* to all particles of a given *isospin universally*. The *coupling* to the *nucleon* would be the same as the *coupling* to the *kaons*.

[*Kaon* (or *K meson*) *composite particles* have proved to be a copious source of information on the nature of fundamental interactions since their discovery in cosmic rays in 1947. They were essential in establishing the foundations of the *Standard Model* of particle physics, such as the *quark model* of *hadrons* and the theory of *quark* mixing.

In 1947, the *kaon*, the first *strange* particle, was co-discovered by George Dixon Rochester, and Clifford Charles Butler, two British physicists at the University of Manchester. They published two cloud chamber photographs of cosmic ray-induced events, one showing what appeared to be a neutral particle decaying into two charged *pions*, and one which appeared to be a charged particle decaying into a charged *pion* and something neutral.

*Kaons* have also played a distinguished role in our understanding of fundamental conservation laws: *CP violation*, a phenomenon generating the observed *matter–antimatter asymmetry* of the universe [???], was discovered in the *kaon* system in 1964. [Christenson, J. H., Cronin, J. W., Fitch, V. L. & Turlay, R. (July 1964). Evidence for the 2π Decay of the K20 Meson. *Phys. Rev. Let*., 13, 4, 138–40; https://doi.org/10.1103/PhysRevLett.13.13.] This was acknowledged by the award of the 1980 Nobel Prize in Physics jointly to James Watson Cronin and Val Logsdon Fitch "for the discovery of violations of fundamental symmetry principles in the decay of neutral K-mesons".

Direct *CP violation* was discovered in the *kaon* decays in the early 2000s by the NA48 experiment at CERN and the KTeV experiment at Fermilab. As the *quark model* shows, assignments that *the kaons form two doublets of isospin*; that is, they belong to the fundamental representation of SU(2) called the **2**. One doublet of strangeness +1 contains the $K^+$ and the $K^0$. The antiparticles form the other doublet (of strangeness −1).]

A *kaon*, also called a *K meson*, is a *composite particle*; any of a group of four *mesons* distinguished by a *quantum number* called *strangeness*. In the *quark model* they are understood to be bound states of a *strange quark* (or *antiquark*) and an *up or down antiquark* (or *quark*).

The four *kaons* are:

$K^-$, *negatively charged* (containing a *strange quark* – *mass* 87 MeV/$c^2$ – and an *up antiquark* – *mass* 1.9 MeV/$c^2$ [???]) has *mass* 493.677 ± 0.013 MeV/$c^2$ (0.53 times the *mass* of a *proton*) and *mean lifetime* (1.2380 ± 0.0020) × $10^{-8}$ s;

$K^+$ (*antiparticle* of above) *positively charged* (containing an *up quark* and a *strange antiquark*) must (by CPT invariance) have *mass* and *lifetime* equal to that of $K^-$. Experimentally, the *mass* difference is 0.032 ± 0.090 MeV, consistent with zero; the difference in lifetimes is $(0.11 \pm 0.09) \times 10^{-8}$ s, also consistent with zero;

$K^0$, *neutrally charged* (containing a *down quark* – *mass* 4.4 MeV/$c^2$ – and a *strange antiquark* – *mass* 87 MeV/$c^2$) has *mass* 497.648 ± 0.022 MeV (0.53 times the *mass* of a *proton*). It has mean squared *charge radius* of − 0.076 ±0.01 $fm^2$;

$K^0$, *neutrally charged* (*antiparticle* of above) (containing a *strange quark* and a *down antiquark*) has the same *mass*.

The *quark model* shows that the *kaons* form two doublets of *isospin*; that is, they belong to the fundamental representation of SU(2) called the **2**. One doublet of *strangeness* +1 contains the $K^+$ and the $K^0$. The *antiparticles* form the other doublet (of *strangeness* −1). The coupling to the *pions* would be the same as the self-coupling of the *vector bosons* to themselves.

When Yang and Mills proposed their theory, there was no candidate *vector boson*, but in 1960 J. J. Sakurai predicted that there should be a *massive vector boson* coupled to *isospin*, that would show *universal couplings*. The *rho mesons* were discovered a short time later, and were quickly identified as Sakurai's *vector bosons*.

The *rho meson* is a short-lived hadronic *composite particle* that is an *isospin triplet* whose three *states* are denoted as $\rho^+$, $\rho^0$ and $\rho^-$. Along with *pions* and *omega mesons*, the *rho meson* carries the *nuclear force* within the *atomic nucleus*. After the *pions* and *kaons*, the *rho mesons* are the lightest *strongly interacting* particle, with a *mass* of 775.45 MeV/$c^2$ (0.83 times the *mass* of a *proton*) for all three *states*.

The *rho mesons* have a very short *lifetime* and their decay width is about 145 MeV with the peculiar feature that the decay widths are not described by a Breit–Wigner form. The principal decay route of the *rho mesons* is to a pair of *pions* with a branching rate of 99.9%.

After several false starts, the ρ *meson* and the ω *meson* were discovered at Lawrence Berkeley Laboratory in 1961.

The *rho mesons* can be interpreted as a *bound state* of a *quark* and an *anti-quark* and is an excited version of the *pion*. Unlike the *pion*, the *rho meson* has spin $j = 1$ (a *vector meson*) and a much higher value of the *mass*. This *mass difference* between the *pions* and *rho mesons* is attributed to a large *hyperfine interaction* between the *quark* and *anti-quark*.

The couplings of the *rho* to the *nucleons* and to each other were verified to be universal, as best as experiment could measure. The fact that the diagonal *isospin current* contains part of the *electromagnetic current* led to the prediction of *rho-photon* mixing and the concept of *vector meson* dominance, ideas which led to successful theoretical pictures of GeV-scale *photon-nucleus scattering*.

## Mean lifetimes of composite particles (hadrons)

| | *Mean lifetime* | *Mass ($MeV/c^2$)* | *Proton masses* |
|---|---|---|---|
| **Mesons:** | | | |
| Pions (pi mesons): | | | |
| $\pi^+$ | $2.60 \times 10^{-8}$ s | 139.6 | 0.15 |
| $\pi^-$ | $2.60 \times 10^{-8}$ s | 139.6 | 0.15 |
| $\pi^0$ | $8.50 \times 10^{-17}$ s | 139.6 | 0.15 |
| Kaons (K mesons): | | | |
| $K^+$ | $1.24 \times 10^{-8}$ s | 493.7 | 0.53 |
| $K^-$ | $1.24 \times 10^{-8}$ s | 493.7 | 0.53 |
| $K^0$ | $1.24 \times 10^{-8}$ s | 497.6 | 0.53 |
| Rho mesons | | | |
| $\rho^+$ | $4.42 \times 10^{-24}$ s | 775.5 | 0.83 |
| $\rho^-$ | $4.42 \times 10^{-24}$ s | 775.5 | 0.83 |
| $\rho^0$ | $4.45 \times 10^{-24}$ s | 775.5 | 0.83 |
| **Baryons:** | | | |
| Proton | stable | 938.3 | 1.00 |
| Anti-proton | claimed to be stable but short-lived | | 1.00 |
| Neutron | $9.9 \times 10^2$ s | 939.6 | 1.001 |
| Delta ($\Delta$) baryons | | | |
| $\Delta^{++}$ | $5.63 \times 10^{-24}$ s | 1,232 | 1.31 |
| $\Delta^+$ | $5.63 \times 10^{-24}$ s | 1,232 | 1.31 |
| $\Delta^-$ | $5.63 \times 10^{-24}$ s | 1,232 | 1.31 |
| $\Delta^0$ | $5.63 \times 10^{-24}$ s | 1,232 | 1.31 |

# PART II The foundational assumptions of New Physics.

## (1) The universe is composed of elementary particles.

Ideally, New Physics should confine itself to *elementary particles* that can be *observed*. On this basis both *quarks* and *gluons*, and possibly the $W^+$ *boson*, $W^-$ *boson*, $Z^0$ *boson*, *Higgs boson*, which are theoretical constructs based on *relativistic quantum field theory*, should be removed from this list.

This restores the *proton* and the *neutron* to the list of *elementary particles*. Although the *proton* and *neutron* can be considered as different *isospin quantum states* of the same particle, they will be treated as separate *elementary particles*.

### Antimatter and antiparticles.

#### Elimination of notion of *antimatter* and supposed problem of asymmetry of *matter* and *antimatter* in the visible universe.

Contrary to Dirac's speculation (below) and to the assertion in the *Standard Model*, *particles* and its corresponding *antiparticles* do not have identical *masses* and *decay lifetimes*. *Antiparticles* are simply the less stable *particles* of similar *mass* but opposite *electric charge*; the *positron* is the less stable *antiparticle* of the *electron*, and the *antiproton* is the less stable *antiparticle* of the *proton*. Then, since *antimatter* comprises the less stable forms of *matter*, this provides a simple explanation for the asymmetry of *matter* and *antimatter* in the visible universe.

**Paul Dirac – 1933 Nobel Lecture, December 12, 1933. *Theory of electrons and positrons***, pp. 324-5: Dirac speculated that there was a *complete symmetry in elementary particles between positive and negative charge*:

"The theory of *electrons* and *positrons* which I have just outlined is a self-consistent theory which fits the experimental facts so far as is yet known. One would like to have an equally satisfactory theory for *protons*. One might perhaps think that the same theory could be applied to *protons*. This would require the possibility of existence of negatively charged *protons* forming a mirror-image of the usual positively charged ones. There is, however, some recent experimental evidence obtained by Stern about the *spin magnetic moment* of the *proton*, which conflicts with this theory for the *proton*. As the *proton* is so much heavier than the *electron*, it is quite likely that it requires some more complicated theory, though one cannot at the present time say what this theory is.

In any case I think it is probable that negative *protons* can exist, since as far as the theory is yet definite, *there is a complete and perfect symmetry between positive and negative electric charge*, and if this *symmetry* is really fundamental in nature, it must be possible to reverse the *charge* on any kind of particle. The negative *protons* would of course be much harder to produce experimentally, since a much larger *energy* would be required, corresponding to the larger *mass*.

If we accept the view of *complete symmetry between positive and negative electric charge* so far as concerns the fundamental laws of Nature, we must regard it rather as an accident that the Earth (and presumably the whole solar system), contains a preponderance of negative *electrons* and positive *protons*. It is quite possible that for some of the stars it is the other way about, these stars being built up mainly of *positrons* and negative *protons*. In fact, there may be half the stars of each kind. The two kinds of stars would both show exactly the same spectra, and there would be no way of distinguishing them by present astronomical methods."

## Preliminary list of elementary particles in New Physics.

These changes would result in the following *preliminary list of observed elementary particles*, each *charged particle* with its *antiparticle* (unstable particle of opposite electric charge). Omitting the *quarks*, *gluons*, $W^+$, $W^-$, $Z^0$ and the *Higgs bosons*, this amounts to a total of 14 *elementary particles* and *antiparticles* in *New Physics* of which the *electron*, *photon* and *proton* (and the *neutron* in the nucleus) are stable.

| **Fermions:** | | **Particle half-life** |
|---|---|---|
| Leptons: | | |
| 1-2 | Electron | stable (6.6 × $10^{28}$ years) |
| 3 | Electron neutrino | ? |
| 4-5 | Muon | 2.2 x $10^{-6}$ s |
| 6 | Muon neutrino | ? |
| 7-8 | Tau | 2.9 x $10^{-13}$ s |
| 9 | Tau neutrino | ? |
| Nucleons: | | |
| 10-11 | Proton | stable (1 x $10^{32}$ years) |
| 12 | Neutron (free standing) | 8.9 x $10^2$ s |
| | Neutron (in nucleus) | stable |
| **Bosons:** | | |
| Gauge bosons: | | |
| 13 | Photon | stable |
| 14 | Graviton | |

## **Paul Dirac – 1933 Nobel Lecture, December 12, 1933.** ***Theory of electrons and positrons.***

[https://www.nobelprize.org/prizes/physics/1933/dirac/lecture/.]

> Dirac shared the 1933 Nobel Prize in Physics with Erwin Schrödinger "for the discovery of new productive forms of atomic theory".
>
> During the intense period of 1925-26 quantum theories were proposed that accurately described the energy levels of electrons in atoms. These equations needed to be adapted to Einstein's theory of relativity, however. In 1928 Paul Dirac formulated a fully *relativistic quantum theory*. The equation gave solutions that he interpreted as being caused by a particle equivalent to the *electron*, but with a positive *charge*. This particle, the *positron*, was later confirmed through experiments. [Paul A.M. Dirac – Facts. NobelPrize.org. https://www.nobelprize.org/prizes/physics/1933/dirac/facts/.]

In his Nobel Lecture, Dirac described the current state of his theory of *electrons* and *positrons*. Dirac recognized that "*there exists no relativistic quantum mechanics (that is, one valid for large velocities) which can be applied to particles with arbitrary properties*", but used this in his prediction of the *positron*. "Thus, when one subjects quantum mechanics to *relativistic* requirements, one imposes restrictions on the properties of the particle. In this way one can deduce information about the particles from purely theoretical considerations, based on general physical principles."

---

> Matter has been found by experimental physicists to be made up of small particles of various kinds, the particles of each kind being all exactly alike. Some of these kinds have definitely been shown to be composite, that is, to be composed of other particles of a simpler nature. But there are other kinds which have not been shown to be composite and which one expects will never be shown to be composite, so that one considers them as elementary and fundamental.
>
> From general philosophical grounds one would at first sight like to have as few kinds of elementary particles as possible, say only one kind, or at most two, and to have all matter built up of these elementary kinds. It appears from the experimental results, though, that there must be more than this. In fact, the number of kinds of *elementary particle* has shown a rather alarming tendency to increase during recent years.
>
> The situation is perhaps not so bad, though, because on closer investigation it appears that the distinction between *elementary* and *composite* particles cannot be made rigorous. To get an interpretation of some modern experimental results one must suppose that particles

can be created and annihilated. Thus, *if a particle is observed to come out from another particle, one can no longer be sure that the latter is composite. The former may have been created.* The distinction between elementary particles and composite particles now becomes a matter of convenience. This reason alone is sufficient to compel one to give up the attractive philosophical idea that all matter is made up of one kind, or perhaps two kinds of bricks.

I should like here to discuss the simpler kinds of particles and to consider what can be inferred about them from purely theoretical arguments. The simpler kinds of particle are:

(i) the *photons* or light-quanta, of which light is composed;

(ii) the *electrons*, and the recently discovered *positrons* (which appear to be a sort of mirror image of the *electrons*, differing from them only in the sign of their *electric charge*);

(iii) the heavier particles - *protons* and *neutrons*.

Of these, I shall deal almost entirely with the *electrons* and the *positrons* - not because they are the most interesting ones, but because in their case the theory has been developed further. There is, in fact, hardly anything that can be inferred theoretically about the properties of the others. The *photons*, on the one hand, are so simple that they can easily be fitted into any theoretical scheme, and the theory therefore does not put any restrictions on their properties. The *protons* and *neutrons*, on the other hand, seem to be too complicated and no reliable basis for a theory of them has yet been discovered.

The question that we must first consider is how theory can give any information at all about the properties of elementary particles. There exists at the present time a general quantum mechanics which can be used to describe the motion of any kind of particle, no matter what its properties are. The general quantum mechanics, however, is valid only when the particles have small velocities and fails for velocities comparable with the velocity of light, when effects of *relativity* come in. *There exists no relativistic quantum mechanics (that is, one valid for large velocities) which can be applied to particles with arbitrary properties. Thus, when one subjects quantum mechanics to relativistic requirements, one imposes restrictions on the properties of the particle. In this way one can deduce information about the particles from purely theoretical considerations, based on general physical principles.*

This procedure is successful in the case of *electrons* and *positrons*. It is to be hoped that in the future some such procedure will be found for the case of the other particles. I should like here to outline the method for *electrons* and *positrons*, showing how one can deduce the *spin* properties of the *electron*, and then how one can infer the existence of *positrons* with similar *spin* properties and with the possibility of being annihilated in collisions with *electrons*.

We begin with the equation connecting the *kinetic energy* W and *momentum* $p_r$ (r = 1, 2, 3), of a particle in *relativistic* classical mechanics

$$W^2/c^2 - p_r^2 - m^2c^2 = 0. \quad (1)$$

From this we can get a *wave equation* of quantum mechanics, by letting the left-hand side operate on the wave function ψ and understanding W and $p_r$ to be the operators $i\hbar\partial/\partial t$ and $-i\hbar\partial/\partial x_r$. With this understanding, the *wave equation* reads

$$[W^2/c^2 - p_r^2 - m^2c^2]\,\psi = 0 \quad (2)$$

Now it is a general requirement of quantum mechanics that its *wave equations* shall be linear in the operator W or $\partial/\partial t$ so this equation will not do. We must replace it by some equation linear in W, and in order that this equation may have *relativistic invariance* it must also be linear in the p's.

We are thus led to consider an equation of the type

$$[W/c - \alpha_r p_r - \alpha_0 mc]\,\psi = 0 \quad (3)$$

This involves four new variables $\alpha_r$ and $\alpha_0$ which are operators that can operate on ψ. We assume they satisfy the following conditions,

$$\alpha_\mu^2 = I \qquad \alpha_\mu\alpha_\nu + \alpha_\nu\alpha_\mu = 0$$

for

$$\mu \neq \nu \text{ and } \mu, \nu = 0, 1, 2, 3$$

and also, the α's commute with the p's and W. These special properties for the α's make Eq. (3) to a certain extent equivalent to Eq. (2), since if we then multiply (3) on the left-hand side by $W/c + \alpha_r p_r + \alpha_0 mc$ we get exactly (2).

The new variables α, which we have to introduce to get a *relativistic wave equation* linear in W, give rise to the *spin* of the *electron*. From the general principles of quantum mechanics one can easily deduce that these variables a give the *electron* a *spin angular momentum* of half a quantum and a *magnetic moment* of one Bohr magneton in the reverse direction to the *angular momentum*. These results are in agreement with experiment. They were, in fact, first obtained from the experimental evidence provided by spectroscopy and afterwards confirmed by the theory.

The variables α also give rise to some rather unexpected phenomena concerning the motion of the *electron*. These have been fully worked out by Schrödinger. *It is found that an electron which seems to us to be moving slowly, must actually have a very high frequency oscillatory motion of small amplitude superposed on the regular motion which appears to us. As a result of this oscillatory motion, the velocity of the electron at any time equals the velocity of light*. This is a prediction which cannot be directly verified by experiment, since the frequency of the oscillatory motion is so high and its amplitude is so small. But one must believe in this consequence of the theory, since other consequences of the theory which are inseparably bound up with this one, such as the law of scattering of light by an *electron*, are confirmed by experiment.

There is one other feature of these equations which I should now like to discuss, *a feature which led to the prediction of the positron*. If one looks at Eq. (1), one sees that it allows the *kinetic energy* W to be either a positive quantity greater than $mc^2$ or a negative quantity less than $-mc^2$. This result is preserved when one passes over to the quantum equation (2) or (3). These quantum equations are such that, when interpreted according to the general scheme of quantum dynamics, they allow as the possible results of a measurement of W either something greater than $mc^2$ or something less than $-mc^2$.

Now in practice the *kinetic energy* of a particle is always positive. We thus see that our equations allow of two kinds of motion for an *electron*, only one of which corresponds to what we are familiar with. *The other corresponds to electrons with a very peculiar motion such that the faster they move, the less energy they have, and one must put energy into them to bring them to rest.*

One would thus be inclined to introduce, as a new assumption of the theory, that only one of the two kinds of motion occurs in practice. But this gives rise to a difficulty, since we find from the theory that if we disturb the *electron*, we may cause a transition from a positive-energy *state* of motion to a negative-energy one, so that, even if we suppose all the electrons in the world to be started off in positive-energy *states*, after a time some of them would be in negative-energy *states*.

Thus, in allowing negative-energy states, the theory gives something which appears not to correspond to anything known experimentally, but which we cannot simply reject by a new assumption. We must find some meaning for these *states*.

An examination of the behavior of these *states* in an *electromagnetic field* shows that they *correspond to the motion of an electron with a positive charge instead of the usual negative one* - what the experimenters now call a *positron*. One might, therefore, be inclined to assume that *electrons* in negative-energy *states* are just *positrons*, but this will not do, because *the observed positrons certainly do not have negative energies*. We can, however, establish a connection between *electrons* in negative-energy *states* and *positrons*, in a rather more indirect way.

We make use of the exclusion principle of Pauli, according to which there can be only one *electron* in any state of motion. We now make the assumptions that in the world as we know it, nearly all the *states* of negative *energy* for the *electrons* are occupied, with just one *electron* in each *state*, and that a uniform filling of all the negative-energy *states* is completely unobservable to us. Further, any unoccupied negative-energy *state*, being a departure from uniformity, is observable and is just a *positron*.

An unoccupied negative-energy *state*, or *hole*, as we may call it for brevity, will have a positive *energy*, since it is a place where there is a shortage of negative *energy*. *A hole is, in fact, just like an ordinary particle, and its identification with the positron seems the most reasonable way of getting over the difficulty of the appearance of negative energies in our equations*. On this view *the positron is just a mirror-image of the electron*, having exactly the same *mass* and opposite *charge*. This has already been roughly confirmed by experiment. The *positron* should also have similar *spin* properties to the *electron*, but this has not yet been confirmed by experiment.

From our theoretical picture, we should expect an ordinary *electron*, with positive *energy*, to be able to drop into a hole and fill up this hole, the energy being liberated in the form of *electromagnetic radiation*. This would mean a process in which an *electron* and a *positron* annihilate one another. The converse process, namely the creation of an *electron* and a *positron* from *electromagnetic radiation*, should also be able to take place. Such processes appear to have been found experimentally, and are at present being more closely investigated by experimenters.

## Fermi, E. (March, 1934) Versuch einer Theorie der β-Strahlen. I. (Attempt at a theory of β rays. I.)[1]

*Zeit. Phys.,* 88, 161–77; https://doi.org/10.1007/BF01351864; also at https://www.nssp.uni-saarland.de/lehre/Vorlesung/Kernphysik_SS19/History/Papers/Fermi_1.pdf (German); translation by T. G. Underwood; also in Underwood, T. G. (2024). *The Standard Model*, Part I, pp. 140-7.

Rome.

[1] Cf. the preliminary communication: (1933). *La Ricerca Scientifica*, 2, 12.

With 3 illustrations.

Received January 16, 1934.

---

***Abstract***

Fermi proposed the first theory of the *weak interaction*, known as *Fermi's interaction*, in which four fermions directly interact with one another. A quantitative theory of *β decay* is proposed, in which the existence of the *neutrino* is assumed, and the emission of *electrons* and *neutrinos* from a nucleus in the β case is treated with a method similar to that of the emission of a quantum of light from an excited atom in radiation theory. Formulas for the lifetime and for the shape of the emitted continuous β radiation spectrum are derived and compared with experience. Tree Feynman diagrams describe the interaction remarkably well. Unfortunately, loop diagrams cannot be calculated reliably because Fermi's interaction is *not renormalizable*.

> [The existence of the *neutrino* had been proposed by Wolfgang Pauli in 1930 to explain the apparent violation of conservation of energy in *beta decay*. In a letter of 4 December to Lise Meitner et al., beginning, "Dear radioactive ladies and gentlemen", he proposed the existence of a hitherto unobserved neutral particle with a small mass, no greater than 1% the mass of a proton, to explain the continuous spectrum of beta decay. In 1934, Enrico Fermi incorporated the particle, which he called a *neutrino*, "little neutral one" in Fermi's native Italian, into his theory of beta decay. [Fermi, E. (1934). Radioattività indotta da bombardamento di neutroni. Ricerca Scientifica, 5, 1, 283; reprinted in Collected Papers (*Note e Memorie*), vol. 1 Italy 1921–1938 (Chicago: University of Chicago Press; Rome: Accademia Nazionale dei Lincei, 1962), 645–46; translated as "Radioactivity Induced by Neutron Bombardment. – I," in Collected Papers, 674–75. Also in Wilson, F. L.

(1968). Fermi's Theory of Beta Decay. *Am. J. of Physics*, 36, 12, 1150–1160; https://doi.org/10.1119/1.1974382: (includes complete English translation of Fermi's 1934 paper published in *Zeitschrift für Physik* in 1934; https://pubs.aip.org/aapt/ajp/article-abstract/36/12/1150/1047952/Fermi-s-Theory-of-Beta-Decay?redirected From=fulltext.

In 1956, the *electron neutrino* was detected by Frederick Reines and Clyde Cowan. [F. Reines, F. & Cowan, C.L. (1956). The Neutrino. *Nature*, 178, 4531, 446–9; https://doi.org/10.1038/178446a0.] At the time it was simply referred to as the *neutrino* since there was only one known *neutrino*. Frederick Reines (March 16, 1918–August 26, 1998) and Clyde Cowan (December 6, 1919–May 24, 1974) were both American physicists. The Nobel Prize in Physics 1995 was awarded "for pioneering experimental contributions to *lepton* physics" jointly with one half to Martin L. Perl "for the discovery of the *tau lepton* [in 1975]" and with one half to Frederick Reines "for the detection of the *neutrino* [in 1956]".]

---

***1. Basic assumptions of the theory.***

In the attempt to build up a theory of nuclear *electrons* and *β emission*, one encounters two visions. The first is due to the continuous β radiation spectrum. If the law of conservation of energy is to remain valid, one must assume that a fraction of the energy released during the β decay escapes our previous observation possibilities. According to W. Pauli's proposal, one can assume, for example, that not only an electron but also a new particle, the so-called "neutrino" (mass of the order of magnitude or less than the electron mass; no electric charge) is emitted during β decay. In the present theory, we will use the hypothesis of the neutrino.

Another difficulty for the theory of nuclear electrons is that the current relativistic theories of light particles (electrons or neutrinos) are not able to explain in a conclusive way how such particles can be bound in orbits of nuclear dimensions.

It therefore seems more appropriate to assume with Heisenberg[2] that a nucleus consists only of heavy particles, protons and neutrons.

[2] Heisenberg, W. (January, 1932). Über den Bau der Atomkerne. I. (About the construction of atomic nuclei. I.). *Zeit. Phys*., 77, 1–11; https://doi.org/10.1007/BF01342433; see above.

Nevertheless, in order to understand the possibility of *β emission*, we will try to construct a theory of the emission of light particles from a nucleus by analogy to the theory of the emission of a light quantum from an excited atom in the ordinary radiation process. In radiation theory, the total number of light quanta is not a constant: light quanta are created when they are emitted by an atom and disappear when they are absorbed. By analogy to this, we want to base the *β radiation* theory on the following assumptions:

a) The total number of *electrons*, as well as *neutrinos*, is never necessarily constant. *Electrons* (or *neutrinos*) can form and disappear. This possibility has no analogy with the emergence or disappearance of a pair of an *electron* and a positron; if one interprets the positron as Dirac's "hole", this last process can indeed be understood as a quantum leap of an *electron* between a *state* with negative energy and a *state* with positive energy with conservation of the total (infinitely large) number of *electrons*.

b) The heavy particles, *neutrons* and *protons*, can be regarded as two *inner quantum states* of the heavy particle, as in Heisenberg. We formulate this by introducing an inner coordinate ρ of the visual particle, which can only take two values: ρ = 1 if the particle is a *neutron*; ρ = – 1 if the particle is a *proton*.

c) The Hamilton function of the system consisting of heavy and light particles must be chosen in such a way that each transition from *neutron* to *proton* is associated with the formation of an *electron* and a *neutrino*. The reverse process, the transformation of a *proton* into a *neutron*, on the other hand, is said to be associated with the disappearance of an *electron* and a *neutrino*. It should be noted that this ensures the preservation of the *charge*.

### *2. The operators that occur in theory.*

A mathematical formalism of the theory in accordance with these three requirements can be most easily constructed with the help of the Dirac-Jordan-Klein method[1] of the "*second quantization*".

[1] Cf. e.g. Jordan, B. P. & Klein, O. (November, 1927). Zum Mehrkörperproblem der Quantentheorie. (On the multibody problem of quantum theory.) *Zeit. Phys.*, 45, 751–65; https://doi.org/10.1007/BF01329553; Heisenberg, W. (1931). Zum Paulischen Ausschließungsprinzip. (On the Paulian exclusion principle.) *Ann. Phys.*, 10, 888; https://doi.org/10.1002/andp.19314020710.

We will therefore consider the *probability amplitudes* ψ and φ of *electrons* and *neutrinos* as well as the *complex conjugated quantities* ψ* and φ* as *operators*; for the description of the heavy particles, on the other hand, we will use the usual *representation* in the *configuration space*, whereby of course ρ must also be counted as a coordinate.

We first introduce two *operators* Q and Q*, which act on the functions of the two-valued variable ρ as the linear substitutions.

$$Q = \begin{vmatrix} 0 & 1 \\ 0 & 0 \end{vmatrix}; \qquad Q^* = \begin{vmatrix} 0 & 0 \\ 1 & 0 \end{vmatrix} \qquad (1)$$

It is easy to see that Q corresponds to a transition from *proton* to *neutron* and Q* to a transition from *neutron* to *proton.*

…

### *3. Setting up the Hamiltonian function.*

The *energy* of the entire system, consisting of heavy and light particles, is the sum of the *energies* $H_{heavy}$ of the heavy particles + $H_{light}$ of the light particles + the *interaction energy* H between heavy and light particles.

We write the first member, considering for the time being only a single visible particle, in the form

$$H_{heavy} = (1 + \rho)/2\ N + (1 - \rho)/2\ P, \qquad (6)$$

where N and P represent the energy operators of the *neutron* and *proton*, respectively. For $\rho = 1$ (*neutron*), (6) is indeed reduced to N; for $\rho = -1$ (*proton*) (6) reduces to P.

The *energy* $H_{light}$ of the light particles assumes the simplest form if one takes as *quantum states* $\psi_1\psi_2$ .. $\psi_s$ ... and $\varphi_1\varphi_2$ .. $\varphi_u$ ... as the *stationary states* for the *electrons* or *neutrinos*. For the *electrons*, for example, the *stationary states* in the *Coulomb field* of the *nucleus* should be chosen, taking into account the *electron displacement*. For *neutrinos*, one can simply assume plane de Broglie waves, since the forces acting on the *neutrinos* do not play a significant role. Let $H_1H_2$ ... Hs... and $K_1K_2$ ... $K_\sigma$ ... the *energies* of the *stationary states* of *electrons* and *neutrinos*; Then we have:

$$H_{light} = \Sigma s\ H_sN_s + \Sigma_\sigma K_\sigma M_\sigma. \qquad (7)$$

All that remains to be written is the *energy of interaction*. First, this consists of the *Coulomb energy* between *proton* and *electrons*; in heavy *nuclei*, however, the attraction of a single *proton* plays only a subordinate role[1] and in no case contributes to the process of *β decay*.

[1] The Coulombian effect of the numerous other *protons* must of course be taken into account as a *static field.*

For the sake of simplicity, therefore, we will not consider this link. On the other hand, we must add to the Hamiltonian function a term that satisfies the condition c) for digit 1.

…

### *4. The perturbation matrix.*

The *theory of β decay* can be carried out with the help of the established Hamiltonian function in full analogy to radiation theory. In the latter, as is well known, the Hamiltonian function consists of the sum: *energy of the atom + energy of the pure radiation field + coupling energy*. This last link is understood as a disturbance of the other two. By analogy to this, in our case the sum

$$H_{heavy} + H_{light} \qquad (16)$$

as an unperturbed Hamiltonian function; in addition, there is the disturbance represented by the *coupling element* (13).

[$H = g[Q\psi^{\sim *}\delta\varphi + Q^*\psi^{\sim}\delta\varphi^*]$, (13)

where ψ and φ are to be written as vertical matrix columns.]

The *quantum states* of the unperturbed system can be numbered as follows:

$$(\rho, n, N_1N_2 \ldots N_S \ldots M_1M_2 \ldots M_\sigma \ldots), \qquad (17)$$

where the first number ρ takes one of the two values ± 1 and indicates whether the heavy particle is a *neutron* or a *proton*. The second number n numbers the *quantum state* of the *neutron* or *proton*. For ρ = 1 (*neutron*), let the corresponding *eigenfunction*

$$\upsilon_n(x), \qquad (18)$$

where x represents the coordinates of the heavy particles, except for ρ. For ρ = – 1 (*proton*) let the *eigenfunction* be

$$v_n(x). \qquad (19)$$

The remaining numbers $N_1N_2 \ldots N_S \ldots M_1M_2 \ldots M_\sigma \ldots$ are only capable of the two values 0 and 1 and indicate whether the respective state of the *electron* or the *neutrino* is occupied.

If we now consider the general form (9)

[$H = Q\,\Sigma_{s\sigma}\, c_{s\sigma}\, a_s\, b_\sigma + Q^*\,\Sigma_{s\sigma}\, c^*_{s\sigma}\, a^*_s\, b^*_\sigma$ (9)

where $c_{s\sigma}$ and $c^*_{s\sigma}$ represent quantities that depend on the coordinates, momenta, etc. of the heavy particle.]

of the *perturbation energy*, we see that it has different elements from zero only for such *transitions* in which either the heavy particle changes from a *neutron* to a *proton state* and at the same time an *electron* and a *neutrino* are produced, or vice versa.

With the help of (1), (3), (5), (9), (18), (19) you can easily find the matrix element in question

$$H_{-1mN1N2\ldots 1s\ldots M1M2\ldots 1\sigma\ldots}^{1nN1N2\ldots 0s\ldots M1M2\ldots 0\sigma\ldots} = \pm \int v^*_m c^*_{s\sigma} \upsilon_n \, d\tau \qquad (20)$$

where the integration must be extended over the *configuration space* of the heavy particle (except for the coordinate ρ). The ± sign means more precisely

$$(-1)^{N1 + N2 + \ldots + Ns-1 + M1 + M2 + \ldots + M\sigma-1}$$

and, by the way, will fall out of the following calculations. The opposite *transition* corresponds to a *complex conjugate matrix element*.

If you introduce the value (15) for $c^*_{s\sigma}$, you get

$$H_{-1m1s1\sigma}^{1n0s0\sigma} = \pm\, g \int v^*_m \, \upsilon_n \, \tilde{\psi}_s \, \delta \, \varphi^*_\sigma \, d\tau \qquad (21)$$

where, because of the brevity, all the constant indices have been omitted in the first term.

### *5. Theory of β decay.*

A *β decay* consists of a process in which a *nuclear neutron* turns into a *proton* and at the same time an *electron*, which is observed as a *β ray*, and a *neutrino* are emitted with the mechanism described. In order to calculate the probability of this process, let us assume that at the time t = 0 a *neutron* exists in a *nuclear state* with *eigenfunction* $\upsilon_n(x)$ and $N_s = M_\sigma = 0$, i.e. the *electron state* s and the *neutrino state* σ are empty. Then for t = 0 is the *probability amplitude* of the *state* $(1, n, 0_s, 0\sigma)$

$$a_1 n 0_s 0_\sigma = 1 \qquad (22)$$

and that of the *state* $(-1, m, 1_s, 1_\sigma)$ where the *neutron* has changed into a *proton* with the *eigenfunction* $v_m(x)$ under emission of an *electron* and a *neutrino*, equal to zero.

…

### *6. Determinants of the transition probability.*

(31) indicates the probability that a *β decay* with the transition of the *electron* to the *state* s will take place during time t. As it is, this probability is proportional to the time t (t has been assumed to be small in terms of lifetime); the coefficient of t indicates the *transition probability* for the process described.

…

### *7. The mass of the neutrino.*

The *transition probability* (32)

$$[P_s = 8\pi^3 g^2/h^4 \,|\, \int \upsilon^*_m u_n \, d\tau \,|^2 \, p_\sigma^2/\upsilon_\sigma \, (\psi\tilde{}_s \psi_s - \mu c^2/K_\sigma \, \psi\tilde{}_s \beta \psi_s). \qquad (32)]$$

determines the shape of the continuous β spectrum. We first want to discuss how this form depends on the *rest mass* μ the *neutrino* in order to determine this constant by comparing it with the empirical curves. The *mass* μ is contained in the factor $p_\sigma^2/v_\sigma$. The dependence of the shape of the *energy distribution curve* on μ is most pronounced near the end point of the *distribution curve*. If $E_0$ is the *limiting energy* of the *β rays*, it is not difficult to see that the *distribution curve* for *energies* E in the vicinity of $E_0$ behaves as follows, except for one factor independent of E.

$$p_\sigma^2/v_\sigma = 1/c^3 \, (\mu c^2 + E_0 - E) \sqrt{\{(E_0 - E)^2 + 2\mu c^2(E_0 - E)\}} \qquad (36)$$

…

### *8. Lifetime and shape of the distribution curve for "permitted" transitions.*

From (39) one can derive a formula which indicates how many *β transitions* take place in the unit of time for which the β particle receives *momentum* between mcη and mc (η + dη). To do this, one must derive a formula for the sum of $\psi\tilde{}_s\psi_s$ at the location of the *nucleus* over all *quantum states* of the respective interval in the continuous spectrum.

It should be noted that the *relativistic eigenfunctions* in the *Coulomb field* for the *states* with j = ½ ($^2s_{1/2}$ and $^2p_{1/2}$) for r = 0 become *infinitely large*. Now, however, the nuclear attraction for the *electrons* obeys *Coulomb's law* only up to r > ρ, where ρ here means the *nuclear radius*. A rough calculation shows that if one makes plausible assumptions about the course of the *electric field* within the *nucleus*, the value of $\psi\tilde{}_s\psi_s$ at the center has a value that is very close to the value that $\psi\tilde{}_s\psi_s$ would assume in the case of *Coulomb's law* at the distance ρ from the center.

…

The reciprocal *lifetime* is obtained from (44) by integrating η = 0 to η = $\eta_0$; one finds:

$$1/\tau = 1.75 \,.\, 10^{95} \, g^2 \,|\, \int \upsilon^*_m u_n \, d\tau \,|^2 \, F(\eta_0), \qquad (45)$$

where $F(\eta_0)$ = … . (46)

### *9. The forbidden transitions.*

Before we move on to the comparison with experience, let's discuss some characteristics of the forbidden *β transitions*.

As already mentioned, a *transition* is forbidden if the corresponding matrix element (35) disappears. If the *representation* of the *nucleus* with individual *quantum states* of the *neutrons* and *protons* is a good approximation, $Q^*_{mn}$ always disappears for symmetry reasons, if not

$$i = i', \qquad (47)$$

where i and i' represent the *momentum* moments (in units h/2π) of the *neutron state* $\upsilon_n$ and the *proton state* $v_m$ .

The *selection rule* (47), if the individual *states* are not a good approximation, corresponds to the more general

$$l = l', \qquad (48)$$

where $l$ and $l'$ mean the *momentum* moments of the *nucleus* before and after the *β decay*.

...

***10. Comparison with experience.***

Formula (45)

$$[1/\tau = 1.75 \,.\, 10^{95}\, g^2 \,|\, \textstyle\int \upsilon^*_m\, u_n\, d\tau\, |^2\, F(\eta_0), \qquad (45)]$$

gives a relationship between the maximum *momentum* of the emitted *β-rays* and the *lifetime* of the β-radiating substance: In this relationship, an unknown element still occurs, namely the integral

$$\int v^*_m\, \upsilon_n\, d\tau, \qquad (50)$$

for the evaluation of which a knowledge of the *eigenfunctions* of the *proton* and the *neutron* in the nucleus would be necessary. However, in the case of *permitted transitions*, (50) is of order 1. So, you can expect that the product

$$\tau F\,(\eta_0) \qquad (51)$$

has the same order of magnitude for all *permitted transitions*. However, if the *transition* in question is *prohibited*, the *lifetime* is about 100 times longer than in the normal case and the product (51) is also correspondingly larger.

Table 2 lists the products (51) for the radioactive elements for which sufficient data are available on the continuous β spectrum.

...

To sum up, it should be said that the theory as given here is in agreement with the experimental data, which are not always particularly accurate. If, moreover, a closer comparison of theory and experience should lead to contradictions, it would still be possible to modify the theory without touching its conceptual foundations. One could keep equation (9) and make a different choice of the $c_{s\sigma}$. In particular, this could lead to a modification of the selection rule (48) and result in a different form of the *energy distribution curve* and the dependence of the *lifetime* on the *maximum energy*. Whether such a change will be necessary, however, can only be shown by a further development of the theory and possibly also by a tightening of the experimental data.

## (2) The speed of light in a vacuum is constant relative to the emitter.

### Avoidance of *length contraction* and *time dilation* for a moving observer by replacing Einstein's theory of Special Relativity with Walter Ritz's theory that the speed of light in a vacuum is constant relative to the emitter.

**Underwood, T. G. (2023). *Special Relativity*,** Conclusion, p. 381, 1st paragraph: "There is no evidence, based directly or indirectly, on the observation of the speed of electromagnetic radiation in a vacuum emitted by an inertial body, or as observed by an inertial observer, moving in a straight line and not involving mirrors. However, by now it may be possible to achieve this in a laboratory experiment in a vacuum without mirrors, using electromagnetic radiation emitted by two sources of the same frequency, one stationary and the other moving at a constant velocity in a straight line; either directly, or by measuring the observed frequency of the radiation. …"

**Walter Ritz's theory that the speed of light in a vacuum is constant relative to the emitter.**

*Walter Ritz's emission theory*, also called emitter theory or ballistic theory of light, was a competing theory for the *special theory of relativity*, explaining the results of the Michelson–Morley experiment of 1887. Emission theories obey the principle of relativity by having no preferred frame for light transmission, but say that *light is emitted at speed "c" relative to its source* instead of applying the invariance postulate. Thus, emitter theory combines electrodynamics and mechanics with a simple Newtonian theory. Although there are still proponents of this theory outside the scientific mainstream, this theory is considered to be conclusively discredited by most scientists. [???]

**Walther Heinrich Wilhelm Ritz (February 22, 1878 – July 7, 1909).**

> [Includes material from Martinez, A. A. (2004). Ritz, Einstein, and the Emission Hypothesis. *Phys. Perspect.*, 6, 4-28; https://doi.org/10.1007/s00016-003-0195-6]

Ritz was a Swiss theoretical physicist. He is most famous for his work with Johannes Rydberg on the Rydberg–Ritz combination principle. He is also famous for his *emission theory* of electromagnetic radiation. Ritz is also known for the variational method named after him, the Ritz method.

Walter Ritz's father Raphael Ritz was born in Valais and was a well-known painter. His mother, born Nördlinger, was the daughter of an engineer from Tübingen. Ritz was a particularly gifted student and attended the municipal lyceum in Sion. From an early age he exhibited a disposition for science and mathematics. Yet, also from an early age, his studies were hampered by recurring ill health. He first began to suffer from respiratory ailments at the age of nineteen, following a traumatic experience in September 1897, "Climbing Mont Pleureur with friends, he looked back to see a group of them slip on fresh snow and plunge over a cliff; the emotional stress was compounded by physical overexertion and overexposure in the rescue efforts." [Paul Forman, "Ritz, Walter," in Charles Coulston Gillispie, ed., *Dictionary of Scientific Biography*, Vol. 11 (New York: Charles Scribner's Sons, 1975), p. 475.] Nevertheless, that fall he took and passed the entrance examination to the Zurich Polytechnikum (later the Eidgenössische Technische Hochschule), to study engineering. Soon, he found out that he could not live with the approximations and compromises associated with engineering, and so he switched to "pure science," that is, to theoretical physics.

Albert Einstein, a year younger than Ritz, also was a student at the Zurich Polytechnikum. Einstein studied in the same section as Ritz, but had entered a year earlier, in 1896. The two registered for some courses with some of the same professors. In 1900, Ritz contracted tuberculosis, possibly also pleurisy, which he later died from. Ritz had made a better impression at Zurich than Einstein. Einstein graduated in 1900, while Ritz left in 1901 after severe illness, and moved to Göttingen for health reasons, to study further at the University of Göttingen, where he was influenced by Woldemar Voigt and David Hilbert.

Ritz wrote a dissertation on spectral lines of atoms and received his doctorate with *summa cum laude*. The theme later led to the Ritz combination principle and in 1913 to the atomic model of Ernest Rutherford and Niels Bohr.

In early 1903 he went to the University of Leiden in the company of his close friend Paul Ehrenfest to attend H.A. Lorentz's lectures on the theory of the electron. In June 1903 he was in Bonn at the Heinrich Kayser Institute, where he found in potash a spectral line that he had predicted in his dissertation. In November 1903, he was in Paris at the Ecole Normale Supérieure. There he worked on infrared photo plates.

In July 1904 his illness worsened and he moved back to Zurich. The disease prevented him from publishing further scientific publications until 1906. In September 1907 he moved to Tübingen, the place of origin of his mother, and in 1908 again to Göttingen, where he became a private lecturer at the university.

In 1908, Ritz found empirically the Ritz combination principle named after him. After that, the sum or difference of the frequencies of two spectral lines is often the frequency of another line. Which of these calculated frequencies is actually observed was only explained later by selection rules, which follow from quantum mechanical calculations.

Unlike many theorists at the time, Ritz was not impressed by Lorentz's approach to problems in electrodynamics. He became increasingly antagonistic to Maxwell's theory in general and to Lorentz's electrodynamics in particular.

Ritz set out on a research program consisting of two parts. First, he undertook a critical study of the contemporary theories of electrodynamics and identified their essential problems and inadequacies. He then sought to devise an alternate synthesis of optics with new electrodynamics that would account better for the experimental facts and provide a foundation on which to advance further.

In 1908 Ritz published his analyses in a 130-page criticism of Maxwell–Lorentz electromagnetic theory, [Ritz, W. (1908). Recherches critiques sur l'Électrodynamique Générale. (Critical Research on General Electrodynamics.) *Annales de Chimie et de Physique*, 13, 145-275; reprinted in (1911). Société Suisse de Physique, ed., *Gesammelte Werke Walther Ritz Œuvres*. Gauthier-Villars, Paris, pp. 317-426.] in which he contended that the theory's connection with the luminescent ether made it "essentially inappropriate to express the comprehensive laws for the propagation of electrodynamic actions." He followed up this work with a series of papers in which he recapitulated and elaborated his arguments.

Ritz pointed out seven problems with Maxwell–Lorentz electromagnetic field equations:

(1) Electric and magnetic forces really express relations about space and time and should be replaced with non-instantaneous elementary actions.

(2) Advanced potentials don't exist (and their erroneous use led to the Rayleigh–Jeans ultraviolet catastrophe).

(3) Localization of energy in the ether is vague.

(4) It is impossible to reduce gravity to the same notions.

(5) The unacceptable inequality of action and reaction is brought about by the concept of absolute motion with respect to the ether.

(6) Apparent relativistic mass increase is amenable to different interpretations.

(7) The use of absolute coordinates, if independent of all motions of matter, requires throwing away the time-honored use of Galilean relativity and our notions of rigid ponderable bodies.

Instead, he indicated that light is not propagated (in a medium) but is projected.

*According to Ritz, the essential difficulties in electrodynamics were rooted in the field equations of electromagnetism*. He stressed that Maxwell's equations admitted far too many possible solutions, infinitely many in principle, and that this plethora of solutions involved absurd physical consequences. He argued that *advanced potentials* were devoid of physical significance; he denied the plausibility of convergent spherical waves; and he complained that Maxwell's equations allowed for the existence of a *perpetuum mobile. To avoid the ambiguous multiplicity of solutions of Maxwell's equations, Ritz claimed that retarded potentials had to be taken as fundamental.* These equations embodied a delay required for electromagnetic effects to traverse distances in space. By allowing only *retarded potentials*, only past states of a system could determine its present state, and *energy could be radiated only from matter*, rather than, say, be drawn out infinitely from a surrounding ether.

Ritz complained that the fundamental electric and magnetic fields were not directly observable, and he argued, like Henri Poincaré before him, that *their physical interpretation involving the hypothesis of a stationary ether violated the principle of action and reaction.* He disdained the ether as a "mathematical phantom," quite undeserving the wide acceptance it had gained. Likewise, he regarded the electric and magnetic force vectors as playing the role of mathematical constructs useful only in particular cases, and he questioned their exact physical significance. Like Heinrich Hertz and others, Ritz deemed only relations of space, time, and matter as fundamental and therefore complained that electrodynamics was based on forces. He concluded that Maxwell's field equations, or more generally partial-derivative equations, were fundamentally inadequate to describe exactly the laws of propagation of physical actions.

Ritz directed his criticisms mainly at Lorentz's electrodynamics, although he was well aware of the related contributions of Poincaré, and of Einstein's theory. Like most other physicists during the first decade of the twentieth century, Ritz regarded Einstein's theory essentially as a generalized reformulation of Lorentz's. He distinguished between the two theories, but thought they both led to identical consequences. Ritz's papers suggest that he appreciated Einstein's theory somewhat, however, since he repeatedly turned to it to undermine Lorentz's. For example, although the Lorentz Maxwell theory involved a stationary ether, Ritz pointed out that Einstein had shown that Lorentz's equations were independent of the concept of absolute motion, and hence of the ether.

*Ritz also argued that Einstein had proven that the FitzGerald-Lorentz contraction was not a true physical effect, but merely an appearance, a consequence of an arbitrary definition*, that is, of the procedure for determining the simultaneity of events. Nonetheless, to Ritz, Einstein's work was basically a refinement of a fundamentally inadequate theoretical program. In renouncing classical mechanics, Einstein had paid too high a price to resolve the difficulties at issue; and however radical, his theory stopped short of altering Maxwell's equations. In general, Ritz just did not care for Einstein's theory. It not only preserved the core of the Maxwell-Lorentz electrodynamics, it also seemed to preserve a vestige of the ether by postulating the constancy of the velocity of light. To Ritz, Einstein had renounced too hastily key parts of classical mechanics, which seemed to be immensely less problematic than electrodynamics and optics.

In February 1909 Ritz completed his Habilitationsschrift at Göttingen, and in early March he gave his inaugural lecture as Privatdozent on the *principle of relativity* in optics. At that time, his reputation was such that a faculty committee at the University of Zurich considered Ritz to be the foremost of nine candidates to become their first professor of theoretical physics, noting that in the opinion of the Zurich physicist Alfred Kleiner, Ritz exhibited "an exceptional talent, bordering on genius". Ritz, however, had to be excluded from consideration because he was too ill to carry the workload, so the job went to Einstein instead.

In 1909 Ritz also developed a direct method to find an approximate solution for boundary value problems. It converts the often-insoluble differential equation into solution of a matrix equation. This method is also known Ritz's variation principle and the Rayleigh-Ritz principle.

As a student, friend or colleague, Ritz had contacts with many contemporary scholars such as Hilbert, Andreas Heinrich Voigt, Hermann Minkowski, Lorentz, Aimé Cotton, Friedrich Paschen, Henri Poincaré and Albert Einstein.

Ritz died in Göttingen on July 7, 1909, at the age of 31, after seven weeks in the Göttingen medical clinic, and was buried in the Nordheim cemetery in Zurich. The family tomb was lifted on November 15, 1999. His tombstone is in section 17 with the grave number 84457.

In his last year and a half, he had published a total of about four hundred pages of articles in the areas of theoretical spectroscopy, the foundations of electrodynamics, the problem of gravitation, and a method for the numerical solution of boundary-value problems. His collected works (541 pages), in German and French, were published two years later, in 1911. [Société Suisse de Physique, ed., *Gesammelte Werke Walther Ritz Œuvres*. Gauthier-Villars, Paris.]

**Ritz, W. (August, 1908) Recherches critiques sur les theories electrodynamiques de Cl. Maxwell et de H. A. Lorentz. (Critical research on the electrodynamic theories of Cl. Maxwell and H. A. Lorentz.)**

*Archives des Sciences phys. et nat.*, 4, 26, 209-36; reprinted in (1911). Société Suisse de Physique, ed., *Gesammelte Werke Walther Ritz Œuvres*. Gauthier-Villars, Paris, pp. 427-46; https://ia904708.us.archive.org/3/items/gesammeltewerkeo00ritzuoft/gesammeltewerkeo00ritzuoft.pdf.; translation by T. G. Underwood; also in Underwood, T. G. (2023). *Special Relativity*, Part I, pp. 281-3.

In this paper Ritz presents a summary of his criticisms of the Maxwell and Lorentz's theories and the resulting experimental uncertainties.

---

The history of the new ideas introduced by Maxwell in the science of electricity and the theories derived from them is certainly one of the most interesting chapters, especially from the psychological point of view, of the history of science. We know with what repugnance the minds, accustomed to the limpid clarity which gave the classical theories of mathematical physics such a high aesthetic value, have admitted these new, disturbing the established order, and which seemed, it must be admitted, strangely confusing at first sight. Maxwell's first publication dates from 1856, "*On Faraday's Lines of Force*"; thirty years later, it took all the authority of a Helmholtz to obtain that the new theory was not accepted, but at least found worthy of some interest. It was the experiments of Hertz and those who followed him that, by demonstrating the identity of light and electrical oscillations, and thus confirming Maxwell's genial views, broke the last resistance and gave this theory the right to cite in Physics. It was then recognized that the origin of the obscurities of Maxwell's work results largely from the fact that we find there two very different tendencies: that of an attempt to explain the electrical actions by the properties of the medium which is the vehicle (explanation which leads Maxwell to various accessory hypotheses and where, despite his efforts, he has completely failed as regards electrostatics), and that of a purely phenomenological description by means of equations with partial derivatives and a hypothesis on electromagnetic energy, and where certain vectors intervene which characterize the electrical and magnetic state of the body. One only has to choose this second method to be faced with many difficulties.

Maxwell's theory, extended by Hertz to moving bodies, does not agree with certain optical experiments (aberration, Fizeau experiment, etc.), nor with those of Eichenwald on the action of mobile dielectrics. The new form that H. A. Lorentz gave to Maxwell's idea, on the contrary, is in perfect agreement with these experiments; moreover, taking up the hypothesis of Fechner and Weber, that all electric current is a convection current, that is to say is due to the transport of electricity, hypothesis which recent research confirms more and more, he considerably simplified the equations; the constitution of the atom which he attributed to electricity allows a clearer and more precise view of the phenomena. Finally, by considering ether as immobile and present even within atoms, he removed an indeterminacy of Maxwell's theory which had not been corrected until then, an indeterminacy resulting from the movements of ether, which was also required, without specifying them sufficiently, by the Hertz theory, but which no experiment had ever been able to evidence. Finally, the reciprocal and complete compatibility of ether and matter explains why bodies move through the ether without experiencing resistance, and that the "ether wind" of 30 km per second which, according to Fresnel and Lorentz, crosses the earth carried away in its motion around the Sun, has never been able to be evidenced even by the most delicate observations.

By thus reducing Maxwell's theory to its simplest expression, and removing many mathematical difficulties, Mr. H. A. Lorentz has bridged the abyss which separates Maxwell's theory from the classical theories, based on the notion of action at a distance, and clarifies the reciprocal relations of Weber and Clausius on the one hand, Maxwell's and his own. This difference is, as we shall see, much narrower than one would have thought at first.

But the theory thus simplified has another advantage: it is that it allows a more rigorous criticism of the principles on which it is built. These principles are of various kinds. These are, first of all, the experimental bases of the theory: evidence, which at first sight seems to have so fully confirmed the theory, has not, without any doubt, always focusing on certain points while leaving others, equally important, in the shadows? What changes could be made to Lorentz's formulas without affecting any evidence?

Second, what is the *true* meaning of the electric force vectors E, magnetic force H, which enter into the equations? And how is the transition from these to the facts of experience that they must represent? Similar questions have arisen, as we know, for Mechanics, and have only recently received their solution. Now, by the notion of electromagnetic mass, through the impotence of the theory is to explain phenomena in terms of the mechanical properties of the ether, modern physics has been induced to conceive inversely an electromagnetic origin of the laws of mechanics, and to make Electrodynamics the pivot of a new conception of nature, replacing the old mechanical design. It is therefore particularly important that there is no cloud obscuring the logical foundations of this vast intellectual edifice.

It is known that among these bases is the hypothesis of an absolute system of coordinates and that the experiments of Michelson and Morley, and others more recently, have on this point gives a formal contradiction to the theory: as in Mechanics, the uniform translation of a system does not seem to have any influence on the optical and electromagnetic phenomena that take place there. Messrs. Lorentz, Einstein, Poincare and others have therefore wondered what new hypotheses should be introduced to justify this fact, *without touching the fundamental equations*. It turned out that it was necessary to abandon the classical concept of universal time; make simultaneity a relative notion, abolish the conception of the invariability of mass, abolish that of rigid bodies and remove the axioms of kinematics, the parallelogram of velocities, etc. When a grain of radium emits β rays in two opposite directions at 250,000 km/sec., it will no longer be said that the relative velocity of these rays is 500,000 km/sec., but it will be equal to 294,000 km/sec. Similarly, two equal times for observer A or two simultaneous events will no longer be equal for observer B, moving relative to A. And, curious and noteworthy, whereas a few years ago, one would have thought it prudent, in order to refute a theory, to show that it entails only one or other of these consequences: at the present time, Maxwell's equations are considered so absolutely intangible, that these consequences do not scare anyone. Rather than concluding that equations need to be modified more or less profoundly, we decided to sacrifice kinematics, the notion of time, etc. After having more or less systematically ignored a theory that was fertile for thirty years, we fall into the extreme opposite. Do these equations really give us such excessive confidence?

The answer is clearly negative, and I propose to present here a summary of the criticisms to which Maxwell and Lorentz's theory gives rise and the experimental uncertainties which it entails. The reader will find in another Memoir[1] the details of the demonstrations.

[1] Ritz, W. (1908). Recherches critiques sur l'Électrodynamique Générale. (Critical Research on General Electrodynamics.) *Annales de Chimie et de Physique*, 13, 145-275; reprinted in (1911). Société Suisse de Physique, ed., *Gesammelte Werke Walther Ritz Œuvres*. Gauthier-Villars, Paris, pp. 317-426; https://ia904708.us. archive.org/3/items/ gesammeltewerkeo00ritzuoft/gesammeltewerkeo00ritzuoft.pdf. See also an article by the author; Ritz, W. (1909). Du rôle de l'éther en physique. (The role of the ether in physics.) *Scientia, 3, 6*; *Gesamm. Werke*, pp. 447-61.

...

## Ritz, W. (December, 1908). Über die Grundlagen der Elektrodynamik und die Theorie der schwarzen Strahlung. (On the basics of electrodynamics and the theory of black body radiation.)

*Phys. Zeit.*, 9, 903-7; reprinted in (1911). Société Suisse de Physique, ed., *Gesammelte Werke Walther Ritz Œuvres*. Gauthier-Villars, Paris, pp. 493–502; https://ia904708.us.archive.org/3/items/gesammeltewerkeo00ritzuoft/gesammeltewerkeo00ritzuoft.pdf; translation by T. G. Underwood; also in Underwood, T. G. (2023). *Special Relativity*, Part I, pp. 284-90.

Initially, even after careful translation from the German text from a PDF of a photocopy of the pages in *Gesammelte Werke*, it was not clear what this was about or why this elicited an immediate response from Einstein. Then during the following night, the penny dropped. The differential equations in the Maxwell-Lorentz formulation of electrodynamics permit infinite solutions, including those with both *retarded* and *advanced* potentials, on which Einstein (1905) relied in deriving the consequences of his two postulates. Ritz shows that *advanced potentials* are inadmissible and how this could be addressed based on *retarded potentials* alone, i.e. by an *emission theory*. He also demonstrates why the role of an ether must be removed from the theory of electrodynamics.

[The italics in this article are the author's, not mine. The *Table des Matieres* at the end of *Gesammelte Werke Walther Ritz Œuvres* on pages 537-41 provides a useful overview of Ritz's publications.]

---

In a recent publication in this journal[1]

[1] Lorentz, H. A. (1908). Zur Strahlungstheorie. (On the theory of radiation.) *Phys. Zeit.*, 9, 562-3; https://www.lorentz.leidenuniv.nl/IL-publications/sources/Lorentz_PZ _1908.pdf.

on the theory of radiation, Mr. H. A. Lorentz concluded that the theory of "black body radiation" developed by Jeans[2] and himself[3]

[2] Jeans, J. (1906). *Proc. Roy. Soc*, 76, p. 296, 545.

[3] Lorentz, H. A. (April, 1908). *Le partage de l'energie entre la mattere ponderable et l'ether*. Conference tenue au congres de Rome (Roma, Tipogralia della R. Accad. dei Lincei).

was incompatible with experience; but since the derivation of the radiation formula in this way could only become obsolete if substantial changes were made to the basic laws of electromagnetics, it was necessary to make such changes, namely, in the sense of Planck's theory, something similar to a *time-energy-atom* can be included in the theory.

In view of the great importance of this question, and in view of the difficulties which, as Mr. H. A. Lorentz shows that even with the introduction of Planck's atom, it is permissible to point out an error in Jeans-Lorentz's theory, which precisely addresses the essential point of the evidence, which makes it obsolete.

*The approach to the electric and magnetic forces given in that proof is too general; it contradicts the formulas of retarded potentials, which any physically permissible solution of the basic equations must* satisfy. However, this condition determines precisely the coordinates of the free ether that Mr. Lorentz (*loc. cit.*) with $q_3$, $q'_3$, are extraordinarily limited in number. But it is these coordinates that ultimately determine the radiation formulas; and the contradiction of theory with observation is caused precisely by the fact that these coordinates, which are infinitely numerous, strive to attract the entire energy of the system on the basis of an analogy to Boltzmann's theorem of the equal distribution of energy over the various degrees of freedom.

As is well known, the basic electromagnetic equations of Lorentz's theory can be brought under the common form by introducing the potentials

$$1/c^2\, \partial^2 f/\partial t^2 - \Delta f = \varphi\, (x, y, z, t). \qquad (1)$$

where c = speed of light, $\varphi$ a given function means of x, y, z, t, and f disappears to infinity. The general solution of this equation, e.g. according to Poisson's method, involves two arbitrary functions of x, y, z, namely the values of f and $\partial f/\partial t$ at the start time $t_0$. Particular solutions of the same are:

$$f_1 = 1/4\pi \int \varphi\, (x', y', z', t - r/c)/r\, dx', dy', dz'$$

and

$$f_2 = 1/4\pi \int \varphi\, (x', y', z', t + r/c)/r\, dx', dy', dz'$$

furthermore, arbitrary linear combinations of $f_1$ and $f_2$ of the form: $f_3 = a_1 f1 + a_2 f_2$ , where $a_1 + a_2 = 1$: finally

$$f_3 = \ldots$$

The solution $f_1$ corresponds to divergent waves, $f_2$ to convergent waves coming from infinity, $f_3$ to both types; still other solutions would correspond to waves that converge towards or diverge from points of pure ether where $\varphi = 0$; in $f_1$ is the resolution in waves, which according to the Kirchhoff's and Poisson's statements can always happen in an infinite number of ways. Experience shows that only the solution $f_1$ can be considered, and Maxwell-Lorentz theory explicitly presupposes this. The necessity of this restriction is already evident from the fact that, just as in the solution $f_1$, a body whose electrons are accelerated radiates energy, so that at a great distance the Poynting vector is directed outwards, at $f_2$ (substitution of c with – c) this vector changes its sign, the body thus receives energy from infinity *without any other body losing a corresponding quantum of energy*. Such a body, which would be able to permanently extract energy from the ether in this way, would have to be called a *perpetual motion* machine and is physically impossible.

In order to eliminate such solutions, which satisfy all conditions — including the conditions at infinity — and which are nevertheless impossible, the initial state, which is still arbitrary at first, must be restricted in an appropriate manner. The necessary and sufficient condition for $f_1$ to be valid in all times is that it is valid at the time $t = t_0$ and $t = t_0 + dt$. However, it is obvious that this initial condition has no reasonable meaning in Maxwell's formulation, and it has therefore been sought to be replaced by others. Assuming, as usually happens, that at time $t_0$ at great distances the field is equal to zero, then the formula $f_1$ follows for later times; *for earlier ones, however, the formula $f_2$ is inadmissible*. Furthermore, the validity of $f_1$ is now subject to a completely unnecessary restriction (field = 0 for $t = t_0$), which, for example, uniform translation does not satisfy. Finally, the character of equation (1) as a hyperbolic differential equation means that if the initial condition is only very approximated (which alone could be stated), it does not follow at all that the formula $f_1$ holds with a similar approximation: it could be, for example, convergent waves, which are very faint and distant at time $t_0$, take on an arbitrary finite value at a given point in space at a later time. Nor do the other additional conditions proposed so far tolerate a precise critique[1];

[1] On these and other weak aspects of Maxwell-Lorentz's theory, see the work of the author: Ritz, W. (1908). Recherches critiques sur l'electrodynamique generale. *Annales de Chimie et de Physique*, 8, t. 13, 143-275; reprinted in (1911). Société Suisse de Physique, ed., *Gesammelte Werke Walther Ritz Œuvres*. Gauthier-Villars, Paris, XVIII, pp. 317-426; https://ia904708.us. archive.org/3/items/gesammeltewerkeo00ritzuoft/ gesammeltewerkeo00ritzuoft.pdf.

the transition from the reversible differential equations to the retarded potentials, *through which irreversibility is introduced into electrodynamics*, can be seen *will not be found* on the basis of the Maxwell's views alone.

It is therefore important to state that the complete expression of the laws of radiation and of Maxwell-Lorentz's theory in general is not the differential equations, but the elementary effects that result from the introduction of the retarded potentials into Lorentz's expression of ponderomotive force. In this form, the electric and magnetic vectors are also eliminated, which can never be observed directly, but only play the role of mathematical auxiliary functions[2],

[2] *Loc. cit.*, p. 3.

while the actual statements of the theory refer only to the quantities space, time and electric charges.

To derive the radiation formula, Jeans and Lorentz now imagine a parallelopipedic cavity bounded by reflective walls, in which there is a body K; the electrical and magnetic forces in the interior are developed as functions of x, y, z in Fourier series, whose coefficients are functions of time and play the role of Lagrangian coordinates, for which the differential equations can be derived from Hamilton's principle. This approach does not take into account the essential condition of representation by retarded potentials; however, as emphasized above, this condition eliminates from the initial states an infinite manifold of states of the ether, which can be represented by two arbitrary functions of xyz and thus by an infinite number of parameters, as inadmissible. In particular, it requires that the forces remain constant when the electric charges are permanently at rest. On the other hand, according to the partial differential equations, this is not necessarily the case: it will always be possible to add a solution to the homogeneous equation $\partial^2/\partial t^2 - c^2 \Delta = 0$, which in the present case must satisfy the boundary conditions on the mirrors and walls, and thus results as the totality of the natural electrical oscillations of the cavity (without the body K). Such solutions appear to be admissible in the Jeans-Lorentz derivative, as Lorentz (*loc. cit.* p. 14) points out, which they should not. For many purposes, the infinitely many parameters they involve (coefficients of the evolution of the general solution of $\partial^2/\partial t^2 - c^2 \Delta = 0$ according to the natural oscillations of the cavity, i.e. in a Fourier series) would have to be permanently zero. But it is precisely these infinitely many parameters of the "pure ether" that, according to the theorem of uniform distribution of energy over the degrees of freedom, strive to absorb all the energy and distribute it to the shortest wavelengths. *Jeans-Loretntz's theory is therefore inadmissible.*

One could argue that the solution just discussed, developed according to Fourier series, can be understood as "retarded" forces originating from the electrons of the reflecting walls. But since perfectly reflecting walls presuppose *an infinite* number of conduction electrons in them, they are to be rejected here as an inadmissible abstraction for this very reason, since in reality the number of degrees of freedom of the body K (or the number of electrons in it) and that of the mirror must not be regarded as infinitely different, and this is precisely what matters. However, if the number of electrons in the mirror is very large, the Jeans-Lorentz approach will remain valid, but only for natural oscillations of the cavity for which the discontinuity of the mirror and the finiteness of the number of electrons and conductivity do not yet come into consideration, i.e. *for long waves or lower temperatures. This is the reason why the approach to such waves, on the other hand, is in no way compatible with the condition of retarded potentials, it represents far too large a variety of solutions.*

However, the condition of retarded potentials seems difficult to include in the statistical analysis, and the question arises whether it is also sufficient to obtain a spectral distribution of the energy from the experimentally given character. Above all, it is necessary to determine how much and what kind of arbitrary constants are involved in the general solution of the equations of motion of a system of electrons if the approach of retarded potentials is used for the forces. It is only to these arbitrary elements that the statistical analysis may extend. In the case of mechanical problems, the question is simplified by the fact that the further course is determined by specifying the coordinates q and the pulses p. This is not the case in electron theory, and perhaps this is another sore point of the same. Even the equations of the force-free motion of the rigid electron, as Mr. Herglotz[1] has shown,

[1] Herglotz, G. (1903). *Götl. Nachr.*, 6; (1904). *Idem.*, 6; (1908). *Math. Ann.* l. 65, 87.

allow for an infinite number of solutions in addition to the uniform translation; at very low speeds, the general solution can be represented as the sum of infinitely many oscillations, with arbitrary amplitudes, at very low speeds, the general solution can be represented as the sum of an infinite number of oscillations, with arbitrary amplitudes, the wavelengths of which are all far beyond the known ultraviolet spectrum, namely at most of the order of magnitude of the electron diameter, and have no lower limit. Since Herglotz's method[2]

[2] *Loc. cit.*, (1904).

remains applicable to more general problems of electron theory and leads to similar integral equations, a similar behavior of the solutions can also be expected in general.

In the last instance, the same is due to the fact that in electron theory the acceleration of an electron is determined by certain *previous* positions, velocities and accelerations of the other electrons or electrons, or charge elements. First of all, we limit ourselves to the case where all occurring functions of the form φ (t – r/c) according to the formula

$$\varphi\,(t - r/c) = \varphi\,(t) - r/c\,\varphi'\,(t) + r2/1.2.c^2\,\varphi''\,(t) - \ldots$$

differential equations of infinitely high order are obtained, the general solution of which depends on an infinite number of constants[1], which in this special case must satisfy certain inequalities imposed by the convergence conditions

[1] Cf. Lalesca, T. (1908). *Sur l'equation de Voltierra*, thesis. Paris.

The investigations of Sommerfeld[2] and P. Hertz[3]

[2] Sommerfeld, A. (1904). *Gott. Nachr.*, p. 363.
[3] Hertz, P. (1908). *Math. Ann.*, l. 65, p. 1.

on the rigid spherical electron show that, given an external force, one can arbitrarily prescribe motion within a time T which is equal to the diameter of the electron divided by the speed of light; especially in the case of a uniform superficial charge, any function with period T satisfies the speed of motion and any such function P(t) can be added to any solution of the problem given external forces.

If the solution is to be analytical, the values of P(t) within a period are not arbitrary, but one can set P = real part of Q where

$$Q = \ldots$$

and the $a_i$ (except for the convergence condition of the series) are arbitrary.

In general, the solution of any system of moving electrons should also require an infinite number of constants to be determined, and allow oscillating solutions of infinitely small wavelengths. These are conditioned precisely by the infinitely many "degrees of freedom of the ether": and it is to be feared that, on the basis of the theorem of uniform distribution of energy, they would eventually want to cause a tendency of radiation to concentrate entirely on the shortest wavelengths, even if the formula might have a shape different from Jeans's. But even without this consideration, the existence, for example, of the force-free oscillations of the electron, which can be superposed to any solution and according to which every solution can be developed, which would therefore have to occur everywhere, must be considered experimentally improbable. Even if a radiation of an extremely short wavelength is inaccessible to our observation methods, a corresponding noticeable energy defect must be shown, which was not perceived anywhere.

One may perhaps conclude from this that, *just as we have already been forced to greatly reduce the variety of solutions of Maxwell's theory by introducing the retarded potentials, a new such restriction is still necessary in order to reduce the number of determining elements (constants) of the solutions to a finite one*.

It is easy to make plausible that among the possible, infinitely many solutions one is always excellent, just as among the solutions of the partial differential equations the retarded potentials would be excellent. Imagine that gravity does not act instantaneously, but according to the laws of electrodynamics. Then, in order to calculate the further motion at given initial values of the coordinates and velocities, the motion according to the classical law will be taken as the first approximation; this solution will then be inserted into the (very small) additional elements introduced by the new law: this will give rise to new second-order differential equivalents, which are integrated with the same initial values, etc. This finally results in a solution in which each coordinate x is a certain analytical function of time t and the initial values $x_{0i}$, $x'_{0t}$, but which only applies to a limited range of these quantities. This solution can then be continued analytically, both as a function of t and as a function of the initial values $x_{0i}$, $x'_{0t}$, and then gives for each time and for arbitrary initial values of the coordinates and velocities a very specific solution to the problem, dependent only on these data, in addition to which, however, there are an infinite number of other solutions, for which the procedure is *never* valid, and which would correspond to most curious planetary systems. In the case of the force-free electron, the simple translation is obtained. If this solution is the only permissible one, then the number of *arbitrary constants, i.e. the manifold of the solution, would not be greater than in mechanics, i.e. equal to twice the number of degrees of freedom of the electrons.*

This can be achieved, for example, by introducing an additional condition in the form of a minimum principle, whereby the variation disappears for all solutions, but a real minimum only for *one* certain solution is likely to occur. The same applies, for example, in the theory of vibrations of strings, diaphragms, etc.[1], where for all natural oscillations, in infinite numbers, the variation disappears, but the minimum is only reached for the fundamental tone.

[1] See, for example, Riemann-Weber (1901). *Parlielle Differentialgleichungen*, t. 2, p. 284, Braunschweig.

Likewise, in addition to the conditions at infinity, it was also possible to introduce conditions for very large t, which could again be derived from the calculus of variations.

Thus, the fundamental difficulties in the theory of black body radiation emphasized by Lorentz *do not lead us both to introducing an energy-time element with Planck, but rather to the demand that the principle of the uniqueness of natural events in the sense of classical mechanics, which has been violated by the current theory of electrons, must be restored by a minimal principle,* so that a certain *finite* number of determinants is sufficient to determine the course of the motion of a system of electrons for all time.

This would mean that the last remnant of what was once called the ether would disappear from the laws of nature. Successively, experience had already compelled us to deny him motion and other properties of matter; from a more or less complicated mechanism, he had become the immutable carrier of electromagnetic phenomena himself. In this reduced field, its existence could still have been demonstrated by solutions of equations which were independent of matter or electrons (the equation $\partial^2/\partial t^2 - c^2 \Delta = 0$ would be sufficient). Experience forces us to reject these solutions. But then the equations of electron theory only express relationships between space and time, the field strengths or "states of the ether" can be completely eliminated. The ether descends to abstraction: it is only an absolute coordinate system and a mathematical construction that introduces an infinite number of constants into the formulas. Experience does not seem to want to leave him either the first or the second of these qualities: it banishes him from physics altogether.

With this assumption, however, one of the essential foundations of Maxwell's description of the processes by partial differential equations, which now have no physical meaning, melts away, as well as the meaning of a mathematical intermediate construction, which, moreover, is insufficient on its own. The belief in their unconditional validity is not strengthened by this, all the more so as it can be shown[1] that the experimental foundations are completely lacking in certain respects.

[1] Ritz, W. *loc.cit.*; Ritz, W. (1908) Recherches critiques sur les theories electrodynamiques de Cl. Maxwell et de H.-A. Lorentz. (Critical research on the electrodynamic theories of Cl. Maxwell and H.-A. Lorentz.) *Archives des Sciences phys. et nat.*, 4 per., t. 26, 209; (1908). Du röle de l'ether en Physique. (Of the röle of the ether in Physics.) *Rivista di Scienza : Scienlia*, 3, 6; (1909). Die Gravitation. (Gravitation.) Rivista di Scienza : Scientia, 5, 10; reprinted in (1911). Société Suisse de Physique, ed., *Gesammelte Werke Walther Ritz Œuvres*. Gauthier-Villars, Paris, XIX, p. 427-46; XX, p. 447-61; XXI, p. 462-77; https://ia904708.us.archive.org/3/items/gesammeltewerkeo00ritzuoft/ gesammeltewerkeo00ritzuoft.pdf.

## Einstein, A. (March, 1909). Zum gegenwärtigen Stand des Strahlungsproblems. (On the Present Status of the Radiation Problem.)

*Phys. Zeit.*, 10, 6, 185-93; reprinted in John Stachel, ed., *The Collected Papers of Albert Einstein.* Vol. 2. The Swiss Years: Writings, 1900–1909 (Princeton: Princeton University Press, 1989), Doc. 56, pp. 542–550; translation at https://ia904704.us.archive.org/27/items/EinsteinOnPresentStatus_201501/Einstein_On_Present_Status_old.pdf; translation also in A. Beck (translator), P. Havas (consultant). (1989). *The Collected Papers of Albert Einstein, Volume 2: The Swiss Years: Writings, 1900-1909*, pp. 357-75; also in Underwood, T. G. (2023). *Special Relativity*, Part I, pp. 291-4.

Received January 23, 1909.

Berne.

Einstein advocated the use of the Maxwell-Lorentz equations, since they yielded expressions for the energy and momentum of a system at any instant of time, while the exclusive use of *retarded* potentials required knowledge of the earlier states of a system to determine any future state. He denied that *retarded* potentials had some fundamental significance; he viewed them merely as auxiliary mathematical formulations. Ritz responded both in print and in person by visiting Einstein in Zurich.

---

This Journal recently published contributions by Messrs. H. A. Lorentz[1], Jeans[2], and Ritz[3]

[1] Lorentz, H. A. (1908). Zur Strahlungstheorie. (On the theory of radiation.) *Phys. Zeit.*, 9, 562-3.

[2] Jeans, J. H. (1908). *Phys. Zeit.* 9, 853-5.

[3] Ritz, W. (December, 1908). Über die Grundlagen der Elektrodynamik und die Theorie der schwarzen Strahlung. (On the basics of electrodynamics and the theory of black body radiation.) *Phys. Zeit.*, 9, 903-7.

which help to simplify the critical interpretation of the present status of this extremely important problem. In the belief that it would be useful if all those who have seriously thought about this matter communicate their views, even if they have not been able to arrive at a final result, I join in with the following contribution.

**1.** The simplest form in which we can express the currently understood laws of electrodynamics is the set of Maxwell-Lorentz partial differential equations. I regard the equations containing *retarded* functions, in contrast to Mr. Ritz,[3] as merely auxiliary mathematical forms. The reason I see myself compelled to take this view is first of all that those forms do not subsume the *energy principle*, while I believe that we should adhere to the strict validity of the *energy principle* until we have found important reasons for renouncing this guiding star. It is certainly correct that Maxwell's equations for empty space, taken by themselves, do not say anything, they only represent an intermediary construct. But, as is well known, exactly the same could be said about Newton's *equations of motion*, as well as about any theory that needs to be supplemented by other theories in order to yield a picture of a complex of phenomena. *What distinguishes the Maxwell-Lorentz differential equations from the forms that contain retarded functions is the fact that they yield an expression for the energy and the momentum of the system under consideration for any instant of time, relative to any unaccelerated coordinate system.* With a theory that operates with *retarded* forces it is not possible to describe the instantaneous state of a system at all without using earlier states of the system for this description. For example, if a light source A had emitted a light complex toward the screen B, but it has not yet reached the screen B, then, according to theories operating with *retarded* forces, *the light complex is represented by nothing except the processes that have taken place in the emitting body during the preceding emission*. Energy and momentum—if one does not want to renounce these quantities altogether—must then be represented as time integrals. *Mr. Ritz certainly claims that experience forces us to abandon these differential equations and introduce the retarded potentials*. However, his arguments do not seem sound to me.

If one defines, in agreement with Ritz,

$$f_1 = 1/4\pi \int \varphi\,(x', y', z', t - r/c)/r\ dx', dy', dz'$$

and

$$f_2 = 1/4\pi \int \varphi\,(x', y', z', t + r/c)/r\ dx', dy', dz'$$

then both $f_1$ and $f_2$ are solutions of the equation

$$1/c^2\ \partial^2 f/\partial t^2 - \Delta f = \varphi\,(x, y, z, t),$$

and hence

$$f_3 = a_1 f1 + a_2 f_2$$

is also a solution if $a_1 + a_2 = 1$. But it is not true that the solution $f_3$ is a more general solution than $f_1$ and that one specializes the theory by setting $a_1 = 1$, $a_2 = 0$. Setting

$$f(x, y, z, t) = f_1$$

*amounts to calculating the electromagnetic effect at the point x, y, z from those motions and configurations of the electric quantities that took place prior to the time point t*. Setting

$$f(x, y, z, t) = f_2,$$

*one determines the electromagnetic effects from the motions and configurations that take place after the time point t*. In the first case the electric field is calculated from the totality of the processes *producing* it, and in the second case from the totality of the processes *absorbing* it. If the whole process occurs in a (finite) space bounded on all sides, then it can be represented in the form

$$f = f_1$$

as well as in the form

$$f = f_2.$$

If we consider a field that is emitted from the finite into the infinite, we can, naturally, only use the form $f = f_1$, exactly because the totality of the absorbing processes is not taken into consideration. But here we are dealing with a misleading paradox of the infinite. Both kinds of representation can always be used, regardless of how distant the absorbing bodies are imagined to be. Thus, one cannot conclude that the solution $f = f_1$ is more special than the solution $a_1 f_1 + a_2 f_2$, where $a_1 + a_2 = 1$.

That a body does not "receive energy from infinity unless some other body loses a corresponding quantity of energy" cannot be brought up as an argument either, in my opinion. First of all, if we want to stick to experience, we cannot speak of infinity but only of spaces lying outside the space considered. Furthermore, *it is no more permissible to infer irreversibility of the electromagnetic elementary processes from the non-observability of such a process than it is permissible to infer irreversibility of the elementary processes of atomic motion from the second law of thermodynamics.*

**2.** Jeans' interpretation can he disputed on the grounds that it might not be permissible to apply the general results of statistical mechanics to cavities filled with radiation. However, the law deduced by Jeans can also be arrived at in the following way[1].

[1] Cf. Einstein, A. (1905) Über einen die Erzeugung und Verwandlung des Lichtes betreffenden heuristischen Gesichtspunkt. (On a Heuristic Viewpoint Concerning the Production and Transformation of Light.) *Ann. Phys.*, 4, 17, 132-48, pp. 133-36.

According to Maxwell's theory, an ion capable of oscillating about an equilibrium position in the direction of the X-axis will, on average, emit and absorb equal amounts of energy per unit time only if the following relation holds between the mean oscillation energy $\underline{E}_\nu$ and the energy density of the radiation $\varrho_\nu$ at the proper frequency $\nu$ of the oscillator:

$$\underline{E}_\nu = c^3/8\pi\nu^2\, \varrho_\nu, \qquad \text{(I)}$$

where c denotes the speed of light. If the oscillating ion can also interact with gas molecules (or, generally, with a system that can be described by means of the molecular theory), then we must necessarily have, according to the statistical theory of heat,

$$\underline{E}_\nu = RT/N \qquad \text{(II)}$$

(R = gas constant, N = number of atoms in one gram-atom, T = absolute temperature), if, on average, no energy is transferred by the oscillator from the gas to the radiation space[2].

[2] Planck, M. (1900). *Ann. Phys*., 1, 69. (ed: original publication has 99 instead of 69 as pg. ref. There is no Planck paper with pg. ref. 99. The intended reference is Planck, M. (1900). Ann. Phys., 1, 69-122.) *Vorlesungen uber die Theorie der Warmestrahlung*. III. Kapitel. (ed: III. Kapitel refers to "Dritter Abschnitt.")

From these two equations we arrive at

$$\varrho_\nu = R/N\ 8\pi/c^3\ \nu^2 T, \qquad \text{(III)}$$

i.e., exactly the same law that has also been found by Messrs. Jeans and H. A. Lorentz[3].

[3] It should be explicitly noted that this equation is an irrefutable consequence of the statistical theory of heat. The attempt, on p. 178 of the book by Planck just cited, to question the general validity of Equation II, is based—it seems to me—only on a gap in Boltzmann's considerations, which has been filled in the meantime by Gibbs' investigations.

**3.** There can be no doubt, in my opinion, that our current theoretical views necessarily lead to the law propounded by Mr. Jeans. However, we can consider it as almost equally well established that formula (III) is not compatible with the facts. Why, after all, do solids emit visible light only above a fixed, rather sharply defined temperature? Why are ultraviolet rays not swarming everywhere if they are indeed constantly being produced at ordinary temperatures? How is it possible to store highly sensitive photographic plates in cassettes for a longtime if they constantly produce short-wave rays? For further arguments I refer to §166 of Planck's repeatedly cited work. Thus, we must say that experience forces us to reject either equation (I), required by electromagnetic theory, or equation (II), required by statistical mechanics, or both equations.

**4.** We must ask ourselves how Planck's radiation theory relates to the theory which is indicated in 2. and which is based on our currently accepted theoretical foundations. In my opinion the answer to this question is made harder by the fact that Planck's presentation of his own theory suffers from a certain logical imperfection. I will now try to explain this briefly.

a) If one adopts the standpoint that the irreversibility of the processes in nature is only apparent, and that the irreversible process consists in a transition to a more probable state, then one must first give a definition of the probability W of a state. …

## Ritz, W. (1909). Zum gegenwärtigen Stand des Strahlungsproblems; Erwiderung auf der Aufsatz des Herrn. A Einstein. (On the Present Status of the Radiation Problem; Response to the essay by Mr. A. Einstein.)

*Phys. Zeit.*, 10, 224-5; reprinted in (1911). Société Suisse de Physique, ed., *Gesammelte Werke Walther Ritz Œuvres*. Gauthier-Villars, Paris, pp. 503-6; https://ia904708.us.archive.org/3/items/gesammeltewerkeo00ritzuoft/gesammeltewerkeo00ritzuoft.pdf; translation by T. G. Underwood; also in Underwood, T. G. (2023). *Special Relativity*, Part I, pp. 295-7.

Sent February, 1909.

Göttingen.

Ritz's response to Einstein (March, 1909).

---

Mr. Einstein is of the opinion[1] that the multiplicity of the integrals of the differential equation emphasized by me[2]

[1] Einstein, A. (March, 1909). Zum gegenwärtigen Stand des Strahlungsproblems. (On the Present Status of the Radiation Problem.) *Phys. Zeit.*, 10, 6, 185-93.
[2] Ritz, W. (December, 1908). Über die Grundlagen der Elektrodynamik und die Theorie der schwarzen Strahlung. (On the basics of electrodynamics and the theory of black body radiation.) *Phys. Zeit.*, 9, 903–907.

$$1/c^2\, \partial^2 f/\partial t^2 - \Delta f = \varphi\, (x, y, z, t)$$

(c = speed of light) does not exist, at least not in the sense that the particulate integrals

$$f_1 = 1/4\pi \int \varphi\, (x', y', z', t - r/c)/r\; dx', dy', dz'$$

and

$$f_2 = 1/4\pi \int \varphi\, (x', y', z', t + r/c)/r\; dx', dy', dz'$$

and finally

$$f_3 = a_1 f_1 + a_2 f_2 \qquad (a_1 + a_2 = 1)$$

do not correspond to the same operations. Rather, the first approach comes down to calculating the field in (x, y, z, t) from certain earlier states, the second from later states, so that the choice between $f_1$ and $f_2$ determines the type of calculation, does not touch on the essence of the process. *This view is quite untenable*. If on a body A at time t associated radiation takes place in a very short period of time, assuming $f_1$, an impulse will act on the electrons of a body B located at a greater distance r at time t + r/c; assuming $f_2$, this will take place on B at time t – r/c (before the process has taken place on A!); assuming $f_3$, on the other hand, there are *two* pulses for B, at times t – r/c and t + r/c. The latter process is therefore substantially different from the other two, and even this is not possible in more general cases by reversing the sign of the time to coincide with each other. This is not a different type of *calculation*, but a completely different *projection*.

In the case of a finitely limited space, Mr. Einstein further believes, one can represent both the processes by $f_1$ and by $f_2$. This is not the case. According to known theorems, a surface integral is added to the integral ($f_1$ or $f_2$) extended over the electrical densities or flows, which does not depend on them. In this form, both earlier times and later times can be used to calculate the field. But Lorentz's assumption consists precisely in the fact that when $f_1$ is applied and large spaces are assumed, the surface integral is omitted, from which it follows that when $f_2$ is applied, it generally *does not* disappear in the same process.

But in addition to these integrals $f_1$, $f_2$, $f_3$, there are an infinite number of others, and it is quite inadmissible to speak at all, as Mr. Einstein does, of *emitted* and *absorbed* fields. If it had only been possible to peel $f_1$, $f_2$ and $f_3$ out of this variety of solutions by some new assumption as the only ones to be considered, it is no longer difficult to negotiate conditions about the direction of the radiation vector at infinity. $f_1$ as the only integral. However, this is a much more difficult question and everything that has been put forward so far to solve it must be rejected as untenable, as I have discussed in detail elsewhere[1].

[1] Ritz, W. (1908). Recherches critiques sur l'Électrodynamique Générale. (Critical Research on General Electrodynamics.) *Annales de Chimie et de Physique*, 13, 145-275; reprinted in (1911). Société Suisse de Physique, ed., *Gesammelte Werke Walther Ritz Œuvres*. Gauthier-Villars, Paris, XVIII, pp. 317-426; https://ia904708.us.archive.org/3/items/gesammeltewerkeo00ritzuoft/gesammeltewerkeo00ritzuoft.pdf.

Mr. Einstein also reproaches the theory operating with *retarded* forces for being able to describe the state (energy and motion) of a system only with reference to earlier states of the system, while the partial differential equations give the instantaneous state. But the question is whether or not this momentary response agrees with the formulas of *retarded potentials*. In the former case, one representation *actually* says exactly the same thing as the other, in a slightly different form; in the second case, on the other hand, the process is one that has in fact *never* been observed. If it has been possible to strictly deduce the fact that a solution that cannot be derived from the *retarded potentials* has never been observed in a fully acceptable way by means of acceptable additional hypotheses from the partial differential equations, then the difficulty will be solved. Until then, I see the root of irreversibility and the second law in the fact that *the retarded forces are the only true integrals of the equations* (against the cold space of the world), and that at a great distance the energy always flows outwards, or at least never inwards. It will not be explained at this point that by replacing the image of the "ether" with certain energetic ideas, one can avoid the difficulties discussed and some others.

I must therefore maintain my conclusion that as long as $f_l$ is still some arbitrary integral of the differential equation

$$1/c^2\, \partial^2 f/\partial t^2 - \Delta f = 0$$

it is necessary (even if one dispenses with limited spaces and perfect mirrors) a radiation formula according to the Jeans-Lorentz method that contradicts experience will result; but that, assuming *retarded potentials*, that method is inadmissible because it operates with an infinite number of electrons (and a perfect mirror).

## Ritz, W. & Einstein, A. (1909). Zum gegenwärtigen Stand des Strahlungsproblems. (On the Present Status of the Radiation Problem.)

*Phys. Zeit.*, 10, 323-4; reprinted in (1911). Société Suisse de Physique, ed., *Gesammelte Werke Walther Ritz Œuvres*. Gauthier-Villars, Paris, XXV, pp. 507-8; https://ia904708.us.archive.org/3/items/gesammeltewerkeo00ritzuoft/gesammeltewerkeo00ritzuoft.pdf; translation in A. Beck (translator), P. Havas (consultant). (1989). *The Collected Papers of Albert Einstein, Volume 2: The Swiss Years: Writings, 1900-1909*, Doc. 57, p. 376; translation by T. G. Underwood; also in Underwood, T. G. (2023). *Special Relativity*, Part I, pp. 298-9.]

Received April 13, 1909.

Zurich.

This led them to publish a concise joint statement of their main differences of opinion in 1909. Whereas Ritz granted physical meaning only to *retarded potentials* in the interest of obtaining irreversibility, Einstein deemed the apparent irreversibility of radiation phenomena to be grounded solely on probabilistic considerations.

> [For a review of this sequence of papers see Frisch, M & Pietsch, W. (August, 2016). Reassessing the Ritz–Einstein debate on the radiation asymmetry in classical electrodynamics. *Studies in History and Philosophy of Science Part B: Studies in History and Philosophy of Modern Physics*, 55, 13-23; https://www.sciencedirect.com/science/article/abs/pii/S1355219815300150:
> "Why do we observe radiation fields coherently diverging from a source but usually not fields coherently converging into a source? In a famous letter to the *Physikalische Zeitschrift*, Albert Einstein and Walter Ritz summarized their opposing views on the origin of the temporal arrow of radiation (Ritz & Einstein, 1909). While Ritz thought that the asymmetry is due to an asymmetry in the fundamental laws governing electromagnetic radiation, Einstein appears to have maintained that the irreversibility of radiation processes can be given a purely probabilistic explanation. This joint letter is frequently cited in philosophical discussions of the radiation asymmetry (see, e.g. Price, 1997; Zeh, 2007; Wheeler, Archibald, & Feynman, 1945; Norton, 2009; Earman, 2011) and almost always in order to appeal to Einstcin's view in support of the idea that the radiation asymmetry ultimately is reducible to the very same statistical considerations that account for the thermodynamic asymmetry. The common view is that Einstein prevailed.
>
> References to the Ritz–Einstein controversy usually do not go beyond a discussion of the joint letter. Yet once we consider further papers by Ritz and Einstein … preceding the joint letter, as well as a paper by Einstein [Einstein, A. (September,

1909). Über die Entwicklung unserer Anschauung über das Wesen und die Konstitution der Strahlung. (On the Development of our View of the Nature and Constitution of Radiation.) *Phys. Zeit.* 10, 817-26, below] published later in the very same year also in the *Physikalische Zeitschrift*, shortly after Ritz's untimely death—a considerably more nuanced picture emerges. Ritz, whose own theory was an *action-at-a-distance* theory, offered several subtle criticisms of attempts to account for the asymmetry within a field-theoretic setting. Moreover, Einstein's last paper on the subject in that year raises a vexing interpretive puzzle concerning what Einstein's view on the asymmetry of radiation in classical electrodynamics were in 1909. One plausible reading of the exchange between Ritz and Einstein—and arguably a more plausible reading than the standard view—is that *by the end of 1909 Ritz had convinced his former classmate Einstein that, within classical radiation theory, the irreversibility has its source at least partly in a fundamental asymmetry of elementary radiation processes.*"]

---

In order to clarify the differences of opinion that came to light in our respective publications[1], we note the following.

[1] Ritz, W. (December, 1908). Über die Grundlagen der Elektrodynamik und die Theorie der schwarzen Strahlung. (On the basics of electrodynamics and the theory of black body radiation.) *Phys. Zeit.*, 9, 903-7; Einstein, A. (March, 1909). Zum gegenwärtigen Stand des Strahlungsproblems. (On the Present Status of the Radiation Problem.) *Phys. Zeit.*, 10, 6, 185-93.

In the special cases where an electromagnetic process *remains restricted to a finite space*, the process can be represented in the form

$$f = f_1 = 1/4\pi \int \varphi\,(x', y', z', t - r/c)/r\; dx', dy', dz'$$

as well as in the form

$$f = f_2 = 1/4\pi \int \varphi\,(x', y', z', t + r/c)/r\; dx', dy', dz'$$

and in other forms possible.

While Einstein believes that one should focus on this case without *substantially* restricting the generality of consideration, Ritz considers this restriction not to be permissible in *principle*. If one takes this standpoint, then experience compels one to consider the representation by *retarded potentials* as the only one possible, if one is inclined to the view that the fact of the *irreversibility* of the radiation processes must already find its expression in the fundamental equations. Ritz considers the restriction to the form of *retarded potentials* as one of the roots of the second law, while Einstein believes that *irreversibility* is exclusively based on probability.

**(3) All elementary particles, including electromagnetic waves (photons), are quantized.**

**Avoidance of the requirement to assume a *point electron*, and to address, through a process of *renormalization*, the still unresolved *divergencies* in the unsuccessful attempt to introduce *special relativity* into quantum electrodynamics.**

Based on a theory of quantum mechanics in which *only relationships among observable quantities occur*. There is no problem of divergencies, nor a renormalization requirement, in *non-relativistic* quantum mechanics and quantum electrodynamics.

**Non-relativistic Quantum Mechanics and Quantum Electrodynamics.**

This development started in 1925 with Heisenberg's introduction of *quantum mechanics in which only relationships among observable quantities occur*, and Dirac's extension of it to *electromagnetic radiation* as *quantum electrodynamics*.

## Heisenberg, W. (July 25, 1925). Über quantentheoretische Umdeutung kinematischer und mechanischer Beziehungen. (Quantum-theoretical re-interpretation of kinematic and mechanical relations.)

*Zeit. Phys.*, 33, 879-93; https://doi.org/10.1007/ BF01328377; (translation (2014) by Luca Doria, Institute of Theoretical Physics, Göttingen; also translation by D. H. Delphenich; https://neo-classical-physics.info/electromagnetism. html); and translation in van der Waerden, B. L., ed. (1968). *Sources of Quantum Mechanics*, 12, 261-76. Dover, New York; also in Underwood, T. G. (2023). *Quantum Electrodynamics – annotated sources*, Volume I, pp. 240-58.

Submitted July 29, 1925.

Göttingen, Institut fur theoretische Physik.

> [This translation is based largely on the December 2014 translation by Luca Doria, though *the notation has been restored to that used by Heisenberg and equation numbering is provided for both versions* (translation in [ ]. [Heisenberg, W. (July, 1925). *On the quantum reinterpretation of kinematical and mechanical relationships*. Werner Heisenberg Institute of Theoretical Physics, Göttingen.]
>
> Most of the comments on this paper are based on Aitchison, A. J. R., MacManus, D. A. & Snyder, T. M. (2004). Understanding Heisenberg's 'Magical' Paper of July 1925: a New Look at the Calculational Details. *arXiv*:quant-ph/0404009; also in (November, 2004). *Am. J. Phys.*, 72, 11, 1370-9. Other comments, as indicated, are from the Appendix to Fedak, W. A. & Prentis, J. J. (2009). The 1925 Born and Jordan paper "On quantum mechanics". *Am. J. Phys.*, 77, 2, 128-139.]

Heisenberg proposes a quantum mechanics in which only relationships among observable quantities occur, not possible to assign to the electron a point in space as a function of time, builds on Kramer's dispersion theory and instead assigns to the electron an *emitted radiation*, substitutes *frequencies* and *amplitudes* of Fourier components of emitted radiation of electron, instead of reinterpreting x(t) as a *sum* over transition components represents position by *set* of transition components, assigns *transition frequencies* and *transition amplitudes* as observables, replaces classical component by *transition* component corresponding to the quantum jump from state $n$ to state $n - \alpha$, translates the old *quantum condition* that fixes the properties of the *states* to a new condition to calculate the amplitude of a *transition* between two states by replacing the differential by a difference, in quantum case *frequencies* do not combine in same way as classical harmonics but in accordance with the *Ritz combination principle* under which spectral lines of any element include frequencies that are either the sum or the difference of the frequencies of two other lines, in quantum case frequencies combine by multiplying *transition amplitudes* (equivalent to matrix multiplication), results in non-commutativity of kinematical quantities, shows simple quantum theoretical connection to Kramers' dispersion theory, the *equation of motion* $\ddot{x} + f(x) = 0$ and the *quantum condition* $h = 4\pi m \sum_{\alpha=-\infty}^{+\infty} \{|a(n, n+\alpha)|^2\omega(n, n+\alpha) - |a(n, n-\alpha)|^2\omega(n, n-\alpha)\}$ together contain if solvable *a complete determination not only of the frequencies and energies but also of the quantum theoretical transition probabilities*.

> [Fedak, W. A. & Prentis, J. J. (2009), p. 128: "The name "*quantum mechanics*" appeared for the first time in the literature in Born, M. (December, 1924). Über Quantenmechanik. *Z. Phys*. 26, 379–395." … For Born and others, *quantum mechanics* denoted a canonical theory of atomic and electronic motion of the same level of generality and consistency as classical mechanics. The transition from classical mechanics to a true quantum mechanics remained an elusive goal prior to 1925.
>
> Heisenberg made the breakthrough in his historic 1925 paper.[2]
>
> > [2] Heisenberg, W. (July, 1925). Über quantentheoretische Umdeutung kinematischer und mechanischer Beziehungen. (On the quantum-theoretical reinterpretation of kinematic and mechanical relations.) *Zeit. Phys*., 33, 879-93.
>
> *Heisenberg's bold idea was to retain the classical equations of Newton but to replace the classical position coordinate with a "quantum-theoretical quantity."* The new position quantity contains information about the measurable line spectrum of an atom rather than the unobservable orbit of the electron."]

---

***Abstract***

In this work we will try to obtain the basis for a quantum mechanics theory which is *based uniquely on relationships between in principle observable quantities*.

***Introduction***

It is known that against the formal rules of the quantum theory used for the calculation of the observable quantities (for example the energy levels of the hydrogen atom) the serious objection can be raised that 1) those calculational rules contain as essential components relationships between quantities that seemingly in principle cannot be observed (like for example the electron position and period of revolution of the electron) and 2) also those rules apparently lack every clear physical basis unless one does not want to remain attached to the hope that those until now unobserved quantities will be made experimentally accessible in the future. This hope might be regarded as justified if the above-mentioned rules were internally consistent and applicable to a clearly defined range of quantum theoretical problems.

Anyway, experience shows that 1) *only the hydrogen atom and its Stark effect fit into those formal rules of quantum theory*, 2) already in the "crossed fields" problem (hydrogen atom in electric and magnetic fields in different directions) fundamental difficulties arise, 3) the reaction of atoms to periodically varying fields surely cannot be described by the mentioned rules and 4) finally an expansion of the quantum rules for the treatment of many-electrons atoms has been proved unfeasible.

It became customary to characterize the failure of the quantum rules (that were already essentially characterized through the application of classical mechanics) as a deviation from classical mechanics. However, this description can hardly be viewed as logical when one considers that already the *Einstein-Bohr frequency condition* represents such a complete departure from classical mechanics or better, from the point of view the wave theory, from the underlying kinematics of this mechanics, that it is absolutely not possible even for the simplest quantum theoretical problem to maintain the validity of classical mechanics.

In this situation, *it is advisable to completely give up any hope about the observation of hitherto unobserved quantities (like the electrons' position and period*) and at the same time acknowledge that 1) the partial agreement with experience of the mentioned quantum rules is more or less an accident and 2) to try to construct a theory of quantum mechanics in which *only relationships among observable quantities occur*. As the most important first steps toward such a theory of quantum mechanics one can refer to the *dispersion theory of* Kramers[1] and following works based on it[2].

[1] Kramers, H. A. (May, 1924). The law of dispersion and Bohr's theory of spectra. *Nature* 113, 673-74 [; derives dispersion (scattering) formula for electromagnetic radiation incident on an atom by assuming incident radiation characterized by train of polarized harmonic waves, positive virtual oscillators correspond to absorption frequencies (same as Ladenburg). Kramers' formula includes an addition term representing negative virtual oscillators that corresponds to emission frequencies].

[2] Born, M. (December, 1924). Über Quantenmechanik, (About quantum mechanics.) *Zeit. Phys*., 26, 379-95 [; the thesis contains an attempt to establish the first step towards quantum mechanics of coupling, which gives an account of the most important properties of atoms (stability, resonance for the jump frequencies, correspondence principle) and arises naturally from the classical laws. This theory contains Kramers' dispersion formula and shows a close relationship to Heisenberg's formulation of the rules of the anomalous Zeeman effect]; Kramers, H. A. & Heisenberg,W. (1925). Über die Streuung von Strahlung durch Atome. (On the scattering of radiation by atoms.) *Zeit. Physik*, 31, 681-708; http://dx.doi.org/10.1007/BF02980624; Born, M. & P. Jordan, P. (1925). Zur Quantenmechanik. (On Quantum Mechanics.) *Zeit. Physik*. (Forthcoming.)

In the following, *we shall try to present some new quantum mechanical relationships and apply them to the detailed treatment of some special problems*. We shall limit ourselves to problems with one degree of freedom.

**§ 1.** In the classical theory, *the radiation of a moving electron* (in the wave-zone **E** ~ **H** ~ 1/r) is not completely given by the expressions

$$\mathbf{E} = e/r^3c^2\ [\mathbf{r}(\mathbf{r}\dot{\mathbf{v}})] \qquad [1]$$
$$\mathbf{H} = e/r^2c^2\ (\dot{\mathbf{v}}\mathbf{r}) \qquad [2]$$

[where **E** and **H** are the fields strengths at a point for the electric field and magnetic field respectively, e is the *electron charge*, **r** is the *distance of the electron from the field point*, and **υ** the *electron velocity*)]

but we have other terms at the next order, e.g. of the form

$$e/rc^3\ (\dot{\mathbf{v}}\mathbf{v}) \qquad [3]$$

that we can denote as quadrupole radiation, and at the next higher order we have terms of the form

$$e/rc^4\ (\dot{\mathbf{v}}\mathbf{v}^2) \qquad [4]$$

and in this way the approximation can be carried out at any desired order.

One can ask himself *how the higher terms look like in the quantum theory.*

Since in the classical theory *the higher orders can be easily calculated when the motion of the electron or its Fourier representation are given respectively, one can expect the same in the quantum theory.* This question does not have to do with electrodynamics but this is - and this seems particularly important to us - of *pure kinematical nature*. We can pose this question as follows: *given instead of the classical quantity x(t) a quantum theoretical one, which quantum theoretical quantity enters in the place of* $x(t)^2$?

Before being able to answer this question, we have to remember that *in the quantum theory it was not possible to assign to the electron a point in space as a function of time through observable quantities*. However surely also in the quantum theory *one can assign to the electron an emitted radiation.* First, *this radiation will be described by frequencies* [ν] which quantum theoretically arise as function of two variables in the form:

$$\nu\ (n, n - \alpha) = 1/h\ \{W(n) - W(n - \alpha)\} \qquad [5]$$

and in the classical theory in the form:

$$\nu\ (n, \alpha) = \alpha\ .\ \nu\ (n,) = \alpha\ 1/h\ dW/dn. \qquad [6]$$

[where the observables are the *energies W(n) of the (Bohr) stationary states*, together with the associated (Einstein-Bohr) *frequencies* $\nu$, which characterize radiation emitted in the transition $n \rightarrow n - \alpha$.]

(From here onwards, we define $n$h = J where J is one of the *canonical constants*).

As characteristic for the comparisons of the classical mechanics to the quantum theory, with regard to the *frequencies* one can write the "*combination relations*"

*Classically*:

$$\nu\,(n, \alpha) + \nu\,(n, \beta) = \nu\,(n, \alpha + \beta) \qquad [7]$$

[where $\nu\,(n, \alpha) = \alpha\ 1/\text{h}\ \text{dW}/\text{d}n$]

*Quantum theoretically*:

$$\nu\,(n, n - \alpha) + \nu\,(n - \alpha, n - \alpha - \beta) = \nu\,(n, n - \alpha - \beta) \qquad [8]$$
$$\nu\,(n - \beta, n - \alpha - \beta) + \nu\,(n, n - \beta) = \nu\,(n, n - \alpha - \beta) \qquad [9]$$

[where $\nu\,(n, n - \alpha) = 1/\text{h}\ \{\text{W}(n) - \text{W}\,(n - \alpha)\}$]

Secondly, besides the *frequencies*, the *amplitudes* are necessary for the description of radiation. The *amplitudes* can be written as complex vectors (each with six independent components) and determine *polarization* and *phase*. They are also function of the two variables $n$ and $\alpha$ so that the corresponding part of the radiation will be represented with

*Quantum theoretically*:

$$\mathbf{R}\{\mathbf{A}(n, n - \alpha)\ e^{i\omega(n,n-\alpha)t}\} \qquad [10](1)$$

*Classically*:

$$\mathbf{R}\{\mathbf{A}_\alpha(n)\ e^{i\omega(n)\alpha t}\} \qquad [11](2)$$

[Heisenberg brings complex numbers and the square root of –1 into quantum theory in the same way as in the classical theory, by expressing the *amplitudes* as *Fourier series* in terms of *exponentials*.

Fourier's original formulation of the *Fourier transform* did not use complex numbers, but rather sines and cosines. Statisticians and others still use this form. An absolutely integrable function $f$ for which Fourier inversion holds can be expanded in terms of genuine frequencies $\lambda$ (avoiding negative frequencies, which are sometimes considered hard to interpret physically) by

$$f(\text{t}) = \int_0^\infty \{a(\lambda) \cos (2\pi\lambda \text{t}) + b(\lambda) \sin (2\pi\lambda \text{t})\}.]$$

First of all, the *phase* (contained in **A**) appears to have no meaning in the quantum theory since in this theory the *frequencies* are not in general commensurable with their harmonics. However, *we will immediately see that also in the quantum theory the phase has a precise meaning which has an analog in the classical theory.*

Let us consider now a particular quantity x(t) in the classical theory such that it can be regarded as represented by the totality of quantities of the form

$$\mathbf{A}_\alpha (n)\, e^{i\omega(n)\alpha t} \qquad [12]$$

which depending on the motion being periodic or not, represents x(t) with a sum or an integral

$$x(t) = \Sigma_{\alpha=-\infty}^{+\infty}\, \mathbf{A}_\alpha (n)\, e^{i\omega(n)\alpha t} \qquad [13]$$

or $$x(t) = \int_{\alpha=-\infty}^{+\infty} \mathbf{A}_\alpha (n)\, e^{i\omega(n)\alpha t}\, d\alpha \qquad [14](2a)$$

A similar combination of the corresponding quantum-theoretical quantities seems to be impossible in a unique manner and therefore not meaningful in view of the equal weight of the quantities *n* and *n* – α [i.e. in the *amplitude* A(n, n – α) and *frequency* ω(n, n – α)]. *However, one may readily regard the ensemble of quantities*

$$\boldsymbol{A}(n, n-\alpha)\, e^{i\omega(n,n-\alpha)t} \qquad [15]$$

*as a representation of the quantity x(t)* and then try to answer the question posed before: *how would the quantity* $x(t)^2$ *be represented?*

> [Aitchison, MacManus & Snyder. (2004). Understanding Heisenberg's 'Magical' Paper of July 1925: a New Look at the Calculational Details, pages 3-4: "An example of something he wishes to exclude from the new theory is the time-dependent *position* coordinate x(t). In considering what might replace it, he turns to the *probabilities for transitions between stationary states*. Consider a simple one-dimensional model of an atom consisting of an electron undergoing periodic motion. For a state characterized by the label *n*, fundamental *frequency* ω(*n*) and *coordinate* x(*n*, t), one can represent x(*n*, t) as a Fourier series
>
> $$x(n, t) = \Sigma_{\alpha=-\infty}^{+\infty}\; A_\alpha(n)\, e^{i\omega(n)\alpha t},$$
>
> where $A_\alpha$ is the amplitude of the α th harmonic.
>
> According to classical theory, the *energy emitted per unit time* (that is, the power) in a transition corresponding to the α th harmonic ν(*n*)α is
>
> $$-(dE/dt)_\alpha = e^2/3\pi\varepsilon_0 c^3\, [\nu(n)\alpha]^4\, |A_\alpha(n)|^2 .$$

In the quantum theory, however, the *transition frequency* corresponding to the classical '$\nu(n)\alpha$' is in general not a simple multiple of a fundamental frequency, but is given by $\omega\,(n, n - \alpha) = 1/h\ \{E(n) - E\,(n - \alpha)\}$, thus $\nu(n)\alpha$ is replaced by $\omega\,(n, n - \alpha)$. Correspondingly, Heisenberg introduces the quantum analogue of $A\alpha(n)$, which he writes as $\mathbf{A}(n, n - \alpha)$. Further, $-\ (dE/dt)_\alpha$ has, in the quantum theory, to be replaced by the *product of the transition probability per unit time*, $P(n, n - \alpha)$, and the *emitted energy* $h\omega(n, n - \alpha)$; resulting in

$$P(n, n - \alpha) = e^2/3\pi\varepsilon_0 hc^3\ [\omega(n, n - \alpha)]^3\ |A(n, n - \alpha)|^2.$$

This equation refers, however, to only one specific transition. For a full description of atomic dynamics (as then conceived), one will need to consider all the quantities $A(n, n - \alpha)\ e^{i\omega(n, n - \alpha)t}$. In the classical case, the terms $A\alpha(n)\ e^{i\omega(n)\alpha t}$ may be combined to yield x(t) using $x(n, t) = \Sigma_{\alpha = -\infty}^{+\infty}\ A_\alpha(n)\ e^{i\omega(n)\alpha t}$.

It is the *transition amplitudes* $\mathbf{A}(n, n - \alpha)$ which Heisenberg fastens upon as being satisfactorily 'observable'; like the *transition frequencies*, they depend on two discrete variables.

... This is the first of Heisenberg's 'magical jumps' - and certainly a very large one. Representing x(t) in this way seems to be the sense in which he considered himself to be 're-interpreting the kinematics'."]

[Fedak, W. A. & Prentis, J. J. (2009), *Appendix,* p. 136: "*Reinterpretation 1: Position.* Heisenberg considered one-dimensional periodic systems. The classical motion of the system (in a stationary state labeled *n*) is described by the time-dependent position $x(n, t)$. Heisenberg represents this periodic function by the Fourier series

$$x(n, t) = \Sigma_{\alpha = -\infty}^{+\infty}\ a_\alpha\,(n)\ e^{i\omega(n)\alpha t} \qquad (A1) \qquad [14](2a)$$

The $\alpha$th Fourier component related to the *n*th stationary state has amplitude $a_\alpha(n)$ and frequency $\alpha\varpi(n)$. According to the *correspondence principle*, the $\alpha$th Fourier component of the classical motion in the state *n* corresponds to the quantum jump from state *n* to state $n - \alpha$. Motivated by this principle, Heisenberg replaced the classical component $a(n)\ e^{i\omega(n)\alpha t}$ by the transition component $a(n, n - \alpha)\ e^{i\omega(n,\ n-\alpha)t}$. ... Unlike the sum over the classical components in Eq. (A1), Heisenberg realized that a similar sum over the transition components is meaningless. Such a quantum *Fourier series* could not describe the electron motion in one stationary state (*n*) because each term in the sum describes a transition process associated with two states (*n* and $n - \alpha$).

Heisenberg's next step was bold and ingenious. Instead of reinterpreting x(t) as a *sum* over transition components, he represented the position by the *set* of transition components. We symbolically denote Heisenberg's reinterpretation as

$$x \rightarrow \{a(n, n-\alpha)\, e^{i\omega(n,\, n-\alpha)t}. \qquad \text{(A2)}$$

*Equation (A2) is the first breakthrough relation.*"]

Classically, the answer is obviously

$$\mathbf{B}_\beta(n)\, e^{i\omega(n)\beta t} = \Sigma_{-\infty}^{+\infty}\, \mathbf{A}_\alpha\, \mathbf{A}_{\beta-\alpha}\, e^{i\omega(n)(\alpha+\beta-\alpha)t} \qquad [16](3)$$

or

$$= \int_{-\infty}^{+\infty} \mathbf{A}_\alpha\, \mathbf{A}_{\beta-\alpha}\, e^{i\omega(n)(\alpha+\beta-\alpha)t}\, d\alpha \qquad [17](4)$$

so that

$$x(t)^2 = \Sigma_{\beta=-\infty}^{+\infty}\, \mathbf{B}_\beta\, e^{i\omega(n)\beta t} \qquad [18](5)$$

or, respectively

$$= \int_{-\infty}^{+\infty} \mathbf{B}_\beta\, e^{i\omega(n)\beta t} d\beta \qquad [19](6)$$

[from $x(t) = \Sigma_{\alpha=-\infty}^{+\infty}\, \mathbf{A}_\alpha(n)\, e^{i\omega(n)\alpha t}$,

$x(t)^2 = \Sigma_{\alpha=-\infty}^{+\infty}\, \Sigma_{\beta-\alpha=-\infty}^{+\infty}\, \mathbf{A}_\alpha\, \mathbf{A}_{\beta-\alpha}\, e^{i\omega(n)(\alpha+\beta-\alpha)t}$,

$x(t)^2 = \Sigma_{\alpha=-\infty}^{+\infty}\, \Sigma_{\beta-\alpha=-\infty}^{+\infty}\, \mathbf{A}_\alpha\, \mathbf{A}_{\beta-\alpha}\, e^{i\omega(n)\beta t}$,

or $x(t)^2 = \Sigma_{\beta=-\infty}^{+\infty}\, B_\beta\, e^{i\omega(n)\beta t}$

where $B_\beta(n)\, e^{i\omega(n)\beta t} = \Sigma_{-\infty}^{+\infty}\, A_\alpha\, A_{\beta-\alpha}\, e^{i\omega(n)(\alpha+\beta-\alpha)t}$.]

[Aitchison, MacManus & Snyder. (2004), p. 4-5: "The crucial difference in the quantum case is that the *frequencies* do not combine in the same way as the classical harmonics, but rather in accordance with the *Ritz combination principle*:

$$\omega(n, n-\alpha) + \omega(n-\alpha, n-\beta) = \omega(n, n-\beta),$$

which is of course consistent with $\omega(n, n-\alpha) = 1/h\,\{W(n) - W(n-\alpha)\}$.

[The *Rydberg–Ritz combination principle* is an empirical generalization proposed by Walther Ritz in 1908 to describe the relationship of the spectral lines for all atoms. The principle states that the spectral lines of any element include frequencies that are either the sum or the difference of the frequencies of two other lines.]

Thus in order to end up with the particular frequency $\omega(n, n-\beta)$, it seems 'almost necessary' (in Heisenberg's words) to combine the quantum *amplitudes* in such a way as to ensure the *frequency* combination above; that is, as

$$B(n, n-\beta)\, e^{i\omega(n,\, n-\beta)t} = \Sigma_\alpha\, A(n, n-\alpha)\, e^{i\omega(n,\, n-\alpha)t}\, A(n-\alpha, n-\beta)\, e^{i\omega(n-\alpha,\, n-\beta)t}.$$

Cancelling the exponentials on both sides, we are left with

$$B(n, n-\beta) = \Sigma_\alpha A(n, n-\alpha)\, A(n-\alpha, n-\beta),$$

which is Heisenberg's law for multiplying *transition amplitudes* together."]

[Fedak, W. A. & Prentis, J. J. (2009), *Appendix,* p. 136: "*Reinterpretation 2: Multiplication.* To calculate the energy of a harmonic oscillator, Heisenberg needed to know the quantity $x^2$. *How do you square a set of transition components*? Heisenberg posed this fundamental question twice in his paper. His answer gave birth to the algebraic structure of quantum mechanics. … Heisenberg answered this question by reinterpreting the square of a Fourier series with the help of the *Ritz principle*. He evidently was convinced that quantum multiplication, whatever it looked like, must reduce to Fourier-series multiplication in the classical limit. … In the new quantum theory Heisenberg replaced

$$x^2(n, t) = \Sigma_{\beta=-\infty}^{+\infty}\, b_\beta(n)\, e^{i\omega(n)\beta t} \qquad \text{(A3)}$$

… with

$$x^2 \rightarrow \{b(n, n-\beta)\, e^{i\omega(n,\, n-\beta)t}, \qquad \text{(A5)}$$

where the $n \rightarrow n-\beta$ *transition amplitude* is

$$b(n, n-\beta) = \Sigma_{\alpha=-\infty}^{+\infty}\ a(n, n-\alpha)\, a(n-\alpha, n-\beta) \qquad \text{(A6)}$$

In constructing Eq. (A6) Heisenberg uncovered the symbolic algebra of atomic processes. … Eq. (A6) allowed Heisenberg to algebraically manipulate the transition components."]

It seems that quantum theoretically the easiest and most natural assumption is to replace Eqs. (3), (4)

$$[\mathbf{B}_\beta(n)\, e^{i\omega(n)\beta t} = \Sigma_{-\infty}^{+\infty}\ \mathbf{A}_\alpha\, \mathbf{A}_{\beta-\alpha}\, e^{i\omega(n)(\alpha+\beta-\alpha)t} \qquad [16](3)$$

$$\text{or} \qquad = \int_{-\infty}^{+\infty} \mathbf{A}_\alpha\, \mathbf{A}_{\beta-\alpha}\, e^{i\omega(n)(\alpha+\beta-\alpha)t}\, d\alpha \qquad [17](4)]$$

with

$$\mathbf{B}(n, n-\beta)\, e^{i\omega(n,n-\beta)t} = \Sigma_{\alpha=-\infty}^{+\infty}\ \mathbf{A}(n, n-\alpha)\, \mathbf{A}(n-\alpha, n-\beta)\, e^{i\omega(n,n-\beta)t} \qquad [20](7)$$

$$\text{or} \qquad = \int_{-\infty}^{+\infty} \mathbf{A}(n, n-\alpha)\, \mathbf{A}(n-\alpha, n-\beta)\, e^{i\omega(n,n-\beta)t}\, d\alpha \qquad [21](8)$$

and indeed, this way of combination follows almost inevitably from the *frequency combination relation*. If we accept the assumptions (7), (8) one recognizes also that the *phases* of the quantum theoretical **A** have the same relevant physical significance as in the classical theory: only the beginning time and hence a phase constant common to all the **A** is arbitrary and without physical meaning but the *phase* of every single A enters in the quantity **B** [1].

[1] Compare also to Kramers, H.A., & Heisenberg, W. (1925). [Über die Streuung von Strahlung durch Atome. (On the scattering of radiation by atoms.) *Zeit. Physik*, 31, 681-

708; https://doi.org/ 10.1007/BF02980624.] In the expressions used there for the induced scattering momentum, the phases are essentially contained.

A geometric interpretation of these quantum theoretic *phase* relationships in analogy to the classical theory seems at first not possible.

We ask now about how to represent the quantity $x(t)^3$ and we find without difficulty:

*Classically*:

$$\mathbf{C}(n, \gamma) = \Sigma_{\alpha=-\infty}^{+\infty} \Sigma_{\beta=-\infty}^{+\infty} \mathbf{A}_{\alpha}(n)\, \mathbf{A}_{\beta}(n)\, \mathbf{A}_{\gamma-\alpha-\beta}(n) \qquad [22](9)$$

*Quantum theoretically*:

$$\mathbf{C}(n, n-\gamma) = \Sigma_{\alpha=-\infty}^{+\infty} \Sigma_{\beta=-\infty}^{+\infty} \mathbf{A}(n, n-\alpha)\, \mathbf{A}(n-\alpha, n-\alpha-\beta)\, \mathbf{A}(n-\alpha-\beta, n-\gamma) \qquad [23](10)$$

or the corresponding formulae with integrals.

In a similar way, all the quantities of the form x(t) n can be expressed quantum theoretically and when a function f[x(t)] is given, one can always obviously find the quantum theoretical analog if it is possible to expand this function in powers of x. *A substantial difficulty arises when we consider two quantities x(t), y(t) and we ask about the product x(t)y(t).*

Let be x(t) characterized with **A** and y(t) with **B** so the representation of x(t)y(t) results:

*Classically*:

$$\mathbf{C}_{\beta} = \Sigma_{\alpha=-\infty}^{+\infty} \mathbf{A}_{\alpha}(n)\, \mathbf{B}_{\beta-\alpha}(n) \qquad [24]$$

*Quantum theoretically*:

$$\mathbf{C}(n, n-\beta) = \Sigma_{\alpha=-\infty}^{+\infty} \mathbf{A}(n, n-\alpha)\, \mathbf{B}(n-\alpha, n-\beta). \qquad [25]$$

[Aitchison, MacManus & Snyder. (2004), p. 5: "Born recognized

$$C(n, n-\beta) = \Sigma_{\alpha=-\infty}^{+\infty} A(n, n-\alpha)\, B(n-\alpha, n-\beta)$$

as matrix multiplication (something unknown to Heisenberg in July 1925), and he and Jordan rapidly produced the first paper to state the fundamental commutation relation (in modern notation …)

$$\mathbf{xp} - \mathbf{px} = i\hbar. \qquad (11)$$

Dirac's paper followed soon after, and then the 'three-man' paper of Born, Heisenberg and Jordan. [Born, M., Heisenberg, W. & Jordan, P. (August, 1926). Zur Quantenmechanik II. (On Quantum Mechanics II.) *Zeit. Phys*., 35, 557-615."]

[Aitchison, MacManus & Snyder. (2004), pp. 6-7: "It took Born only a few days to show that Heisenberg's *quantum condition* (16) was in fact the diagonal matrix element of

$\Sigma_{\alpha=-\infty}^{+\infty}$ $\mathbf{A}(n, n-\alpha)\ \mathbf{B}(n-\alpha, n-\beta)$. (7)

or in modern notation, of **xp** − **px**, and to guess that the off-diagonal elements of **xp** − **px** were zero, a result which was shown to be compatible with the *equations of motion* in Born and Jordan's paper. [Born, M. & Jordan, P. (December, 1925). Zur Quantenmechanik. (On Quantum Mechanics.) *Zeit. Phys.*, 34, 858-88."]

… Heisenberg's *transition amplitude* $\mathbf{A}(n, n-\alpha)$ is indeed precisely the same as the quantum-mechanical matrix element $(n-\alpha|x|n)$, where $|n)$ is the exact eigenstate with energy $W(n)$. The relation of (16)

$$[h = 4\pi m\ \Sigma_{\alpha=0}^{+\infty} \{|a(n, n+\alpha)|^2\omega(n, n+\alpha) - |a(n, n-\alpha)|^2\omega(n, n-\alpha)\} \qquad [34](16)]$$

to the *fundamental commutator* (11)

$$[\ddot{x} + f(x) = 0 \qquad [26](11)]$$

is briefly recalled in Appendix A.

…

*Appendix A: The quantum condition and* ***xp*** − ***px*** = iħ.

Consider the $(n, n)$ element of $(x\dot{x} - \dot{x}x)$. This is

$$\Sigma_\alpha\, a(n, n-\alpha)\ i\omega(n-\alpha, n)\ a(n-\alpha, n) - \Sigma_\alpha\, i\omega(n, n-\alpha)\ a(n, n-\alpha)\ a(n-\alpha, n).$$

$$[\Sigma_\alpha\, a(n, n-\alpha)\ .\ d/dt\ a(n-\alpha, n) - \Sigma_\alpha\, d/dt\ a(n, n-\alpha)\ .\ a(n-\alpha, n),$$
$$\text{where } a(n, n-\alpha) = a_\alpha(n, n-\alpha)\ e^{i\alpha\omega(n,\, n-\alpha)t}]$$

[The square root of − 1 (i) is introduced into $\mathbf{x\dot{x}} - \mathbf{\dot{x}x}$ by differentiating the Fourier series $x = \Sigma_{\alpha=-\infty}^{+\infty}\ a_\alpha(n)\ e^{i\alpha\omega(n)t}$ expressed in terms of exponentials with respect to t in $\dot{x}$.]

In the first term, the sum over $\alpha > 0$ may be re-written as

$$-\, i\ \Sigma_{\alpha>0}\ \omega(n, n-\alpha)\ |a(n, n-\alpha)|^2$$

using $\omega(n, n-\alpha) = -\ \omega(n-\alpha, n)$ from $\nu(n, n-\alpha) = 1/h\ \{W(n) - W(n-\alpha)\}$ in paragraph 1(original page 881), and $a_\alpha(n-\alpha, n) = a_\alpha(n, n-\alpha)$ from the quantum-theoretical analogue of $a_{-\alpha}(n) = a_\alpha(n)$ on original page 885, assuming as Heisenberg did that the $a_\alpha$'s are chosen to be real, while the sum over $\alpha < 0$ becomes, similarly,

$$i\ \Sigma_{\alpha>0}\ \omega(n+\alpha, n)|a(n+\alpha, n)|^2$$

on changing α to − α.

Similar steps in the second term led to the result

$$(\mathbf{x\dot{x}} - \mathbf{\dot{x}x})\ (n, n) = 2i\ \Sigma_{\alpha>0}\ \{\omega(n+\alpha, n)|a(n+\alpha, n)|^2 - \omega(n, n-\alpha)|a(n, n-\alpha)|^2\} = 2ih/(4\pi m),$$

where the last step follows from the '*quantum condition*' (16). Setting $\mathbf{p} = m\mathbf{\dot{x}}$ [in $m(x\dot{x} - \dot{x}x)\ (n, n) = ih/4\pi$] we find

$(\mathbf{xp} - \mathbf{px})(n, n) = i\hbar$

for all values of n, [which is the modern formulation of (16).] This is the result which Born found shortly after reading Heisenberg's paper. *In the further development of the theory the value of the 'fundamental commutator' xp − px, namely iħ times the unit matrix, was taken to be a basic postulate*. The sum rule (16) is then derived by taking the $(n, n)$ matrix element of the relation $[\mathbf{x}, [\mathbf{H}, \mathbf{x}]] = \hbar^2/m$."]

While classically x(t)y(t) always equal to y(t)x(t) is, in general it must not be the case in the quantum theory. In special cases, for example when one considers $x(t)x(t)^2$, the difficulty does not arise.

[Aitchison, MacManus & Snyder. (2004), p. 5: "This 'difficulty' clearly unsettled Heisenberg: but *it very quickly became clear that the non-commutativity (in general) of kinematical quantities in quantum theory was the really essential new technical idea in the paper*."]

As in the question posed at the beginning of this paragraph, when one considers a form like

$v(t)\dot{v}(t)$

one has to substitute $v\dot{v}$ quantum theoretically with $(v\dot{v} + \dot{v}v)/2$ so that $v\dot{v}$ becomes the derivative of $v^2/2$. In a similar way, natural mass quantum-theoretic mean values can always be given, which, however, are hypothetical to an even higher degree than the formulas (7) and (8). *

$$[\mathbf{B}(n, n-\beta)\, e^{i\omega(n,n-\beta)t} = \Sigma_{\alpha=-\infty}^{+\infty}\ \mathbf{A}(n, n-\alpha)\, \mathbf{A}(n-\alpha, n-\beta)\, e^{i\omega(n,n-\beta)t} \qquad [20](7)$$

or

$$= \int_{-\infty}^{+\infty} \mathbf{A}(n, n-\alpha)\, \mathbf{A}(n-\alpha, n-\beta)\, e^{i\omega(n,n-\beta)t}\, d\alpha. \qquad [21](8)]$$

Apart from the difficulty just described, formulas of type (7), (8) were general enough to express the interaction of the electrons in an atom through the characteristic *amplitudes* of the electrons. *

* My translation. These two sentences were omitted from Luca Dora's translation:

"In ahnlicher Weise lassen sich wohl stets naturgemasse quanten-theoretisehe Mittelwerte angeben, die allerdings in noch hoherem Grade hypothetiseh sind als die Formela (7) und (8).

Abgesehen von der eben geschilderten Schwierigkeit durften Formeln vom Typus (7), (8) allgemein genugen, um aueh die Wechselwirkung der Elektronen in einem Atom durch die eharakteristischen Amplituden der Elektronen auszudrucken."

[Original page 884.]

**§ 2.** After these considerations which subject was the kinematic of the quantum theory, we will turn to *mechanical problems aiming at the determination of* ***A****, v,* ***W*** *from the given forces of the system.*

> [Aitchison, MacManus & Snyder. (2004), p. 5: "Having identified the *transition amplitudes* X($n$, $n-\alpha$) and *frequencies* $\omega(n, n-\alpha)$ as the 'observables' with which the new theory should deal, Heisenberg now turns his attention to how they may be determined 'from the given forces of the system' - that is, by the dynamics."]

In the previously presented theory [the Old Quantum Mechanics], this problem is solved in two steps:

1. Integration of the *equations of motion*

$$\ddot{x} + f(x) = 0 \qquad [26](11)$$

[where f(x) is the *force per mass* function.]

2. Determination of the constants arising from periodic motion [through the quantum condition] with

$$\int p dq = \int m\dot{x} dx = J\ (= nh), \qquad [27](12)$$

[where m is the *mass* and the integral is to be evaluated over the period of one period of the motion].

> [Fedak, W. A. & Prentis, J. J. (2009), Appendix, pp 136-7: "*Reinterpretation 3: Motion.* Equations (A2), (A5), and (A6) represent the new "kinematics" of quantum theory—the new meaning of the position x. Heisenberg next turned his attention to the new "mechanics." *The goal of Heisenberg's mechanics is to determine the amplitudes, frequencies, and energies from the given forces.* Heisenberg noted that in the old quantum theory $a_\alpha(n)$ and $\varpi(n)$ are determined by solving the classical *equation of motion*
>
> $$\ddot{x} + f(x) = 0 \qquad (A7) \qquad [26](11)$$
>
> and quantizing the classical solution—making it depend on $n$ - via the quantum condition
>
> $$\int m\dot{x} dx = nh. \qquad (A8) \qquad [27](12)$$
>
> Heisenberg assumed that Newton's second law in Eq. (A7) is valid in the new quantum theory provided that the classical quantity x is replaced by the set of quantities in Eq. (A2),
>
> $$x \rightarrow \{a(n, n-\alpha)\, e^{i\omega(n,\, n-\alpha)t}. \qquad (A2)$$
>
> and f(x) is calculated according to the new rules of amplitude algebra. *Keeping the same form of Newton's law of dynamics, but adopting the new kinematic meaning of x is the third Heisenberg breakthrough.*"]

If one wants to construct a quantum theoretical mechanics which is the possible classical analog, it is probably very close to bring the *equation of motion* Eq. (11)

[$\ddot{x} + f(x) = 0$ (A7) [26](11)]

directly into the quantum theory *where it is only necessary to take over*, for not abandoning the foundation of in principle observable quantities, *instead of the quantities $\ddot{x}$, $f(x)$, their quantum theoretic representations known from § 1.*

> [Aitchison, MacManus & Snyder. (2004), p. 5: "or, as we would say today, by taking matrix elements of the operator equation of motion $\ddot{x} + f(x) = 0$."]

In the classical theory, it is possible to search for a solution of Eq. (11) by first expressing x[(t)] in Fourier series or Fourier integrals with undetermined coefficients (and frequencies); although in general we obtain infinitely many equations with infinitely many unknowns (or integral equations) which can be solved only in special cases with simple recursion formulae for [the Fourier coefficients] X.

However, in the quantum theory, we are dependent on this kind of solution for Eq. (11)

[$\ddot{x} + f(x) = 0$ [26](11)]

which, as discussed before, prevents the definition of direct [quantum-theoretical] analogues of the function x(n, t).

This has as consequence that the quantum theoretical solution of Eq. (11) is feasible at first only in the simplest cases. Before going over these simple examples, we would like to derive quantum theoretically the value of the constant in Eq. (12).

[$\int p\,dq = \int m\dot{x}\,dx = J\ (= nh)$, [27](12)]

We assume also that the (classical) motion is periodic:

$$x = \Sigma_{\alpha=-\infty}^{+\infty}\, a_\alpha(n)\, e^{i\alpha\omega(n)t} \qquad [28](13)$$

> [Note: the complex amplitude vectors for absorption $\mathbf{A}_a$ and emission $\mathbf{A}_e$ transitions are replaced by $a_\alpha$, which are no longer vectors but amplitudes in the Fourier expansion of the coordinate x of the electron.]

then

$$m\dot{x} = m\, \Sigma_{\alpha=-\infty}^{+\infty}\, a_\alpha(n)\, .\, i\alpha\omega(n)\, e^{\alpha\omega(n)t} \qquad [29]$$

and

$$\int m\dot{x}\, dx = \int m\dot{x}^2\, dt = 2\pi m\, \Sigma_{\alpha=-\infty}^{+\infty}\, a_\alpha(n) a_{-\alpha}(n) \alpha^2 \omega(n). \qquad [30]$$

> [where the integral is evaluated over one period of the motion].

Further, since $a_{-\alpha}(n) = a_\alpha(n)$ (x must be real), it follows

$$\int m\dot{x}^2\, dt = 2\pi m\, \Sigma_{\alpha=-\infty}^{+\infty}\, |a_\alpha(n)|^2 \alpha^2 \omega(n). \qquad [31](14)$$

[Aitchison, MacManus & Snyder. (2004), p. 6: "Heisenberg argues that (14) does not sit well with the *Correspondence Principle*, since the latter should only determine J up to an additive constant (times h). Setting (14) equal to *n*h, he converts it to the form

$$h = 2\pi m \Sigma_{\alpha=-\infty}^{+\infty} \alpha \, d/dn \, \{\alpha\omega(n) \cdot |a_\alpha(n)|^2\}$$

which determines the $a_\alpha(n)$'s only to within a constant."]

Until now, this *phase* integral was set to a multiple of h (*n*h); such a condition is not only forced into the classical calculation but it looks arbitrary also from the previous point of view of the *correspondence principle* because correspondence-wise the J is set only up to an additive constant as a multiple integer of h and instead of Eq. (14)

$$[\int m\dot{x}^2 \, dt = 2\pi m \Sigma_{\alpha=-\infty}^{+\infty} |a_\alpha(n)|^2\alpha^2\omega(n). \qquad [31](14)]$$

one should have had

$$d/dn \, (nh) = d/dn \cdot \int m\dot{x}^2 \, dt \qquad [32]$$

which means

$$h = 2\pi m \Sigma_{\alpha=-\infty}^{+\infty} \alpha \, d/dn \, \{\alpha\omega(n) \cdot |a_\alpha(n)|^2\} \qquad [33](15).$$

[Bohr's *correspondence principle* states that the behavior of systems described by the theory of quantum mechanics (or by the old quantum theory) reproduces classical physics in the limit of large quantum numbers. In other words, it says that for large orbits and for large energies, quantum calculations must agree with classical calculations. The principle was formulated by Niels Bohr in 1920, though he had previously made use of it as early as 1913 in developing his model of the atom. [Bohr, N. (October, 1920), Über die Serienspektra der Elemente. (About the serial spectra of the elements.), *Zeit. Physik*, 2, 5, 423-78; https://doi.org/10.1007/BF01329978; translation in Niels Bohr Collected Works (1976). Edited by L. Rosenfeld, J. Rud Nielsen. Volume 3, 241-282.]]

Such a relation though fixes the $a_\alpha$ only up to a constant and this indetermination led empirically to the difficulty of half-integer quantum numbers.

[Aitchison, MacManus & Snyder. (2004), p. 6: "… the summation can alternatively be written as over positive values of α, replacing 2πm by 4πm. *In another crucial jump, Heisenberg now replaces the differential in (15) by a difference*, giving

$$h = 4\pi m \Sigma_{\alpha=0}^{+\infty} \{|a(n, n+\alpha)|^2\omega(n, n+\alpha) - |a(n, n-\alpha)|^2\omega(n, n-\alpha)\}.\text{"]}$$

If we ask for a quantum theoretical relation between observable quantities according to Eq. (14) and (15), the missing unambiguity comes out by itself again.

Indeed *only Eq. (15) has a simple quantum theoretical connection to the Kramer's dispersion theory*:

$$h = 4\pi m \Sigma_{\alpha=0}^{+\infty} \{|a(n, n+\alpha)|^2\omega(n, n+\alpha) - |a(n, n-\alpha)|^2\omega(n, n-\alpha)\} \qquad [34](16)$$

[Aitchison, MacManus & Snyder. (2004), p. 6: "As Heisenberg later recalled, he had noticed that 'if I wrote down this [presumably (15) above] and tried to translate it according to the scheme of dispersion theory, I got the Thomas-Kuhn sum rule [which is equation (16)]. And that is the point. Then I thought, 'That is apparently how it is done'.

By 'the scheme of dispersion theory', Heisenberg is referring to what Jammer called *Born's correspondence rule*, [Born, M. (December, 1924). Über Quantenmechanik, (About quantum mechanics.) *Zeit. Phys.*, 26, 379-95] namely

$$\alpha\, \partial\Phi(n)/\partial n \leftrightarrow \Phi(n) - \Phi(n-\alpha),$$

or rather to its iteration in the form

$$\alpha\, \partial\Phi(n, \alpha)/\partial n \leftrightarrow \Phi(n+\alpha, n) - \Phi(n, n-\alpha)$$

as used in the *Kramers-Heisenberg theory of dispersion*. [Kramers, H. A. & Heisenberg, W. (February, 1925). Über die Streuung von Strahlung durch Atome. (On the scattering of radiation by atoms.) *Zeit. Physik*, 31, 681-708; http://dx.doi.org/10.1007/BF02980624]".]

[The *Kramers-Heisenberg dispersion formula* is an expression for the cross section for scattering of a photon by an atomic electron. It was derived before the advent of quantum mechanics by Hendrik Kramers and Werner Heisenberg in 1925, based on the *correspondence principle* applied to the classical dispersion formula for light.

The quantum mechanical derivation was given by Paul Dirac in 1927. [Dirac, P. A. M. (March, 1927). The quantum theory of the emission and absorption of radiation. *Roy. Soc. Proc., A*, 114, 767, 243-65.]]

[Fedak, W. A. & Prentis, J. J. (2009), *Appendix*: "*Reinterpretation 4: Quantization.* "How did Heisenberg reinterpret the old quantization condition in Eq. (A8)?

$$\int m\dot{x}\, dx = nh. \qquad (A8) \qquad [27](12)$$

Given the Fourier series in Eq. (A1),

$$x(n, t) = \Sigma_{\alpha=-\infty}^{+\infty} a_\alpha(n)\, e^{i\omega(n)\alpha t} \qquad (A1) \qquad [14](2a)$$

the quantization condition, $nh = \int m\dot{x}^2\, dt$, can be expressed in terms of the Fourier parameters $a_\alpha(n)$ and $\omega(n)$ as

$$nh = \int m\dot{x}^2\, dt = 2\pi m \Sigma_{\alpha=-\infty}^{+\infty} |a_\alpha(n)|^2\alpha^2\omega(n). \qquad (A9) \qquad [31](14)$$

For Heisenberg, setting $\int p dx$ (= $\int m\dot{x}\, dx$) equal to an integer multiple of h was an arbitrary rule that did not fit naturally into the dynamical scheme. Because his theory focuses exclusively on transition quantities, Heisenberg needed to translate the old quantum condition that fixes the properties of the *states* to a new condition that fixes the properties of the *transitions between states*. Heisenberg believed that what matters is the difference between ∫pdx evaluated for neighboring states: $[\int pdx]_n - [\int pdx]_{n-1}$. He therefore took the derivative of Eq. (A9) with respect to $n$ to eliminate the forced $n$ dependence and to produce a differential relation that can be reinterpreted as a difference relation between transition quantities. In short, Heisenberg converted

$$h = 2\pi m \sum_{\alpha=-\infty}^{+\infty} \alpha\, d/dn\, (|a_\alpha(n)|^2 \alpha\omega(n)) \qquad \text{(A10)}$$

to

$$h = 4\pi m \sum_{\alpha=0}^{+\infty} \{|a(n, n+\alpha)|^2\omega(n, n+\alpha) - |a(n, n-\alpha)|^2\omega(n, n-\alpha)\} \qquad \text{(A11)} \qquad [34](16)$$

In a sense Heisenberg's "*amplitude condition*" in Eq. (A11) (16) is the counterpart to Bohr's *frequency condition* (Ritz's *frequency combination rule*). Heisenberg's condition relates the *amplitudes* of different lines within an atomic spectrum and Bohr's condition relates the *frequencies*.

*Equation (A11) is the fourth Heisenberg breakthrough.*

Equations (A7) and (A11)

$$[\ddot{x} + f(x) = 0 \qquad [26](11)$$

$$h = 4\pi m \sum_{\alpha=0}^{+\infty} \{|a(n, n+\alpha)|^2\omega(n, n+\alpha) - |a(n, n-\alpha)|^2\omega(n, n-\alpha)\} \qquad \text{(A11)} \qquad [34](16)$$

constitute Heisenberg's new mechanics. In principle, these two equations can be solved to find $a(n, n-\alpha)$ and $\omega(n, n-\alpha)$. *No one before Heisenberg knew how to calculate the amplitude of a quantum jump*. Equations (A2), (A6), (A7), and (A11) define Heisenberg's program for constructing the line spectrum of an atom from the given force on the electron.]

Indeed, this relationship

$$[h = 4\pi m \sum_{\alpha=0}^{+\infty} \{|a(n, n+\alpha)|^2\omega(n, n+\alpha) - |a(n, n-\alpha)|^2\omega(n, n-\alpha)\} \qquad [34](16)]$$

is sufficient for a unique determination of the $a$'s because the initially undetermined constant in the quantities $a$ will be fixed by itself by the condition which should give a normal state where no more radiation is present. Let the normal state be described by $n_0$, then

$$a(n_0, n_0 - \alpha) = 0 \text{ for } \alpha > 0. \qquad [35]$$

The question about half-integer or integer quantization cannot be present in a quantum mechanics where only relations between observable quantities are used.

Eqs. (11) and (16)

$$[\ddot{x} + f(x) = 0 \qquad [26](11)$$

$$h = 4\pi m \sum_{\alpha=-\infty}^{+\infty} \{|a(n, n+\alpha)|^2\omega(n, n+\alpha) - |a(n, n-\alpha)|^2\omega(n, n-\alpha)\} \qquad [34](16)]$$

together contain, if solvable, *a complete determination not only of the frequencies and energies, but also of the quantum theoretical transition probabilities*. However, the actual mathematical procedure succeeds only in the easiest cases. A particular complication comes also from systems like the hydrogen atom: since the solutions represent partly periodic and partly aperiodic motions, it has the consequence that the quantum theoretic series (7), (8)

$$[B(n, n-\beta)\, e^{i\omega(n,n-\beta)t} = \Sigma_{\alpha=-\infty}^{+\infty}\, A(n, n-\alpha)\, A(n-\alpha, n-\beta)\, e^{i\omega(n,n-\beta)t} \qquad [20](7)$$

or

$$= \int_{-\infty}^{+\infty} A(n, n-\alpha)\, A(n-\alpha, n-\beta)\, e^{i\omega(n,n-\beta)t}\, d\alpha. \qquad [21](8)]$$

and Eq. (16) always fall in both the sum and the integral case. *Quantum mechanically, it is not possible to divide "periodic and aperiodic motions"*.

Despite that, one might see Eq. (11) and Eq. (16) at least in principle as a satisfactory solution of the mechanical problem, if it is possible to show that this solution coincides (or is not in contradiction) with the until now known quantum mechanical relationships and that a small perturbation of a mechanical problem gives rise to additional orders in the energies or frequencies respectively which correspond to the expressions found by Kramers and Born (in contrast to which would have led to the classical theory). Further, one must investigate if in general Eq. (11) in the suggested quantum theoretical interpretation corresponds an energy integral $m\, \dot{x}^2/2 + U(x) = \text{const.}$ and if such obtained energy (analogously as classically holds $\nu = \partial W/\partial J$) the relation $\Delta W = h\nu$ is sufficient. A general answer to these questions might demonstrate the coherence of the present experiments and lead to a quantum mechanics which operates only with observable quantities. Apart from a general relationship between the *Kramer's dispersion formula* and Eq. (11) and (16), we can only answer the above stated questions in very special solvable cases through simple recursion.

The general connection between *Kramer's dispersion theory* and our Eqs. (11) and (16)

$$[\ddot{x} + f(x) = 0 \qquad [26](11)$$

$$h = 4\pi m\, \Sigma_{\alpha=-\infty}^{+\infty}\, \{|a(n, n+\alpha)|^2\omega(n, n+\alpha) - |a(n, n-\alpha)|^2\omega(n, n-\alpha)\} \qquad [34](16)]$$

consists in the fact that in Eq. (11) (more precisely from its quantum-theoretical analog) one finds, just as in the classical theory, that the oscillating electron behaves like a free electron when acted upon by light, of much higher frequency than any eigenfrequency of the system. This result follows also from Kramers' dispersion theory when Eq. (16) is taken into account. Indeed, Kramers finds for *moment* induced by a wave of the form $E \cos 2\pi\nu t$

$$M = 2e^2E\cos(2\pi\nu t)/h\, \Sigma_{\alpha=0}^{+\infty}\, [|a(n, n+\alpha)|^2\nu(n, n+\alpha)/\{\nu^2(n, n+\alpha) - \nu^2\} - |a(n, n-\alpha)|^2\nu(n, n-\alpha)/\{\nu^2(n, n-\alpha) - \nu^2\}] \qquad [36]$$

So that for $\nu > \nu(n, n+\alpha)$

$$M = 2e^2E \cos(2\pi\nu t)/\nu^2 h \, \Sigma_{\alpha=0}^{+\infty} \, [|a(n, n+\alpha)|^2 \nu(n, n+\alpha) - |a(n, n-\alpha)|^2 \nu(n, n-\alpha)] \qquad [37]$$

which using Eq. (16) becomes

$$M = -\, e^2E \cos(2\pi\nu t)/\nu^2 4\pi^2 m. \qquad [38]$$

**§ 3.** In the following, as the simplest example, the anharmonic oscillator will be treated:

$$\ddot{x} + \omega_0^2\, x + \lambda x^2 = 0 \qquad [39](17)$$

Classically, this equation can be satisfied by an Anzatz for the solution of the form:

$$x = \lambda a_0 + a_1 \cos \omega t + \lambda a_2 \cos 2\omega t + \lambda^2 a3 \cos 3\omega t + \ldots + \lambda^{\tau-1}\, a_\tau \cos \tau\omega t \qquad [40]$$

where the *a* are power series in λ, the first terms of which are independent from λ.

[Aitchison, MacManus & Snyder. (2004), pp. 8-9: "… Heisenberg proposes to seek a solution analogous to this, using the 'representation' of x(t) in terms of the quantities $\{a(n, n-\alpha)\, e^{i\omega(n,\, n-\alpha)t}\}$. It seems reasonable to assume that, as the index α increases away from zero, in integer steps, each successive amplitude will (to leading order in λ) be suppressed by an additional power of λ, as in the classical case. Thus, Heisenberg suggests that, in the quantum case, x(t) should be represented by terms of the form

$\lambda a(n, n)$ ; $a(n, n-1) \cos \omega(n, n-1)t$ ; $\lambda a(n, n-2)t$ ; ... $\lambda^{\tau-1} a(n, n-\tau)$ $\cos \omega(n, n-\tau)t$ ...

where,

$a(n, n) = a^{(0)}(n, n) + \lambda a^{(1)}(n, n) + \lambda^2\, a^{(2)}(n, n) + \ldots$

$a(n, n-1) = a^{(0)}(n, n-1) + \lambda a^{(1)}(n, n-1) + \lambda^2\, a^{(2)}(n, n-1) + \ldots$

and so on, and

$\omega(n, n-\alpha) = \omega^{(0)}(n, n-\alpha) + \lambda\omega^{(1)}(n, n-\alpha) + \lambda^2\omega^{(2)}(n, n-\alpha) + \ldots$

As Born and Jordan pointed out*, some use of 'correspondence' arguments has been made here, in assuming that, as λ → 0, only transitions between adjacent states are possible.

* Born, M. & Jordan, P. (December, 1925). Zur Quantenmechanik. (On Quantum Mechanics.) *Zeit. Phys.*, 34, 858-88.

*Heisenberg now simply writes down what he asserts to be the quantum versions.*"]

[For derivation relating the amplitudes $a(n, n-\alpha)$ to the corresponding quantities $\lambda^{\tau-1} a(n, n-\tau)$ see Aitchison, MacManus & Snyder. (2004), pp. 10-11.]

Quantum theoretically, we try to find an analogous expression representing x with terms of the form

$\lambda a(n, n)$ ; $a(n, n-1) \cos \omega(n, n-1)t$ ; $\lambda a(n, n-2)t$ ;
... $\lambda^{\tau-1} a(n, n-\tau) \cos \omega(n, n-\tau)t$ ... [41]

The recursion formulae for the determination of $a$ and $\omega$ (up to order $\lambda$) according to Eq. (3), (4) or Eq. (7), (8) are:

*Classically*

… [Original page 888.]

> [Aitchison, MacManus & Snyder. (2004). Understanding Heisenberg's 'Magical' Paper of July 1925: a New Look at the Calculational Details. § 4. *Conclusion*, pages 18-19: "The fact is, Heisenberg's 'amplitude calculus' works: at least for the simple one-dimensional problems on which he tried it out, it is an eminently practical procedure, requiring no sophisticated mathematical knowledge to implement. Since it uses the correct equations of motion, and incorporates the fundamental commutator (11)
>
> [$\mathbf{xp} - \mathbf{px})(n, n) = ih$ (11)]
>
> via the '*quantum condition*' (16)
>
> [$h = 4\pi m \sum_{\alpha=-\infty}^{+\infty} \{|a(n, n+\alpha)|^2 \omega(n, n+\alpha) - |a(n, n-\alpha)|^2 \omega(n, n-\alpha)\}$ [34](16)]
>
> the answers obtained are completely correct, in the sense of agreeing with conventional quantum mechanics.
>
> … The multiplication law (10)
>
> [$B(n, n-\beta) = \Sigma_\alpha A(n, n-\alpha) A(n-\alpha, n-\beta)$ (10)]
>
> has a convincing physical rationale, even for those who (like Heisenberg) do not recognize it as matrix multiplication … . The simple examples of this introduce the fundamental quantum idea that a transition from one state to the other occurs via all possible intermediate states, something which can take time to emerge in the traditional wave-mechanical approach. … Finally, the type of perturbation theory employed here … [is] more easily related to the classical analysis than is conventional quantum-mechanical perturbation theory ... .It is of course true that many important problems in quantum mechanics are much more conveniently handled in the wave-mechanical formalism … [in which] the … 'matrix elements' are the elements of Heisenberg's matrices."]

…

[Original page 890 ff; translation page 13 ff:]

… Furthermore, the energy calculated from Eq. 27 satisfies the relation (cf. Eq. 24):

$\omega(n, n-1)/2\pi = 1/h \,.\, [W(n) - W(n-1)]$, [62]

which can be regarded as a necessary condition for the possibility of a determination of the transition probabilities according to Eqs. 11 and 16

$$[\ddot{x} + f(x) = 0 \qquad [26](11)$$

$$h = 4\pi m \sum_{\alpha=-\infty}^{+\infty} \{|a(n, n+\alpha)|^2\omega(n, n+\alpha) - |a(n, n-\alpha)|^2\omega(n, n-\alpha)\} \qquad [34](16)]$$

…

Whether a method to determine quantum-theoretical data using relations between observable quantities as proposed here, can be regarded as satisfactory in principle, or whether this method indeed after all represents a too rough approach to the physical problem of constructing a theoretical quantum mechanics, an obviously very involved problem at the moment, can be decided only by a deeper mathematical investigation of the method which has been very superficially employed here.

## Dirac, P. A. M. (March, 1927). The quantum theory of the emission and absorption of radiation.

*Roy. Soc. Proc., A*, 114, 767, 243-65; https://doi.org/10.1098/rspa.1927.0039; also in Underwood, T. G. (2023). *Quantum Electrodynamics – annotated sources*, Volume I, pp. 471-88.

Communicated by N. Bohr, For. Mem. R.S.

Received February 2, 1927.

St. John's College, Cambridge, and Institute for Theoretical Physics, Copenhagen.

Addresses *non-relativistic quantum electrodynamics*, treats problem of an assembly of similar systems satisfying the Einstein-Bose statistical mechanics which interact with another different system by obtaining a Hamiltonian function to describe the motion, theory of system in which *forces are propagated with velocity of light* instead of instantaneously, time counted as a c-number instead of being treated symmetrically with the space co-ordinates, addition of *interaction term*, production of electromagnetic field (emission of radiation) by moving electron, reaction of radiation field on emitting system, applies to the interaction of an assembly of *light-quanta* with an atom, shows that it leads to *Einstein's laws for the emission and absorption of radiation*, the interaction of an atom with *electromagnetic waves* is then considered, treats *field* of radiation as a dynamical system whose interaction with an ordinary atomic system may be described by a Hamilton function, dynamical variables specifying the *field* are the *energies* and *phases* of the harmonic components of the waves, shows that if one takes the *energies* and *phases* of the waves to be *q-numbers* satisfying the proper quantum conditions instead of *c-numbers* the Hamiltonian function for the interaction of the *field* with an atom takes the same form as that for the interaction of an assembly of *light-quanta* with the atom, provides a complete formal reconciliation between the wave and light-quantum point of view, leads to the correct expressions for Einstein's A's and B's, radiative processes of the more general type considered by Einstein and Ehrenfest in which more than one light-quantum take part simultaneously are not allowed on the present theory, the mathematical development of the theory made possible by Dirac's *general transformation theory* of the quantum matrices [Dirac (January, 1927). The Physical Interpretation of the Quantum Dynamics].

---

### § 1. *Introduction and Summary.*

The new quantum theory, based on the assumption that the dynamical variables do not obey the commutative law of multiplication, has by now been developed sufficiently to form a fairly complete theory of *dynamics*. One can treat mathematically the problem of any dynamical system composed of a number of particles with instantaneous forces acting between them, provided it is describable by a Hamiltonian function, and one can interpret the mathematics physically by a quite definite general method. On the other hand, hardly anything has been done up to the present on *quantum electrodynamics*.

> [*Quantum electrodynamics* deals with the *electromagnetic field* and its interaction with electrically charged particles.]

*The questions of the correct treatment of a system in which the forces are propagated with the velocity of light instead of instantaneously, of the production of an electromagnetic field by a moving electron, and of the reaction of this field on the electron have not yet been touched.* In addition, *there is a serious difficulty in making the theory satisfy all the requirements of the restricted principle of relativity* [*Theory of Special Relativity*], since a Hamiltonian function can no longer be used. This *relativity* question is, of course, connected with the previous ones, and it will be impossible to answer any one question completely without at the same time answering them all. However, it appears to be possible to build up a fairly satisfactory theory of the *emission of radiation* and of the *reaction of the radiation field on the emitting system* on the basis of a kinematics and dynamics *which are not strictly relativistic*. This is the main object of the present paper. *The theory is non-relativistic only on account of the time being counted throughout as a c-number, instead of being treated symmetrically with the space co-ordinates.* The relativity variation of *mass* with *velocity* is taken into account without difficulty.

The underlying ideas of the theory are very simple. Consider *an atom interacting with a field of radiation*, which we may suppose for definiteness to be confined in an enclosure so as to have only a discrete set of degrees of freedom. Resolving the radiation into its Fourier components, we can consider the *energy* and *phase* of each of the components to be dynamical variables describing the *radiation field*. Thus, if $E_r$ is the *energy* of a component labelled r and $\vartheta_r$ is the corresponding *phase* (defined as the time since the wave was in a standard phase), we can suppose each $E_r$ and $\vartheta_r$ to form a pair of *canonically conjugate* variables. In the absence of any interaction between the field and the atom, the whole system of field plus atom will be describable by the Hamiltonian

$$H = \Sigma_r E_r + H_0 \qquad (1)$$

equal to the total *energy*, $H_0$ being the Hamiltonian for the atom alone, since the variables $E_r$, $\vartheta_r$ obviously satisfy their *canonical equations of motion*

$$E^{\cdot}_r = \delta H/\delta\vartheta_r = 0, \qquad \vartheta^{\cdot}_r = \delta H/\delta E_r = 1.$$

*When there is interaction between the field and the atom, it could be taken into account on the classical theory by the addition of an interaction term to the Hamiltonian* (1), which would be a function of the variables of the atom and of the variables $E_r$, $\vartheta_r$ that describe the field. This interaction term would give the effect of the radiation on the atom, and also the reaction of the atom on the *radiation field.*

In order that an analogous method may be used on the quantum theory, *it is necessary to assume that the variables $E_r$, $\vartheta_r$ are q-numbers* satisfying the standard *quantum conditions* $\vartheta_r E_r - E_r \vartheta_r = ih$, etc., where h is $(2\pi)^{-1}$ times the usual Planck's constant, like the other dynamical variables of the problem. *This assumption immediately gives light-quantum properties to the radiation**.

* Similar assumptions have been used by Born and Jordan [Born, M. & Jordan, P. (December, 1925). Zur Quantenmechanik. (On Quantum Mechanics.) *Zeit. Phys.*, 34, 858-88] p. 886, for the purpose of taking over the classical formula for the emission of radiation by a dipole into the quantum theory, and by Born, Heisenberg and Jordan [Born, M., Heisenberg, W. & Jordan, P. (August, 1926). Zur Quantenmechanik II. (On Quantum Mechanics II.) *Zeit. Phys.*, 35, 557-615] p. 606, for calculating the energy fluctuations in a field of *black-body radiation.*

For if $\nu_r$ is the *frequency* of the component r, $2\pi\nu_r\vartheta_r$ is an *angle variable*, so that its *canonical conjugate* $E_r/2\pi\nu_r$ can only assume a discrete set of values differing by multiples of h, which means that $E_r$ can change only by integral multiples of the quantum $(2\pi h)\ \nu_r$. If we now add an *interaction term* (taken over from the classical theory) to the Hamiltonian (1), the problem can be solved according to the rules of quantum mechanics, and we would expect to obtain the correct results for the action of the radiation and the atom on one another. It will be shown that we actually get the correct laws for the *emission* and *absorption* of radiation, and the correct values for Einstein's A's and B's.

[*Einstein coefficients* are mathematical quantities which are a measure of the probability of absorption or emission of light by an atom or molecule. The Einstein A coefficients are related to the rate of *spontaneous emission* of light, and the Einstein B coefficients are related to the *absorption* and *stimulated emission* of light.]

In the author's previous theory#,

# Dirac, P. A. M. (October, 1926). On the Theory of Quantum Mechanics. *Roy. Soc. Proc., A*, 112, 762, 661-77, § 5[; *relativistic* treatment of Schrodinger's wave theory in which the time t and its *conjugate momentum* –W are treated from the beginning on the same footing as the other variables, sets $x_4 = ict$ (so that $x_1^2 + x_2^2 + x_3^2 + x_4^2 = 0$ and $x_1^2 + x_2^2 + x_3^2 = c^2t^2$) and $p_4 = iW/c$ where – W is the *momentum* conjugate to t, substitutes $(t - x_1/c)$ for t as *uniformizing variable* in order that its contribution to the exchange of energy with the radiation field may vanish, applies to system containing an atom with two electrons, finds that if the positions of the two electrons are interchanged the new state of the atom is physically indistinguishable from the original one, in order that theory only enables

calculation of *observable quantities* must treat ($mn$) and ($nm$) as only one *state*, must infer that *unsymmetrical* functions of the co-ordinates (and momenta) of the two electrons cannot be represented by matrices, *symmetrical functions* such as the total *polarization* of the atom can be considered to be represented by matrices without inconsistency, these matrices are by themselves sufficient to determine all the physical properties of the system, theory of uniformizing variables introduced by the author can no longer apply, allows two solutions satisfying necessary conditions, one leads to Pauli's exclusion principle that not more than one electron can be in any given orbit, the other leads to the Einstein-Bose statistical mechanics, accounts for the *absorption* and stimulated *emission* of radiation by an atom, elements of matrices representing total *polarization* determine *transition probabilities*, *cannot be applied to spontaneous emission*; applies to theory of ideal gas and to problem of an atomic system subjected to a perturbation from outside (e.g., an incident electromagnetic field) which can vary with time in an arbitrary manner, *with neglect of relativity mechanics* accounts for the absorption and stimulated emission of radiation and shows that the elements of the matrices representing the total polarization determine the *transition probabilities*].

*where the energies and phases of the components of radiation were c-numbers*, only the B's could be obtained, and *the reaction of the atom on the radiation could not be taken into account*.

It will also be shown that the Hamiltonian which describes the *interaction of the atom and the electromagnetic waves* can be made identical with the Hamiltonian for the problem of the *interaction of the atom with an assembly of particles moving with the velocity of light and satisfying the Einstein-Bose statistics, by a suitable choice of the interaction energy for the particles*. The number of particles having any specified *direction of motion* and *energy*, which can be used as a dynamical variable in the Hamiltonian for the particles, is equal to the number of quanta of *energy* in the corresponding wave in the Hamiltonian for the waves. *There is thus a complete harmony between the wave and light-quantum descriptions of the interaction*. We shall actually build up the theory from the light-quantum point of view, and show that the Hamiltonian transforms naturally into a form which resembles that for the waves.

The mathematical development of the theory has been made possible by the author's *general transformation theory* of the quantum matrices[$].

[$] Dirac, P. A. M. (January, 1927). The Physical Interpretation of the Quantum Dynamics. *Roy. Soc. Proc., A*, 113, 765, 621-41[; *non-relativistic* matrix mechanics, Heisenberg's original matrix mechanics assumed that the elements of the diagonal matrix that represents the energy are the *energy levels* of the system, and the elements of the matrix that represents the total polarization, which are periodic functions of the time, determine the *frequencies* and *intensities* of the spectral lines in analogy to classical theory, in *Schrodinger's wave representation* physical results are based on assumption that the square of the *amplitude* of the wave function can be interpreted as a probability, enables probability of a *transition* being produced in a system by an arbitrary external perturbing force to be worked out, this paper provides a *general theory of obtaining physical results from quantum theory*, it shows all the physical information that one can hope to get from quantum dynamics and provides a general method for obtaining it, replaces special assumptions previously used, requires a theory of the more general schemes of matrix representation in which the rows and columns refer to any set of constants of integration that commute and of the laws of transformation from one such scheme to another, *does not take relativity mechanics into account*, counts time variable wherever it occurs as a parameter (a c-number), *transformation equations* that satisfy *quantum conditions* and *equations of motion*, *eigenfunctions* of Schrodinger's wave equation as *transformation functions* that enable transformation from scheme of matrix representation to scheme in which Hamiltonian is a diagonal matrix, dynamical variables represented by matrices whose rows and columns refer to the initial values of the *action variables* or to the *final values*, coefficients that enable transformation from one set of matrices to the other are those that determine the *transition probabilities*].

An essentially equivalent theory has been obtained independently by Jordan [Jordan, P. (November, 1927; received October 18, 1926). Über eine neue Begründung der Quantenmechanik. (On a new justification for quantum mechanics.) *Zeit. Phys*., 40, 809-38; https://doi.org/10.1007/BF01390903]. See also, London, F. (1926). Winkelvariable und Kanonische Transformationen in der Undulationsmechanik. *Zeit. Phys*., 40, 193-210.

Owing to the fact that we count the time as a *c-number*, we are allowed to use the notion of the value of any dynamical variable at any instant of time. This value is a *q-number*, capable of being represented by a generalized "matrix" according to many different matrix schemes, some of which may have continuous ranges of rows and columns, and may require the matrix elements to involve certain kinds of infinities (of the type given by the S functions). A matrix scheme can be found in which any desired set of constants of integration of the dynamical system that commute are represented by diagonal matrices, or in which a set of variables that commute are represented by matrices that are diagonal at a specified time*.

* One can have a matrix scheme in which a set of variables that commute are at all times represented by diagonal matrices if one will sacrifice the condition that the matrices must satisfy the *equations of motion*. The transformation function from such a scheme to one in which the *equations of motion* are satisfied will involve the time explicitly. See p. 628 in Dirac (January, 1927). The Physical Interpretation of the Quantum Dynamics [*loc. cit.*]

The values of the diagonal elements of a diagonal matrix representing any *q-number* are the *characteristic values* of that *q-number*. A Cartesian *co-ordinate* or *momentum* will in general have all *characteristic values* from $-\infty$ to $+\infty$, while an *action variable* has only a discrete set of *characteristic values*. (We shall make it a rule to use unprimed letters to denote the *dynamical variables* or *q-numbers*, and the same letters primed or multiply primed to denote their *characteristic values*. *Transformation functions* or *eigenfunctions* are functions of the *characteristic values* and not of the *q-numbers* themselves, so they should always be written in terms of primed variables.)

If $f(\xi, \eta)$ is any function of the *canonical variables* $\xi_k$, $\eta_k$, the matrix representing $f$ at any time t in the matrix scheme in which the $\xi_k$ at time t are diagonal matrices may be written down without any trouble, since the matrices representing the $\xi_k$ and $\eta_k$ themselves at time t are known, namely,

$$\xi_k (\xi' \xi'') = \xi'_k \, \delta(\xi' \xi''),$$
$$\eta_k (\xi' \xi'') = - ih \, \delta(\xi'_1 - \xi''_1) \ldots \delta(\xi'_{k-1} - \xi''_{k-1}) \delta(\xi'_k - \xi''_k) \, \delta(\xi'_{k+1} - \xi''_{k+1}) \ldots \quad (2)$$

Thus if the Hamiltonian H is given as a function of the $\xi_k$ and $\eta_k$ we can at once write down the matrix $H(\xi' \xi'')$. We can then obtain the *transformation function*, $(\xi' /\alpha')$ say, which transforms to a matrix scheme $(\alpha)$ in which the Hamiltonian is a diagonal matrix, as $(\xi' /\alpha')$ must satisfy the integral equation

$$\int H(\xi' \xi'') \, d\xi''(\xi''/\alpha') = W(\alpha') \, . \, (\xi'/\alpha'), \quad (3)$$

of which the *characteristic values* $W(\alpha')$ are the *energy levels*. This equation is just Schrodinger's *wave equation* for the *eigenfunctions* $(\xi' /\alpha')$ which becomes an ordinary differential equation when H is a simple algebraic function of the $\xi_k$ and $\eta_k$ on account of the special equations (2) for the matrices representing $\xi_k$ and $\eta_k$. Equation (3) may be written in the more general form

$$\int H(\xi' \xi'') \, d\xi''(\xi''/\alpha') = ih \, \partial(\xi' /\alpha')/\partial t, \quad (3')$$

in which it can be applied to systems for which the Hamiltonian involves the time explicitly.

One may have a dynamical system specified by a Hamiltonian H which cannot be expressed as an algebraic function of any set of *canonical variables*, but which can all the same be represented by a matrix $H(\xi' \xi'')$. Such a problem can still be solved by the present method, since one can still use equation (3) to obtain the *energy* levels and *eigenfunctions*. We shall find that the Hamiltonian which describes the interaction of a *light-quantum* and an atomic system is of this more general type, so that the interaction can be treated mathematically, although one cannot talk about an interaction *potential energy* in the usual sense.

*It should be observed that there is a difference between a light-wave and the de Broglie or Schrodinger wave associated with the light-quanta. Firstly, the light-wave is always real, while the de Broglie wave associated with a light-quantum moving in a definite direction must be taken to involve an imaginary exponential.* A more important difference is that their *intensities* are to be interpreted in different ways. The number of *light-quanta* per unit volume associated with a monochromatic *light-wave* equals the *energy* per unit volume of the wave divided by the *energy* $(2\pi h)\nu$ of a single *light-quantum*. On the other hand, a monochromatic de Broglie wave of *amplitude a* (multiplied into the imaginary exponential factor) must be interpreted as representing $a^2$ *light-quanta* per unit volume for all *frequencies*. This is a special case of the general rule for interpreting the matrix analysis*,

* *Loc. cit.* Dirac, P. A. M. (January, 1927). The Physical Interpretation of the Quantum Dynamics, §§ 6, 7.

according to which, if $(\xi' / \alpha')$ or $\psi_{\alpha'}(\xi_k')$ is the *eigenfunction* in the variables $\xi_k$ of the state $\alpha'$ of an atomic system (or simple particle), $|\psi_{\alpha'}(\xi_k')|^2$ is the probability of each $\xi_k$ having the value $\xi_k'$, [or $|\psi_{\alpha'}(\xi_k')|^2 d\xi_1' d\xi_2' \ldots$ is the probability of each $\xi_k$ lying between the values $\xi_k'$ and $\xi_k' + d\xi_k'$, when the $\xi_k$ have continuous ranges of characteristic values] on the assumption that all phases of the system are equally probable. *The wave whose intensity is to be interpreted in the first of these two ways appears in the theory only when one is dealing with an assembly of the associated particles satisfying the Einstein-Bose statistics. There is thus no such wave associated with electrons.*

### § 2. *The Perturbation of an Assembly of Independent Systems.*

We shall now consider *the transitions produced in an atomic system by an arbitrary perturbation*. The method we shall adopt will be that previously given by the author, #

\# *Loc. cit.* Dirac, P. A. M. (October, 1926). On the Theory of Quantum Mechanics.

which leads in a simple way to equations which determine the *probability* of the system being in any *stationary state* of the unperturbed system at any time$.

$ The theory has recently been extended by Born [(Born, M. (192[7]). Das Adiabatenprinzip in der Quantenmechanik. *Zeit. Phys.*, 40. 167-192; https://doi.org/10.1007/bf01400360] so as to take into account the adiabatic changes in the stationary states that may be produced by the perturbation as well as the transitions. This extension is not used in the present paper.

This, of course, gives immediately the probable number of systems in that *state* at that time for an assembly of the systems that are independent of one another and are all perturbed in the same way. The object of the present section is to show that the equations for the rates of change of these probable numbers can be put in the Hamiltonian form in a simple manner, which will enable further developments in the theory to be made.

Let $H_0$ be the Hamiltonian for the unperturbed system and V the perturbing *energy*, which can be an arbitrary function of the dynamical variables and may or may not involve the time explicitly, so that the Hamiltonian for the perturbed system is

$H = H_0 + V$. The *eigenfunctions* for the perturbed system must satisfy the *wave equation*

$$ih\ \delta\psi/\delta t = (H_0 + V)\ \psi,$$

where $(H_0 + V)$ is an operator. If $\psi = \Sigma_r\ a_r\psi_r$ is the solution of this equation that satisfies the proper initial conditions, where the $\psi_r$'s are the *eigenfunctions* for the unperturbed system, each associated with one *stationary state* labelled by the suffix *r*, and the $a_r$'s are functions of the time only, then $|a_r|^2$ is the *probability* of the system being in the state at any time. The $a_r$'s must be normalized initially, and will then always remain normalized. The theory will apply directly to an assembly of N similar independent systems if we multiply each of these $a_r$'s by $N^{1/2}$ so as to make $\Sigma_r\ |a_r|^2 = N$. We shall now have that $|a_r|^2$ is the probable number of systems in the *state r*.

The equation that determines the rate of change of the $a_r$'s is*

$$ih\ \dot{a}_r = \Sigma_s\ V_{rs}\ a_s, \qquad (4)$$

where the $V_{rs}$'s are the elements of the matrix representing V.

* *Loc. cit.* Dirac, P. A. M. (October, 1926). On the Theory of Quantum Mechanics, equation (25).

The *conjugate* imaginary equation is

$$ih\ \dot{a}_r^* = \Sigma_s\ V_{rs}^* a_s^* = \Sigma_s\ a_s^* V_{sr}^*, \qquad (4')$$

If we regard $a_r$ and ih $a_r^*$ as *canonical conjugates*, equations (4) and (4') take the Hamiltonian form with the Hamiltonian function $F_1 = \Sigma_{rs}\ a_r^* V_{rs} a_s$, namely,

$$da_r/dt = 1/ih\ \partial F_1/\partial a_r^*, \qquad ih\ da_r^*/dt = -\ \partial F_1/\partial a_r.$$

We can transform to the *canonical variables* $N_r$, $\phi_r$ by the *contact transformation*

$$a_r = N_r^{\frac{1}{2}}\, e^{-i\phi/h}, \qquad a_r^* = N_r^{\frac{1}{2}}\, e^{i\phi/h}.$$

This transformation makes the new variables $N_r$ and $\phi_r$ real, $N_r$ being equal to $a_r\ a_r^* = |a_r|^2$, the probable number of systems in the state r, and $\phi_r/h$ being the *phase* of the *eigenfunction* that represents them. The Hamiltonian $F_1$ now becomes

$$F_1 = \Sigma_{rs}\ V_{rs}\ N_r^{\frac{1}{2}}\ N_s^{\frac{1}{2}}\ e^{i(\phi r - \phi s)/h},$$

and the equations that determine the rate at which transitions occur have the *canonical* form

$$\dot{N}_r = \partial F_1/\partial \phi_r, \qquad \dot{\phi}_r = \partial F_1/\partial N_r.$$

A more convenient way of putting the *transition equations* in the Hamiltonian form may be obtained with the help of the quantities

$$b_r = a_r\, e^{-iW_r t/h}, \qquad b_r{}^* = a_r{}^*\, e^{iW_r t/h},$$

$W_r$ being the *energy* of the *state* r. We have $|b_r|^2$ equal to $|a_r|^2$, the probable number of systems in the *state* r. For $\dot{b}_r$ we find

$$\begin{aligned} ih\,\dot{b}_r &= W_r b_r + ih\,\dot{a}_r\, e^{-iW_r t/h} \\ &= W_r b_r + \Sigma_s V_{rs} b_s\, e^{i(W_r - W_s)\, t/h} \end{aligned}$$

with the help of (4). If we put $V_{rs} = \upsilon_{rs}\, e^{i(W_r - W_s)\, t/h}$, so that $\upsilon_{rs}$ is a constant when V does not involve the time explicitly, this reduces to

$$ih\,\dot{b}_r = W_r b_r + \Sigma_s\, \upsilon_{rs}\, b_s = \Sigma_s\, H_{rs} b_s, \qquad (5)$$

where $H_{rs} = W_r\, \delta_{rs} + \upsilon_{rs}$, which is a matrix element of the total Hamiltonian $H = H_0 + V$ with the time factor $e^{i(W_r - W_s)\, t/h}$ removed, so that $H_{rs}$ is a constant when H does not involve the time explicitly. Equation (5) is of the same form as equation (4), and may be put in the Hamiltonian form in the same way.

It should be noticed that equation (5) is obtained directly if one writes down the Schrodinger equation in a set of variables that specify the *stationary states* of the unperturbed system. If these variables are $\xi_h$, and if $H(\xi'\xi'')$ denotes a matrix element of the total Hamiltonian H in the ($\xi$) scheme, this Schrodinger equation would be

$$ih\, \partial\psi(\xi')/\partial t = \Sigma_{\xi''}\, H(\xi'\xi'')\, \psi(\xi''), \qquad (6)$$

like equation (3'). This differs from the previous equation (5) only in the notation, a single suffix r being there used to denote a *stationary state* instead of a set of numerical values $\xi'_k$ for the variables $\xi'_k$, and $b_r$ being used instead of $\psi(\xi')$. Equation (6), and therefore also equation (5), can still be used when the Hamiltonian is of the more general type which cannot be expressed as an algebraic function of a set of *canonical variables*, but can still be represented by a matrix $H(\xi'\xi'')$ or $H_{rs}$.

We now take $b_r$ and $ihb_r{}^*$ to be *canonically conjugate variables* instead of $a_r$ and $iha_r{}^*$. The equation (5) and its *conjugate imaginary equation* will now take the Hamiltonian form with the Hamiltonian function

$$F = \Sigma_{rs}\, b_r{}^* H_{rs} b_s. \qquad (7)$$

Proceeding as before, we make the *contact transformation*

$$b_r = N_r{}^{1/2}\, e^{-i\theta/h}, \qquad b_r{}^* = N_r{}^{1/2}\, e^{i\theta/h}, \qquad (8)$$

to the new *canonical variables* $N_r$, $\theta_r$, where $N_r$ is, as before, the probable number of systems in the *state* r, and $\theta_r$ is a new *phase*. The Hamiltonian F will now become

$$F = \Sigma_{rs}\, H_{rs}\, N_r{}^{1/2}\, N_s{}^{1/2}\, e^{i(\theta_r - \theta_s)/h},$$

and the equations for the rates of change of $N_r$ and $\theta_r$ will take the *canonical* form

$$\dot{N}_r = \partial F/\partial \theta_r, \qquad \dot{\theta}_r = \partial F/\partial N_r.$$

The Hamiltonian may be written

$$F = \Sigma_r W_r N_r + \Sigma_{rs} \upsilon_{rs} N_r^{1/2} N_s^{1/2} e^{i(\theta_r - \theta_s)/h}. \qquad (9)$$

The first term $\Sigma_r W_r N_r$ is the *total proper energy* of the assembly, and the second may be regarded as the additional *energy* due to the perturbation. If the perturbation is zero, the phases $\theta_r$ would increase linearly with the time, while the previous *phases* $\phi_r$ would in this case be constants.

### §3. *The Perturbation of an Assembly satisfying the Einstein-Bose Statistics.*

According to the preceding section we can describe the effect of a perturbation on an assembly of independent systems by means of *canonical variables* and Hamiltonian *equations of motion*. The development of the theory which naturally suggests itself is to make these *canonical variables q-numbers* satisfying the usual *quantum conditions* instead of *c-numbers*, so that their Hamiltonian *equations of motion* become true *quantum equations*. The Hamiltonian function will now provide a Schrodinger *wave equation*, which must be solved and interpreted in the usual manner. The interpretation will give not merely the probable number of systems in any *state*, but the *probability* of any given distribution of the systems among the various *states*, this *probability* being, in fact, equal to the square of the modulus of the normalized solution of the *wave equation* that satisfies the appropriate initial conditions. We could, of course, calculate directly from elementary considerations the *probability* of any given distribution *when the systems are independent*, as we know the probability of each system being in any particular *state*. We shall find that the *probability* calculated directly in this way *does not agree with that obtained from the wave equation except in the special case when there is only one system in the assembly.* In the general case it will be shown that *the wave equation leads to the correct value for the probability of any given distribution when the systems obey the Einstein-Bose statistics instead of being independent.*

We assume the variables $b_r$, $ihb_r^*$ of §2 to be *canonical* q-numbers satisfying the *quantum conditions*

$$b_r \,.\, ihb_r^* - ihb_r^* \,.\, b_r = ih,$$
$$b_r b_r^* - b_r^* b_r = 1,$$

and

$$b_r b_s - b_s b_r = 0, \qquad b_r^* b_s^* - b_s^* b_r^* = 0,$$
$$b_r b_s^* - b_s^* b_r = 0 \qquad (s \neq r).$$

The *transformation equations* (8) must now be written in the *quantum form*

$$b_r = (N_r + 1)^{½} e^{-i\theta/h} = e^{-i\theta/h} N_r^{½}$$
$$b_r^* = N_r^{½} e^{i\theta/h} = e^{i\theta/h} (N_r + 1)^{½}, \qquad (10)$$

in order that the $N_r$, $\theta_r$ may also be *canonical variables*. These equations show that the $N_r$ can have only integral *characteristic values* not less than zero, #

> # See § 8 of the author's paper, p. 281. [Dirac, P. A. M. (May, 1926). The elimination of the nodes in quantum mechanics. *Roy. Soc. Proc., A*, 111, 757, 281–305 [; the laws of classical mechanics must be generalized when applied to atomic systems, *the commutative law of multiplication* as applied to dynamical variables is replaced by certain *quantum conditions* which are just sufficient to enable one to evaluate xy – yx when x and y are given, it follows that the dynamical variables cannot be ordinary numbers expressible in the decimal notation (which numbers will be called *c-numbers*), but may be considered to be numbers of a special kind (which will be called *q-numbers*), whose nature cannot be exactly specified, but which can be used in the algebraic solution of a dynamical problem in a manner closely analogous to the way the corresponding classical variables are used, the object of this paper is to simplify the *non-relativistic* quantum treatment by the introduction of *quantum variables*, in the classical treatment of the dynamical problem of a number of particles or electrons moving in a central field of force and disturbing one another one always begins by making the initial simplification known as the *elimination of the nodes*, this consists in obtaining a *contact transformation* from the Cartesian co-ordinates and momenta of the electrons to a set of canonical variables of which all except three are independent of the orientation of the system as a whole while these three determine the orientation, introduces *action variables and their canonical conjugate angle variables*, *transformation equations*, substitutes set of *c-numbers* for *action variables* to fix *stationary state* and obtain physical results, applies to *anomalous Zeeman effect*, showed that *non-relativistic* theory gave the correct g-formula for *energy* of stationary states and Kronig's results for the relative intensities of the lines of a multiplet and their components in a weak magnetic field].

which provides us with a justification for the assumption that the variables are *q-numbers* in the way we have chosen. The numbers of systems in the different *states* are now ordinary quantum numbers.

The Hamiltonian (7)

$$[F = \Sigma_{rs}\, b_r^* H_{rs} b_s. \qquad (7)]$$

now becomes

$$F = \Sigma_{rs}\, b_r^* H_{rs} b_s = \Sigma_{rs}\, N_r^{½} e^{i\theta r/h} H_{rs}(N_s + 1)^{½} e^{-i\theta s/h}$$
$$= \Sigma_{rs}\, H_{rs} N_r^{½} (N_s + 1 - \delta_{rs})^{½} e^{i(\theta r-\theta s)/h} \qquad (11)$$

in which the $H_{rs}$ are still *c-numbers*. We may write this F in the form corresponding to (9)

$$[F = \Sigma_r\, W_r N_r + \Sigma_{rs}\, \upsilon_{rs} N_r^{½} N_s^{½} e^{i(\theta r-\theta s)/h}. \qquad (9)]$$

$$F = \Sigma_r W_r N_r + \Sigma_{rs} \upsilon_{rs} N_r^{1/2} (N_s + 1 - \delta_{rs})^{1/2} e^{i(\theta r - \theta s)/h} \qquad (11')$$

in which it is again composed of a *proper energy* term $\Sigma_r W_r N_r$ and an *interaction energy* term.

...

*Consider first the case when there is only one system in the assembly.* The probability of its being in the *state* q is determined by the *eigenfunction* $\psi(N_1', N_2', ...)$ in which all the N's are put equal to zero except $N'_q$, which is put equal to unity. This *eigenfunction* we shall denote by $\psi\{q\}$. When it is substituted in the left-hand side of (13), all the terms in the summation on the right-hand side vanish except those for which r = q, and we are left with

$$ih\, \partial/\partial t\, \psi\{q\} = \Sigma_r H_{qs}\, \psi\{s\},$$

which is the same equation as (5)

$$[ih\, b^{\cdot}_r = W_r b_r + \Sigma_s \upsilon_{rs} b_s = \Sigma_s H_{rs} b_s, \qquad (5)]$$

with $\{q\}$ playing the part of $b_q$. This establishes the fact that the present theory is equivalent to that of the preceding section when there is only one system in the assembly.

*Now take the general case of an arbitrary number of systems in the assembly, and assume that they obey the Einstein-Bose statistical mechanics.* This requires that, in the ordinary treatment of the problem, only those *eigenfunctions* that are *symmetrical* between all the systems must be taken into account, these *eigenfunctions* being by themselves sufficient to give a complete quantum solution of the problem[#].

[#] *Loc. cit.* Dirac, P. A. M. (October, 1926). On the Theory of Quantum Mechanics, § 3.

We shall now obtain the equation for the rate of change of one of these symmetrical *eigenfunctions*, and show that it is identical with equation (13). ...

... We have thus established that the Hamiltonian (11)

$$[F = \Sigma_{rs} b_r^* H_{rs} b_s = \Sigma_{rs} N_r^{1/2} e^{i\theta r/h} H_{rs}(N_s + 1)^{1/2} e^{-i\theta s/h}$$
$$= \Sigma_{rs} H_{rs} N_r^{1/2} (N_s + 1 - \delta_{rs})^{1/2} e^{i(\theta r - \theta s)/h} \qquad (11)]$$

describes the effect of a perturbation on an assembly satisfying the *Einstein-Bose statistics*.

## § 4. *The Reaction of the Assembly on the Perturbing System.*

Up to the present we have considered only perturbations that can be represented by a perturbing *energy* V added to the Hamiltonian of the perturbed system, V being a function only of the dynamical variables of that system and perhaps of the time. The theory may readily be extended to the case when the perturbation consists of interaction with a perturbing dynamical system, *the reaction of the perturbed system on the perturbing system being taken into account*. (The distinction between the perturbing system and the perturbed system is, of course, not real, but it will be kept up for convenience.)

We now consider a perturbing system, described, say, by the canonical variables $J_k$, $w_k$, the J's being its first integrals when it is alone, interacting with an assembly of perturbed systems with no mutual interaction, that satisfy the *Einstein-Bose statistics*. The total Hamiltonian will be of the form

$$H_T = H_P(J) + \Sigma_n H(n),$$

where $H_P$ is the Hamiltonian of the perturbing system (a function of the J's only) and H(n) is equal to the *proper energy* $H_0(n)$ plus the *perturbation energy* V(n) of the nth system of the assembly. H(n) is a function only of the variables of the nth system of the assembly and of the J's and ϖ's, and does not involve the time explicitly.

The Schrodinger equation corresponding to equation (14) is now …

…

This is the Schrodinger equation corresponding to the Hamiltonian function

$$F = H_P(J) + \Sigma_{rs} H_{rs}N_r^{½} (N_s + 1 - \delta_{rs})^{½} e^{i(\theta r-\theta s)/h}, \qquad (19)$$

in which $H_{rs}$ is now a function of the J's and ϖ's, being such that when represented by a matrix in the (J) scheme its (J' J") element is H ($J'_r$; $J''_s$). (It should be noticed that $H_{rs}$ still commutes with the N's and θ's.)

Thus, *the interaction of a perturbing system and an assembly satisfying the Einstein-Bose statistics can be described by a Hamiltonian of the form (19)*

$$[F = H_P(J) + \Sigma_{rs} H_{rs}N_r^{½} (N_s + 1 - \delta_{rs})^{½} e^{i(\theta r-\theta s)/h}. \qquad (19)]$$

We can put it into the form corresponding to (11')

$$[F = \Sigma_r W_rN_r + \Sigma_{rs} \upsilon_{rs} N_r^{½} (N_s + 1 - \delta_{rs})^{½} e^{i(\theta r-\theta s)/h} \qquad (11')]$$

by observing that the matrix element H($J'_r$; $J''_s$) is composed of the sum of two parts, a part that comes from the *proper energy* $H_0$, which equals $W_r$ when $J''_k = J'_k$ and s = r and vanishes otherwise, and a part that comes from the *interaction energy* V, which may be denoted by υ($J'_r$; $J''_s$). Thus, we shall have

$$H_{rs} = W_r \delta_{rs} + \upsilon_{rs},$$

where $\upsilon_{rs}$ is that function of the J's and ϖ's which is represented by the matrix whose (J' J") element is υ($J'_r$; $J''_s$), and so (19) becomes

$$F = H_P(J) + \Sigma_r W_rN_r + \Sigma_{rs} \upsilon_{rs}N_r^{½} (N_s + 1 - \delta_{rs})^{½} e^{i(\theta r-\theta s)/h}, \qquad (20)$$

*The Hamiltonian is thus the sum of the proper energy of the perturbing system $H_P(J)$, the proper energy of the perturbed systems $\Sigma_r W_rN_r$ and the perturbation energy* $\Sigma_{rs} \upsilon_{rs}N_r^{½} (N_s + 1 - \delta_{rs})^{½} e^{i(\theta r-\theta s)/h}$.

## § 5. *Theory of Transitions in a System from One State to Others of the Same Energy.*

Before applying the results of the preceding sections to *light-quanta*, we shall consider the solution of the problem presented by a Hamiltonian of the type (19)

$$[F = H_P(J) + \Sigma_{rs} H_{rs} N_r^{1/2} (N_s + 1 - \delta_{rs})^{1/2} e^{i(\theta r - \theta s)/h}. \qquad (19)]$$

The essential feature of the problem is that it refers to a dynamical system which can, under the influence of a perturbation *energy* which does not involve the time explicitly, make transitions from one state to others of the same *energy*. The problem of collisions between an atomic system and an electron, which has been treated by Born*, is a special case of this type.

* Born, M. (November, 1926) Quantenmechanik der Stoßvorgänge. (Quantum mechanics of collision processes.) *Zeit. Phys.*, 38, 803-27; https://doi.org/10.1007/BF01397184.

Born's method is to find a *periodic* solution of the *wave equation* which consists, in so far as it involves the *co-ordinates* of the colliding electron, of plane waves, representing the incident electron, approaching the atomic system, which are scattered or diffracted in all directions. The square of the *amplitude* of the waves scattered in any direction with any *frequency* is then assumed by Born to be the *probability* of the electron being scattered in that direction with the corresponding *energy*.

*This method does not appear to be capable of extension in any simple manner to the general problem of systems that make transitions from one state to others of the same energy*. Also, there is at present no very direct and certain way of interpreting a periodic solution of a *wave equation* to apply to a non-periodic physical phenomenon such as a collision. (The more definite method that will now be given shows that Born's assumption is not quite right, it being necessary to multiply the square of the *amplitude* by a certain factor.)

*An alternative method of solving a collision problem is to find a solution of the wave equation which consists initially simply of plane waves moving over the whole of space in the necessary direction with the necessary frequency to represent the incident electron.* In course of time waves moving in other directions must appear in order that the *wave equation* may remain satisfied. The *probability* of the electron being scattered in any direction with any *energy* will then be determined by the rate of growth of the corresponding harmonic component of these waves. The way the mathematics is to be interpreted is by this method quite definite, being the same as that of the beginning of §2.

We shall apply this method to the general problem of a system which makes transitions from one *state* to others of the same *energy* under the action of a perturbation. Let $H_0$ be the Hamiltonian of the unperturbed system and V the *perturbing energy*, which must not involve the time explicitly. If we take the case of a continuous range of *stationary states*, specified by the first integrals, $\alpha_k$ say, of the unperturbed motion, then, following the method of § 2, we obtain

$$ih\, \dot{a}\,(\alpha') = \int V\,(\alpha'\alpha'')\, d\alpha''.\, a\,(\alpha''), \qquad (21)$$

corresponding to equation (4)

$$[ih\, \dot{a}_r = \Sigma_s V_{rs}\, a_s. \qquad (4)]$$

…

This result differs by the factor $(2\pi h)^2/2mE'.\ P'/P^0$ from Born's* [where $P^0$ refers to the *initial momentum*, and E' and P' refer, respectively, to the *resultant energy and momentum* of the scattered electron].

* In a more recent paper [Born, M. (1926). *Nachr. Gesell. d. Wiss., Gottingen*, 146] Born has obtained a result in agreement with that of the present paper for *non-relativity* mechanics, by using an interpretation of the analysis based on the conservation theorems. I am indebted to Prof. N. Bohr for seeing an advance copy of this work.

The necessity for the factor $P'/P^0$ in (26) could have been predicted from the *principle of detailed balancing*, as the factor $|\, v\,(p';\, p^0)\,|^2$ is symmetrical between the direct and reverse processes#.

# See Klein, O., Rosseland, S. (March, 1921).Über Zusammenstöße zwischen Atomen und freien Elektronen. (About collisions between atoms and free electrons.) *Zeit. Phys.,* 4, 46-51; https://doi.org/10.1007/BF01328041, eq. (4).

### § 6. *Application to Light-Quanta*$.

[$ Dirac, P. A. M. (May, 1927). The quantum theory of dispersion. *Roy. Soc. Proc., A*, 114, 769, 710-28, page 711, ff: "In Dirac (March, 1927)., § 6, *it was in error assumed that* $V_{mn}$ *caused transitions from state m to state n*, and consequently the information there obtained about an absorption (or emission) process in terms of the number of *light-quanta* existing *before the process* should really apply to an emission (or absorption) process in terms of the number of *light-quanta* in existence *after the process*. This change, of course, does not affect the results (namely the proof of Einstein's laws) which can depend on $|V_{mn}|^2 = |V_{nm}|^2$."]

We shall now apply the theory of §4 to the case when the systems of the assembly are *light-quanta*, the theory being applicable to this case since *light-quanta* obey the *Einstein-Bose statistics* and have no mutual interaction. A *light-quantum* is in a *stationary state* when it is moving with constant *momentum* in a straight line. Thus, a *stationary state r* is fixed by the three components of *momentum* of the *light-quantum* and a variable that specifies its state of *polarization*. We shall work on the assumption that there are a finite number of these *stationary states*, lying very close to one another, as it would be inconvenient to use continuous ranges. The interaction of the *light-quanta* with an atomic system will be described by a Hamiltonian of the form (20)

$$[F = H_P(J) + \Sigma_r W_r N_r + \Sigma_{rs} \upsilon_{rs} N_r^{1/2} (N_s + 1 - \delta_{rs})^{1/2} e^{i(\theta r - \theta s)/h}, \qquad (20)]$$

in which $H_P(J)$ is the Hamiltonian for the *atomic system alone*, and the coefficients $\upsilon_{rs}$ are for the present unknown. *We shall show that this form for the Hamiltonian, with the $\upsilon_{rs}$ arbitrary, leads to Einstein's laws for the emission and absorption of radiation.*

*The light-quantum has the peculiarity that it apparently ceases to exist when it is in one of its stationary states, namely, the zero state, in which its momentum, and therefore also its energy, are zero.* When a *light-quantum* is absorbed, it can be considered to jump into this *zero state*, and when one is emitted, it can be considered to jump from the *zero state* to one in which it is physically in evidence, so that it appears to have been created. Since there is no limit to the number of *light-quanta* that may be created in this way, we must suppose that there are an infinite number of *light-quanta* in the *zero state, so that the $N_0$ of the Hamiltonian (20) is infinite*. We must now have $\theta_0$, the variable *canonically conjugate* to $N_0$, a constant, since

$$\theta^{\cdot}_0 = \partial F/\partial N_0 = W_0 + \text{terms involving } N_0^{-1/2} \text{ or } (N_0 + 1 - \delta_{rs})^{-1/2}$$

and $W_0$ is zero. *In order that the Hamiltonian (20) may remain finite it is necessary for the coefficients $\upsilon_{r0}$, $\upsilon_{0r}$ to be infinitely small.* We shall suppose that they are infinitely small in such a way as to make $\upsilon_{r0}N_0^{1/2}$ and $\upsilon_{0r}N_0^{1/2}$ finite, in order that the transition probability coefficients may be finite. Thus, we put

$$\upsilon_{r0} (N_0 + 1)^{1/2} e^{-i\theta_0/h} = \upsilon_r, \qquad \upsilon_{r0} N_0^{1/2} e^{i\theta_0/h} = \upsilon_r^*,$$

where $\upsilon_r$ and $\upsilon_r^*$ are finite and *conjugate imaginaries*. We may consider the $\upsilon_r$ and $\upsilon_r^*$ to be functions only of the J's and ϖ's of the *atomic system*, since their factors $(N_0 + 1)^{1/2} e^{-i\theta_0/h}$ and $N_0^{1/2} e^{i\theta_0/h}$ are practically constants, the rate of change of $N_0$ being very small compared with $N_0$. The Hamiltonian (20)

$$[F = H_P(J) + \Sigma_r W_r N_r + \Sigma_{rs} \upsilon_{rs} N_r^{1/2} (N_s + 1 - \delta_{rs})^{1/2} e^{i(\theta r - \theta s)/h}, \qquad (20)]$$

now becomes

$$F = H_P(J) + \Sigma_r W_r N_r + \Sigma_{r\neq 0} [\upsilon_r N_r^{1/2} e^{i\theta r/h} + \upsilon_r^*(N_r + 1)^{1/2} e^{-i\theta r/h}]$$
$$+ \Sigma_{r\neq 0} \Sigma_{s\neq 0} \upsilon_{rs} N_r^{1/2} (N_s + 1 - \delta_{rs})^{1/2} e^{i(\theta r - \theta s)/h} \qquad (27)$$

The *probability of a transition* in which a *light-quantum* in the *state* r is absorbed is proportional to the square of the modulus of that matrix element of the Hamiltonian which refers to this transition. This matrix element must come from the term $\upsilon_r N_r^{1/2} e^{i\theta r/h}$ in the Hamiltonian, and must therefore be proportional to $N'_r{}^{1/2}$ where $N'_r$ is the number of *light-quanta* in *state* r *before the process*.

The *probability of the absorption* process is thus proportional to $N'_r$. In the same way the probability of a *light-quantum* in *state* r being *emitted* is proportional to $(N'_r + 1)$, and the *probability* of a *light-quantum* in *state* r being *scattered* into *state* s is proportional to $N'_r (N'_r + 1)$. *Radiative processes of the more general type considered by Einstein and Ehrenfest*[$] *in which more than one light-quantum take part simultaneously, are not allowed on the present theory.*

[$] Einstein, A. & Ehrenfest, P. (December, 1923). Zur Quantentheorie des Strahlungsgleichgewichts. (On the quantum theory of radiation equilibrium.) *Zeit. Phys.*, 19, 301-6; https://doi.org/10.1007/BF01327565.

To establish a connection between the number of *light-quanta* per *stationary state* and the *intensity* of the radiation, we consider an enclosure of finite volume, A say, containing the radiation. The number of *stationary states* for *light-quanta* of a given type of *polarization* whose *frequency* lies in the range $\nu_r$, to $\nu_r + d\nu_r$ and whose direction of motion lies in the solid angle about the direction of motion for *state* r will now be $A\nu_r^2 d\nu_r d\varpi_r/c^3$. The *energy* of the light-quanta in these *stationary states* is thus $N'_r \,.\, 2\pi h\nu_r \,.\, A\nu_r^2 d\nu_r d\varpi_r/c^3$. This must equal $Ac^{-1} I_r d\nu_r d\varpi_r$, where $I_r$ is the *intensity* per unit *frequency* range of the radiation about the *state* r. Hence

$$I_r = N'_r (2\pi h) \nu_r^3/c^2, \qquad (28)$$

so that $N'_r$ is proportional to $I_r$ and $(N'_r + 1)$ is proportional to $I_r + (2\pi h) \nu_r^3/c^2$.

We thus obtain that the *probability* of an *absorption* process is proportional to $I_r$, the incident *intensity* per unit *frequency* range, and that of an *emission* process is proportional to $I_r + (2\pi h)\nu_r^3/c^2$, which are just *Einstein's laws*[*].

[*] The ratio of stimulated to *spontaneous emission* in the present theory is just twice its value in Einstein's. This is because in the present theory either polarized component of the incident radiation can stimulate only radiation polarized in the same way, while in Einstein's the two polarized components are treated together. This remark applies also to the *scattering process*.

In the same way the *probability* of a process in which a *light-quantum* is *scattered* from a state *r* to a *state s* is proportional to $I_r [I_s + (2\pi h)\nu_r^3/c^2)]$, which is *Pauli's law for the scattering of radiation by an electron.*[#]

[#] Pauli, W. (December, 1923). Über das thermische Gleichgewicht zwischen Strahlung und freien Elektronen. (About the thermal equilibrium between radiation and free electrons.) *Zeit. Phys.*, 18, 272-86; https://doi.org/10.1007/BF01327708.

## §7. *The Probability Coefficients for Emission and Absorption.*

We shall now consider the *interaction of an atom and radiation* from the *wave* point of view. We resolve the *radiation* into its *Fourier components*, and suppose that their number is very large but finite. Let each component be labelled by a suffix *r*, and suppose there are $\sigma_r$ components associated with the *radiation* of a definite type of *polarization* per unit solid angle per unit *frequency* range about the component *r*. Each component *r* can be described by a vector potential $\kappa_r$ chosen so as to make the scalar potential zero. The perturbation term to be added to the Hamiltonian will now be, according to the classical theory *with neglect of relativity mechanics*, $c^{-1}\ \Sigma_r\ \kappa_r\ \dot{X}_r$, where $X_r$ is the component of the total *polarization* of the atom in the direction of $\kappa_r$, which is the direction of the *electric vector* of the component *r*.

We can, as explained in § 1, suppose the field to be described by the *canonical variables* $N_r$, $\vartheta_r$, of which $N_r$ is the number of *quanta* of *energy* of the component *r*, and $\vartheta_r$ is its *canonically conjugate phase*, equal to $2\pi h\nu_r$ times the $\vartheta_r$ of § 1. …

…

The Hamiltonian for the whole system of *atom plus radiation* would now be, according to the classical theory,

$$F = H_P(J) + \Sigma_r\ (2\pi h\nu_r)\ N_r + 2^{c-1}\ \Sigma_r\ (h\nu_r/c\sigma_r)^{1/2}\ \dot{X}_r\ N_r^{1/2}\cos\theta_r/h, \qquad (29)$$

*where $H_P(J)$ is the Hamiltonian for the atom alone.* On the quantum theory we must make the variables $N_r$ and $\theta_r$ canonical q-numbers like the variables $J_k$, $\varpi_k$ that describe the *atom*. We must now replace the $N_r^{1/2}\cos\theta_r/h$ in (29) by the real q-number

$$\tfrac{1}{2}\ \{N_r^{1/2}\ e^{i\theta r/h} + e^{-i\theta r/h}\ N_r^{1/2}\} = \tfrac{1}{2}\ \{N_r^{1/2}\ e^{i\theta r/h} + (N_r + 1)^{1/2} e^{-i\theta r/h}\}$$

so that the Hamiltonian (29) becomes

$$F = H_P(J) + \Sigma_r\ (2\pi h\nu_r)\ N_r + h^{1/2}c^{-3/2}\ \Sigma_r\ (\nu_r/\sigma_r)^{1/2}\ \dot{X}_r\ \{N_r^{1/2}\ e^{i\theta r/h} + (N_r + 1)^{1/2} e^{-i\theta r/h}]\ (30)$$

This is of the form (27)

$$[F = H_P(J) + \Sigma_r\ W_r N_r + \Sigma_{r\neq 0}\ [\upsilon_r N_r^{1/2}\ e^{i\theta r/h} + \upsilon_r^*(N_r + 1)^{1/2} e^{-i\theta r/h}]$$
$$+ \Sigma_{r\neq 0}\ \Sigma_{s\neq 0}\ \upsilon_{rs} N_r^{1/2}\ (N_s + 1 - \delta_{rs})^{1/2}\ e^{i(\theta r - \theta s)/h}, \qquad (27)]$$

with $\upsilon_r^* = \upsilon_r^* = h^{1/2}c^{-3/2}(\nu_r/\sigma_r)^{1/2}\ \dot{X}_r$,

$\upsilon_{rs} = 0 \qquad (r, s \neq 0). \qquad (31)$

*The wave point of view is thus consistent with the light-quantum point of view and gives values for the unknown interaction coefficient $\upsilon_{rs}$ in the light-quantum theory*. These values are not such as would enable one to express the *interaction energy* as an algebraic function of *canonical variables*. Since the *wave* theory gives $\upsilon_{rs} = 0$ for r, s ≠ 0, *it would seem to* *show that there are no direct scattering processes, but this may be due to an incompleteness in the present wave theory.*

We shall now show that the Hamiltonian (30)
$[F = H_P(J) + \Sigma_r (2\pi h\nu_r) N_r + h^{1/2}c^{-3/2} \Sigma_r (\nu_r/\sigma_r)^{1/2} \dot{X}_r \{N_r^{1/2} e^{i\theta r/h} + (N_r + 1)^{1/2}e^{-i\theta r/h}\} \qquad (30)]$
leads to the correct expressions for *Einstein's A's and B's*. We must first modify slightly the analysis of § 5 so as to apply to the case when the system has a large number of discrete *stationary states* instead of a continuous range. …
…

The present theory, since it gives a proper account of *spontaneous emission*, must presumably give the effect of *radiation reaction* on the emitting system, and enable one to calculate the natural breadths of *spectral lines*, *if one can overcome the mathematical difficulties involved in the general solution of the wave problem corresponding to the Hamiltonian (30)*. Also, the theory enables one to understand how it comes about that there is no violation of the law of the *conservation of energy* when, say, a photo-electron is emitted from an atom under the action of extremely weak incident radiation. The *energy of interaction* of the atom and the radiation is a *q-number* that does not commute with the first integrals of the *motion* of the atom alone or with the *intensity* of the radiation. Thus, one cannot specify this energy by a *c-number* at the same time that one specifies the *stationary state* of the atom and the *intensity* of the radiation by *c-numbers*. *In particular, one cannot say that the interaction energy tends to zero as the intensity of the incident radiation tends to zero*. There is thus always an unspecifiable amount of *interaction energy* which can supply the *energy* for the photo-electron.

***Summary.***

The problem is treated of *an assembly of similar systems satisfying the Einstein-Bose statistical mechanics, which interact with another different system*, a Hamiltonian function being obtained to describe the motion. *The theory is applied to the interaction of an assembly of light-quanta with an ordinary atom, and it is shown that it gives Einstein's laws for the emission and absorption of radiation*. The interaction of an atom with *electromagnetic waves* is then considered, and it is shown that if one takes the *energies* and *phases* of the waves to be *q-numbers* satisfying the proper quantum conditions instead of *c-numbers*, the Hamiltonian function takes the same form as in the *light-quantum* treatment. *The theory leads to the correct expressions for Einstein's A's and B's*.

I would like to express my thanks to Prof. Niels Bohr for his interest in this work, and for much friendly discussion about it.

**(4) All elementary particles have mass, apart from the photon and gluons.**

## Masses of elementary particles in New Physics.

Only the *photon* has no *mass*. *Photons* are their own *anti-particle* and there is no difference between them. The *anti-particle* of the *electron* is the *positrino*. The *mass* of the *electron* is $9.1094 \times 10^{-31}$ kg; the *mass* of the *proton* is $1.6726 \times 10^{-27}$ kg (1,836 times the *mass* of the *electron*); and the *mass* of the *neutron* is $1.6749 \times 10^{-27}$ kg.

**Masses of elementary particles in New Physics.**

| *#* | *Symbol* | *Description* | *Value* (MeV/c²) | *Proton masses* |
|---|---|---|---|---|
| **Fermions:** | | | | |
| Leptons*:* | | | | |
| 1-2 | $m_e$ | Electron mass | 0.511 | 0.00054 |
| 3 | | Electron neutrino mass | 0.511 | 0.00054 |
| 4-5 | $m_\mu$ | Muon mass | 105.7 | 0.113 |
| 6 | | Muon neutrino mass | 105.7 | 0.113 |
| 7-8 | $m_\tau$ | Tau mass | 1,776.9 | 1.894 |
| 9 | | Tau neutrino mass | 1,776.9 | 1.894 |
| Nucleons: | | | | |
| 10-11 | $m_p$ | Proton mass | 938.3 | 1.000 |
| 12 | $m_n$ | Neutron mass | 939.6 | 1.001 |
| **Bosons:** | | | | |
| Gauge bosons: | | | | |
| 13 | $m_v$ | Photon mass | 0.00 | 0.00 |
| 14 | | Graviton | | |

## (5) Elementary and composite particles can have electric charge or be neutral.

## Electric charges of elementary particles in New Physics.

The *electric charge* of the *electron,* e = –1.602 × $10^{-19}$ coulombs. The *anti-particles* of the *elementary particles* have an equal but opposite *electric charge*. The *electric charge* of the *positron* (the anti-electron) is – e (i.e. – 1 *electron charge*, positive).

**Electric charges of elementary particles in New Physics.**

| # | *Description* | *Particle value (electron charges, e)* |
|---|---|---|
| **Fermions:** | | |
| Leptons: | | |
| 1-2 | Electron charge | +1 (negative) |
| 3 | Electron neutrino charge | 0 |
| 4-5 | Muon charge | +1 (negative) |
| 6 | Muon neutrino charge | 0 |
| 7-8 | Tau charge | +1 (negative) |
| 9 | Tau neutrino charge | 0 |
| Nucleons: | | |
| 10-11 | Proton charge | −1 (positive) |
| 12 | Neutron charge | 0 |
| **Bosons:** | | |
| Gauge bosons: | | |
| 13 | Photon charge | 0 |
| 14 | Graviton (hypothetical) | 0 |

## (6) Elementary particles have a quantum state called spin.

**Spin obeys the mathematical laws of angular momentum quantization.**

The specific properties of *spin angular momenta* include:

- *Spin quantum numbers may take either half-integer or integer values.*
- Although the *direction* of its *spin* can be changed, *the magnitude of the spin of an elementary particle cannot be changed.*
- *The spin of a charged particle is associated with a magnetic dipole moment with a g-factor that differs from 1*. (In the classical context, this would imply the internal charge and mass distributions differing for a rotating object.).

The conventional definition of the *spin quantum number* is s = n/2, where n can be any non-negative integer. Hence the allowed values of s are 0, 1/2, 1, 3/2, 2, etc. The value of s for an *elementary particle* depends only on the type of particle and cannot be altered in any known way (in contrast to the spin direction described below). The *spin angular momentum* S of any physical system is quantized. The allowed values of S are

$$S = \hbar\sqrt{s(s+1)} = h/2\pi \sqrt{n/2\,(n+2)/2} = h/4\pi \sqrt{n(n+2)},$$

where h is the Planck constant, and $\hbar = h/2\pi$ is the reduced Planck constant. In contrast, *orbital angular momentum* can only take on integer values of s; i.e., even-numbered values of n.

*Those particles with half-integer spins, such as 1/2, 3/2, 5/2, are known as fermions, while those particles with integer spins, such as 0, 1, 2, are known as bosons*. The two families of particles obey different rules and broadly have different roles in the world around us. A key distinction between the two families is that *fermions obey the Pauli exclusion principle*: that is, *there cannot be two identical fermions simultaneously having the same quantum numbers* (meaning, roughly, having the same position, velocity and spin direction). *Fermions obey the rules of Fermi–Dirac statistics*. In contrast, *bosons obey the rules of Bose–Einstein statistics* and have *no such restriction*, so they may "bunch together" in identical states.

## Spin of elementary particles in New Physics.

**Spin of elementary particles in New Physics.**

| # | *Description* | *Value* (±) |
|---|---|---|
| **Fermions:** | | |
| Leptons: | | |
| 1-2 | Electron spin | 1/2 |
| 3 | Electron neutrino spin | 1/2 |
| 4-5 | Muon spin | 1/2 |
| 6 | Muon neutrino spin | 1/2 |
| 7-8 | Tau spin | 1/2 |
| 9 | Tau neutrino spin | 1/2 |
| Nucleons: | | |
| 10-11 | Proton spin | 1/2 |
| 12 | Neutron spin | 1/2 |
| **Bosons:** | | |
| Gauge bosons: | | |
| 13 | Photon spin | 1 |
| 14 | Graviton spin | 1 |

**Non-relativistic theory.**

In 1925 Wolfgang Pauli introduced a new two-valued quantum number, identified by Samuel Goudsmit and George Uhlenbeck as *electron spin*. From this he formulated his *Exclusion Principle* for electrons, which he later extended to all *fermions* with his *spin–statistics theorem* of 1940.

> [Pauli, W. (1925). Über den Zusammenhang des Abschlusses der Elektronengruppen im Atom mit der Komplexstruktur der Spektren. *Zeit. Phys.*, 31, 1, 765–83; doi:10.1007/BF02980631.]

*Pauli's Exclusion Principle* states that two or more identical particles with half-integer spins (i.e. *fermions*) cannot simultaneously occupy the same quantum state within a system that obeys the laws of quantum mechanics.

*In the case of electrons in atoms*, the exclusion principle can be stated as follows: in a poly-electron atom *it is impossible for any two electrons to have the same two values of all four of their quantum numbers*, which are: n, the *principal quantum number*; $\ell$, the *azimuthal quantum number*; $m_\ell$, the *magnetic quantum number*; and $m_s$, the *spin quantum number*. For example, if two *electrons* reside in the same orbital, then their values of n, $\ell$, and $m_\ell$ are equal. In that case, the two values of $m_s$ (*spin*) pair must be different. Since the only two possible values for the *spin* projection $m_s$ are +1/2 and −1/2, it follows that one *electron* must have $m_s = +1/2$ and one $m_s = -1/2$.

Particles with an *integer spin* (*bosons*) are not subject to the Pauli exclusion principle. *Any number of identical bosons can occupy the same quantum state.*

A more rigorous statement is: *under the exchange of two identical particles, the total (many-particle) wave function is antisymmetric for fermions and symmetric for bosons.* This means that if the space and spin coordinates of two identical particles are interchanged, then *the total wave function changes sign for fermions, but does not change sign for bosons.*

Pauli, W. (1927). Zur Quantenmechanik des magnetischen Elektrons. (On the quantum mechanics of magnetic electrons.). *Zeit. Phys.*, 43, 601-23: In 1927 Pauli formalized the theory of *spin* using the theory of quantum mechanics invented by Erwin Schrödinger and Werner Heisenberg. He pioneered the use of Pauli matrices as a representation of the spin operators and introduced a two-component spinor wave-function. *Pauli's theory of spin was non-relativistic.*

## **Compton, A. H. (August, 1921). The Magnetic Electron*.**

*Journ. Frankl. Inst.*, 192, 2, 145-55; https://www.semanticscholar.org/paper/The-magnetic-electron-Compton/f602176e15e52ed703a67865f4c87eeb7a83048e; also in Underwood, T. G. (2024). *Electricity & Magnetism*, Part IV, pp. 335-41.

* Based on a paper read before Section B of the American Association for the Advancement of Science, December 27, 1920.

Washington University, St. Louis.

Compton's paper on investigations of ferromagnetic substances with X-rays was the first to introduce the idea of *electron spin*. Compton hypothesized that the electron's *magnetic moment* was intrinsically connected to the electron's *spin* and pointed out the possible bearing of this idea on the origin of the natural unit of magnetism.

---

The evidence brought forward by the speakers who have preceded me has shown that many magnetic phenomena find a satisfactory explanation on the hypothesis that *matter contains a large number of minute elementary magnets*. The theories of *para-* and *ferro-magnetism* as developed by Langevin, Weiss and others, though based upon the hypothesis of such ultimate magnetic particles, make no assumptions concerning their nature. The explanation of *diamagnetism*, on the other hand, is based upon the view that this effect owes its origin to the circulation of electricity in resistanceless paths. The success of these theories in explaining the principal characteristics of magnetism gives us confidence in the real existence of these magnetic particles. Let us see, therefore, if it is possible to identify these elementary magnets with any of the fundamental divisions of matter.

The original investigations of *ferromagnetism* which led to the hypothesis of an elementary magnetic particle credited molecules with the properties of small permanent magnets. This view finds some support in the profound effect of heating, mechanical jarring, etc., on the ease of magnetization of iron. The dependence of magnetic permeability upon the chemical condition of a substance suggests the same view. But perhaps the strongest argument that has been brought forward in support of the idea of molecular magnets has been the discovery of the Heusler alloys, in which by melting together elements which arc only slightly magnetic an alloy with ferromagnetic properties is produced. It is, however, difficult to imagine what mechanism could reasonably give to a group of atoms, such as the chemical molecule, the properties of a single magnetic particle. Moreover, if on magnetization such a group of atoms should actually turn around within a crystal, as the elementary magnets are supposed to do, the resulting change in the positions of the atoms composing the molecule should produce a change in the crystal form; since, as we know, the form of the crystal is dependent upon the arrangement. of its component atoms. It is, however, a matter of common observation that a magnetic field effects no such change in the form of a magnetic crystal.

*Perhaps the most natural, and certainly the most generally accepted view of the nature of the elementary magnet, is that the revolution of electrons in orbits within the atom give to the atom as a whole the properties of a tiny permanent magnet.* Support of this view is found in the quantitative explanation which it affords of the *Zeeman effect*. It seems but a step from the explanation of this effect to Langevin's explanation of *diamagnetism* as another result of the induced electronic currents within the atom. On Langevin s view the electronic orbits act as resistanceless circuits in which an external magnetic field induces changes of current. By Lenz's law these induced currents will always be in the direction to give the electronic orbit a magnetic polarity opposite to the applied field, thus accounting for the atom's *diamagnetic* properties. This theory offers a satisfactory qualitative explanation of *diamagnetism*, and accounts for the fact that *diamagnetism* is independent of temperature. But quantitatively it is inadequate. For, in order to explain the magnitude of the observed *diamagnetic* susceptibility on this view, one must suppose either that the atom possesses a number of electrons equal to several times its atomic number, or the distance between the electrons in the atom must be several times as great as is estimated by more direct methods. Moreover, the experiments of Barnett[1] and J. Q. Stewart[2] show that the *ratio of charge to mass* of the elementary magnet, though of the same order of magnitude, *is appreciably greater than one would expect if the magnetic moment is due solely to electrons revolving in orbits.*

[1] Barnett, S. J. (1915). *Phys. Rev.*, 6, 240.
[2] Stewart, J. Q. (1918). *Phys. Rev.*, 11, 100.

But perhaps a more serious difficulty with the usual electron theory of *diamagnetism* is that the induced change in magnetic moment of the electronic orbit involves also a change in its angular momentum. It is obvious, *according to the classical electrical theory, that any electron revolving in an orbit will soon radiate its energy.* Any angular momentum induced by an applied magnetic field will, on this theory, therefore, rapidly disappear so that *diamagnetism* should be merely a transient effect. *Let us then assume with Bohr that if each electron has some definite angular momentum such as $h/2\pi$, no radiation occurs.* On this view the electrons in the normal atom will all possess the requisite angular momentum, and when an external magnetic field is applied the induced change in angular momentum will put the electrons in an unstable condition. *On this view also, therefore, the additional rotational energy induced by an applied magnetic field will not be permanent, but will soon be dissipated.* In fact, the theory of atomic structure has yet to be proposed according to which *diamagnetism*, accounted for by the induced magnetic moment of electrons revolving in orbits, can be more than a transient phenomenon.

*Besides the molecule and the atom, we have the other two fundamental divisions of matter, the atomic nucleus and the electron.* The sign of the Richardson-Barnett effect indicates that *it is negative electricity which is chiefly responsible for magnetic effects*, which makes the view that the positive nucleus is the elementary magnet difficult to defend. On the other hand, many of the magnetic properties of matter receive a satisfactory explanation on Parson's hypothesis[3], that the electron is a continuous ring of negative electricity spinning rapidly about an axis perpendicular to its plane, and therefore possessing a magnetic moment as well as an electric charge.

[3] Parson, A. L. (1915). *Smithsonian Misc. Collections.*

Thus, for example, the fact that such a ring can rotate without radiating enables this hypothesis to account for *diamagnetism* as a permanent instead of a transient effect. While retaining Parson's view of a magnetic electron of comparatively large size, we may suppose with Nicholson that instead of being a ring of electricity, the electron has a more nearly isotropic form with a strong concentration of electric charge near the center and a diminution of electric density as the radius increases. It is natural to suppose that the mass of such an electron is concentrated principally near its center and that the ratio of the charge to the mass of its external portions will be greater than that for the electron as whole. While the explanation of the inertia of such a charge of electricity is perhaps not obvious, it is at least consistent with our usual conceptions and it has the advantage of offering an explanation for the large value of e/m observed in Barnett and Stewart's experiments. It also makes possible an explanation of the relatively large induced currents required to account for *diamagnetism* without introducing the assumption of a prohibitively large radius for the electric charge. A series of experiments has recently been performed, designed to determine which of these fundamental divisions of matter is identical with the elementary magnet in ferromagnetic substances. The first of these, due to K. T. Compton and E. A. Trousdale[4], had for its object the detection of any displacement of the atoms of a substance on magnetization.

[4] Compton, K. T. & Trousdale, E. A. (1915). *Phys. Rev.*, 5. 315.

If the elementary magnet consists of a group of atoms such as the chemical molecule, the rotation of this elementary magnet into alignment with an applied external field will cause a displacement of the individual atoms. It is known, however, that the position of the spots on a Lane photograph depends upon the arrangement of the atoms within the crystal employed. If then, such a photograph is taken with a magnetic crystal, the character of the diffraction pattern should change when the direction of magnetization of the crystal is altered. In these experiments, however, no effect of this character was found. The obvious conclusion is that the ultimate magnetic particle does not consist of any group of atoms such as the chemical molecule. The second of these experiments, performed by Mr. Rognley and myself [5],

[5] Compton, A. H. & Rognley, 0. (1920). *Phys. Rev.*, 16, 464.

was based upon the fact that the intensity of reflection of X-rays from the surface of a crystal depends not only upon the arrangement of the atoms within the crystal, but also upon the distribution of the electrons within the atoms. Let us suppose that the atom acts as a tiny magnet due to the orbital motion of its component electrons. Magnetization of the crystal will orient these atomic magnets and in so doing will change the planes of revolution of the electrons. This change in the electronic distribution should, therefore, affect the intensity of reflection of a beam of X-rays from the crystal's surface. An attempt was made to detect such a change in the intensity of X-ray reflection from a crystal of magnetite when strongly magnetized. Apparatus sufficiently sensitive to detect a change in intensity of less than one per cent. was employed, but magnetization of the crystal failed to produce any measurable effect. ... The following table shows in the first column the order of the X-ray spectrum line which was being studied; in the second column the calculated ratio of intensity from the magnetized to that from the unmagnetized crystal, supposing the atom to have the Rutherford form; and the third and fourth columns represent the similar ratios as estimated from a cubic form of atom. ... According to experiment the value of these ratios was always unity, at least within one per cent. It is clear that none of the types of atoms considered could be oriented by a magnetic field without producing a noticeable effect. In fact, it is difficult to imagine any form of magnetic atom which would be so nearly isotropic that it would have given no effect in our experiment. *It is, therefore, difficult to avoid the conclusion that the elementary magnet is not the atom as a whole.*

*Since neither the molecule nor the atom gives a satisfactory explanation of these experiments, the view suggests itself that it is something within the atom, presumably the electron, which is the ultimate magnetic particle.* Let us see then if we can find any positive evidence for the existence of an electron with a magnetic moment.

On the basis of the classical dynamics we should expect the electron, whatever its form, to possess thermal energy of rotational motion, equal on the average to that of a molecule or atom at the same temperature. On Planck's more recent quantum hypothesis, however, which is perhaps the more reasonable view, at the absolute zero of temperature each particle of matter - including the electron - should retain an average amount of energy

½ hv for each degree of freedom for motion. For a rotating system this corresponds to, an angular momentum of h/2π. Thus, whatever view we adopt, the thermal motions of the electron will give to it an appreciable magnetic moment. For a particle of the small moment of inertia of the electron, the frequency of rotation corresponding to an angular momentum h/2π will be exceedingly high, and the corresponding energy ½ hv will be large compared with the additional energy which it may acquire due to an increase in temperature. Thus, the angular momentum, and hence also the magnetic moment of the electron, will be nearly the same at different temperatures - a property characteristic of the elementary magnets. *It is interesting to notice, also, that the magnitude of the magnetic moment of an electron spinning with an angular momentum h/2π is of the proper order to account for ferromagnetic properties, being about one-third the magnetic moment of the iron atom.*

If an electron with such an angular momentum is to have a peripheral velocity which does not approach that of light, it is necessary that the radius of gyration of the electron shall be greater than $10^{-11}$ cm. While such an electron is much larger than the spherical electron of Lorentz, recent experiments on the scattering of X-rays and gamma rays indicate the electron's diameter may be even greater than the minimum value thus required to explain magnetic properties. Experiment shows that the scattering of very high frequency radiation is considerably less than theory demands if the electron is supposed to have negligible dimensions. In the case of hard gamma rays, indeed, I have found the scattering at certain angles to fall below 1/1000, the intensity predicted on the usual theory[6].

[6] Compton, A. H. *Phil. Mag.*, (in printer's hands).

The only adequate explanation of these experiments seems to be that interference occurs between the rays scattered from the different parts of the same electron. Such an explanation clearly implies that the diameter of the electron is comparable with the wave-length of the radiation employed, which means that the effective radius of the electron is of the order of $10^{-10}$ cm. *Considerations of the size of the electron, therefore, support rather than oppose the view that the electron may have an appreciable magnetic moment.*

Further evidence that the electron possesses properties other than those of an electric charge of negligible dimensions is afforded by a study of the white X-radiation emitted at the target of an X-ray tube. It was noticed by Kaye that the X-rays emitted in the direction of the cathode ray beam are harder and more intense than those traveling in the opposite direction. The difference in both hardness and intensity of the radiation at different angles is in good accord with the view proposed by D. L. Webster that the particles emitting the radiation are moving in the direction of the cathode-ray beam, giving rise to a Doppler effect. Indeed, it is very difficult to give any other explanation of the difference in wave-length of the radiation in different directions. But, on this view, in order to account for the difference in hardness observed in the case of gamma rays, *the radiating particles must have a velocity of about one-half the speed of light*. Since the highest known speeds at which atoms travel is only about one-tenth the velocity of light, as observed in the case of alpha particles, the swiftly moving radiators giving rise to this high-frequency X-radiation must therefore be free electrons. If this view is correct, it follows, as Webster has pointed out, that *the electron must be a system capable of emitting radiation*, and is therefore, not a mere charge of electricity of negligible dimensions. *On the present view we may well suppose that the electron is spinning like a gyroscope and on traversing matter is set into mutational oscillations*, resulting in the observed radiation.

Strong evidence that *the electron possesses a magnetic moment* is afforded by H. S. Allen's recent explanation of the rotation of the plane of polarization by optically active substances[7].

[7] Allen, H. S. (1920). *Phil. Mag.*, 40, 426.

You will remember in Drude's classical work it is found that optical rotation may be explained if the electrons, when made to oscillate by a passing electric wave, do not move exactly in the plane of the electric vector. *He supposes rather that there is a component of motion at right angles to the electric vector and finds that such a motion will account for the observed rotation*. Allen shows that the motion perpendicular to the electric vector which Drude assumes is *a natural consequence of the view that the electron is magnetic and has an appreciable diameter*. It would take us too far afield to discuss the details of this work, but the significance of the result is obvious, since it has heretofore been difficult to give a reasonable account of the type of motion postulated by Drude.

Finally, I wish to discuss a phenomenon, first noticed by C. T. R. Wilson and brought to my attention by Mr. Shimizu, which, if its obvious explanation is correct, gives *direct evidence that free electrons possess magnetic polarity*. Suppose that a magnetic electron is placed in a homogeneous *paramagnetic* medium. Every part of the medium will be slightly magnetized in the direction of the lines of force, and the magnetic field at the electron due to the magnetic moment of each portion of the medium will have a positive component in the direction of the electron's magnetic axis. Thus, the magnetization induced in the surrounding medium will give rise to a magnetic force at the electron in the direction of its own magnetic axis. The case is exactly analogous to placing a bar magnet in a field of iron filings. The iron filings will be magnetized by induction in the direction of the *lines of force* and if the bar magnet is removed, there still exists a magnetic field where the magnet was because of the magnetization of the surrounding iron filings. If now the electron is in motion, this induced magnetic field will produce the same effect as would an externally applied field of the same intensity. That is, the force due to the magnetic field from the surrounding medium acting on the moving electric charge will make it follow a curved instead of a straight path. If, because of its gyroscopic action, the axis of the electron does not change its direction, the induced magnetic field will always be in the same direction, and the electron will describe a helical orbit. In any actual medium, composed of discreet particles and therefore not homogeneous on an electronic scale, this spiral motion will be superposed upon an irregular motion due to collisions, and the axis of the electron will not remain fixed in direction. Thus, any spiral motion that may appear should be rather broken. A rough calculation, assuming an electron to be projected into air with a speed corresponding to a drop through 10,000 volts, which is about that of the secondary cathode rays produced by ordinary X-rays, and having a magnetic moment corresponding to the angular momentum $h/2\pi$, indicates that the induced magnetic field at the electron should be of the order of 3000 gauss, if the permeability of the medium is that of ordinary air. This field is strong enough to produce a very decided curvature in the electron's path, so in spite of the irregularities in the electron's motion we might hope to observe experimentally the predicted helical tracks.

Below are a few of C. T. R. Wilson's photographs of the tracks of secondary cathode rays and beta particles. ...

...

Let us then review the different lines of evidence that have given us information concerning the nature of the elementary magnet. In the first place, the Richardson-Barnett effect shows that *magnetism is due chiefly to the circulation of negative electricity whose ratio of charge to mass is not greatly different from that of the electron*. In the second place, *experiments on the diffraction of X-rays by magnetic crystals* indicate that the elementary magnet is not any group of atoms, such as the chemical molecule, nor even the atom itself; but *lead rather to the view that it is the electron rotating about its own axis which is responsible for the ferro-magnetism*. And finally, *positive evidence in favor of the hypothesis of some form of magnetic electron is supplied by a consideration of the curvature of the tracks of beta rays through air*. May I then conclude that *the electron itself, spinning like a tiny gyroscope, is probably the ultimate magnetic particle*.

**Uhlenbeck, G. E. & Goudsmit, S. (November, 1925). Ersetzung der Hypothese vom unmechanischen Zwang durch eine Forderung bezuglich des inneren Verhaltens jedes einzelnen Elektrons. (Replacement of the hypothesis of unmechanical coercion by a requirement regarding the internal behavior of each individual electron.)**

*Naturw.*, 13, 47, 953-4; https://doi.org/10.1007/BF01558878; translation by T. G. Underwood; also in Underwood, T. G. (2023). *Quantum Electrodynamics - annotated sources*, Volume I, pp. 282-6.

October 17, 1925.

Instituut voor Theoretische Natuurkunde, Leiden.

The idea of a *quantized spinning of the electron* was put forward for the first time by Compton in August 1921. Without being aware of Compton's suggestion Uhlenbeck and Goudsmit noted doublets in the alkali spectra that did not conform to current models of the atom. They proposed applying the model of the *spinning electron* to interpret a number of features of the quantum theory of the *anomalous Zeeman effect*, and applied the classical formula for spherical rotating electron with finite radius and surface charge.

---

**§ 1.** As is well known, the structure and magnetic behavior of the spectra can be described in detail with the help of *Landé's vector model* R, K, J and m[1].

[1] See Back, E. & Landé, A. (1925). *Zeemaneffekt und Multiplettstruktur der Spektrallinien.* (Zeeman effect and Multiplet structure of the spectral lines.) Berlin: Verlag von Julius Springer).

Here, R denotes the *momentum* moment of the atomic remnant ~ i.e. of the atom without the luminous electron - K the *momentum* moment of the luminous electrons, J their resultant and m the projection of J on the direction of an *external magnetic field*, all expressed in the branch quantum units:

(a) that for the rest of the atom the behavior of the *magnetic moment* to the *mechanical* is *twice as large as you would expect classically.*

(b) that in the formulae, where $R^2$, $K^2$, $J^2$ occurs, you can do this by using these expressions $R^2 - ¼$, $K^2 - ¼$, $J^2 - ¼$. [The Heisenberg Averaging[2])].

[2] Heisenberg, W. (1924). Über eine Änderung der formalen Regeln der Quantentheorie in einem Problem anomaler Zeeman-Effekte. (On an alteration to the formal rules of quantum theory in a problem of anomalous Zeeman effect.) *Zeit. Phys.*, 26, 291-307.

This model has shown itself to be very robust and has, among other things, fought to unravel the most complicated spectra.

**§ 2.** However, one starts to encounter difficulties as soon as one tries to connect *Landé's vector model* to our ideas about the formation of the atom from electrons. E.g.:

a) Pauli[3] has already shown that in the case of the alkali atoms, the atomic radical must be magnetically ineffective, otherwise the influence of the *relativity correction* would cause a dependency of the Zeeman effect on the nuclear charge, which is not perceived in these spectra.

[3] Pauli Jr., W. (1925). Über den Einfluss der Geschwindigkeitsabhängigkeit der Elektronenmasse auf den Zeemaneffekt. (On the influence of the velocity dependence of the electron mass on the Zeeman effect.) *Zeit. Phys.*, 31, 373.

b) In *Lande's model*, one must not identify the *momentum moment* of the atomic radical with that of the positive ions, as one would expect it according to the definition of the atomic radical. [Landé-Heisenberg branching theorem[4] — unmechanical coercion].

[4] See Back, E. & Landé, A. (1925). *Zeemaneffekt und Multiplettstruktur der Spektrallinien. Loc. cit.*, pages 55ff.

c) For some spectra recently analyzed with the help of Lande's scheme (e.g. vanadium, titanium), the K of the basic term did not correspond at all with the values expected from the Bohr-Stone periodic system.

**§ 3.** The above-mentioned difficulties point all in the same direction, namely that *the meaning of which is attributed to Lande's vectors is probably not correct*. Pauli[5] has already embarked on a new path, which is particularly difficult.

[5] Pauli Jr., W. (1925). Über den Zusammenhang des Abschlusses der Elektronengruppen im Atom mit der Komplexstruktur der Spektren. (On the relationship between the completion of the electron groups in the atom and the complex structure of the spectra.) *Zeit. Phys.*, 31, 765.

From this he concluded that in the case of alkali spectra, all quantum numbers must be written to the luminous *electron* alone. According to Pauli, each *electron* in the *magnetic field* then gets 4 independent quantum numbers. With the help of Bohr's construction principle and a few general sentences, he was then able to achieve the same results as Landé in a simple way[6].

[6] Compare: Goudsmit, S. (December, 1925). Über die Komplexstruktur der Spektren. *Zeit. Phys.*, 32, 1, 794-98; https://doi.org/10.1007/BF01331715; Heisenberg, W. (1925). Quantentheorie der multiplen Struktur und des abnormalen Zeeman-Effekts. (Quantum theory of multiple structure and the abnormal Zeeman effect.) *Zeit. Phys.*, 32, 841-60;

Hund, F. (1925). Zur Deutung verwickelter Spektren, insbesondere der Elemente Scandium bis Nickel. (On the interpretation of entangled spectra, in particular the elements scandium to nickel.) *Zeit. Phys*., 33, 345-71; http://dx.doi.org/ 10.1007/BF01328319.

The difficulties mentioned in § 2 disappear completely in the Pauli procedure. The connection to the Bohr-Stoner periodic system is achieved, and new aspects are still opened[7].

[7] See those in 5) below.

**§ 4.** In both cases, however, the appearance of the so-called *relativistic doublet* in the rontgen and alkali spectra remains an enigma. To explain this fact, one has recently come to the assumption of a classically indescribable ambiguity in the quantum theoretical properties of the *electron*[1].

[1] Heisenberg, W. (1925). Quantentheorie der multiplen Struktur und des abnormalen Zeeman-Effekts. (Quantum theory of multiple structure and the abnormal Zeeman effect.) *Zeit. Phys*., 32, 841-60; *loc. cit.*

**§ 5.** There seems to us to be another way open. Pauli does not bind himself to a model idea. The 4 quantum numbers assigned to each electron have lost their original Landé meaning. It is now obvious to give to each electron with its 4 quantum numbers 4 degrees of freedom. One can then give the quantum numbers, for example, the following meaning: n and k remain as before the main and azimuthal quantum number of the electron in its orbit. *R, however, will be assigned its own rotation of the electron*[2].

[2] Note that the quantum numbers of the *electron* occurring here must be taken from the alkali spectra. R therefore has only the value 1 for each *electron* (in Landé standardization).

The other quantum numbers retain their old meaning. Through our imagination, the conceptions of Landé and Pauli with all their advantages have formally merged with each other[3].

[3] For example, the meaning of the Heisenberg's Scheme III is now becoming more understandable, in which one has to assemble both the R and the K of the *electrons* for an entire atom.

The *electron* must now take over the still misunderstood property (referred to in § I under a), which Landé attributed to the atomic remnant.

The closer quantitative implementation of this idea will probably depend heavily on the choice of the *electron* model. In order to come into line with the facts, the following demands must therefore be made of this model:

a) *The ratio of the magnetic moment of the electron to the mechanical one must be twice as large for the self-rotation as for the orbital motion*[4].

[4] For example, for a spherical rotating electron with surface charge, the Abraham formulas can be used [Abraham, M. (1903). Prinzipien der Dynamik des Elektrons. (Principles of electron dynamics.) *Ann. Phys*., 315, 105-79] read:

Rotational energy $1/9\ e^2 a/c^2\ \dot{\varphi}^2$ ($a$ = electron radius),

also: $p_\varphi = 2/9\ e^2 a/c^2\ \dot{\varphi}$

Magnetic moment: $\Phi = 1/3\ e a^2/c\ \dot{\varphi}$

Mass: $m = 2/3\ e^2/c^2 a$

Also: $\Phi/p_\varphi = 3/2\ ac/e = 2 \times e/2mc$,

in fact, twice as much as in the orbital motion. Note, however, that when quantizing this rotational motion, *the peripheral speed of the electron is far from the speed of light*.

b) The different orientations from the R to the orbital plane (or K) of the *electron* must be able to provide the explanation of *relativity-doublets*, perhaps in connection with a Heisenberg -Wentzel averaging rule[5].

[5] Heisenberg, W. *loc. cit*. Wentzel, G. (1925). *Ann. Phys*., 76, 803.

## Pauli, W. (February, 1925). Über den Zusammenhang des Abschlusses der Elektronengruppen im Atom mit der Komplexstruktur der Spektren. (On the connection between the completion of electron groups in an atom and the complex structure of spectra.)

*Zeit. Phys*., 31, 1, 765–83; https://doi.org/10.1007/BF02980631; translation at http://www.fisicafundamental.net/relicario/doc/Pauli_1925.pdf.

Institut für theoretische Physik, Hamburg.

Received January 16, 1925.

Pauli first reviewed the established theories for the energy differences *of the triplet levels of the alkaline earths*, based respectively, on *the anomaly of the relativity correction* of the *optically active electron*, and *the dependence of the interaction between the electron and the atom core on the relative orientation of these two systems*. He noted a serious difficulty with the former is the connection of these ideas with the *correspondence principle*, which was well known to be a necessary means to explain the selection rules for the *quantum numbers* $k_1$, j, and m and the polarization of the Zeeman components, in particular, that *it was necessary that the totality of the stationary states of an atom corresponded to a collection (class) of orbits with a definite type of periodicity properties*. The dynamic explanation of this kind of motion of the *optically active electron*, which was based upon the assumption of deviations of the forces between the *atom* core and the *electron* from central symmetry, *seemed to be incompatible with the possibility to represent the alkali doublet (and thus also the magnitude of the corresponding precession frequency) by relativistic formulae*. Consequently, Pauli, decided to pursue instead the alternative *non-relativistic* theory to the problem of *completion of electron groups in an atom*, in order to draw conclusions only about the *number of possible stationary states* of an *atom* when several equivalent *electrons* are present. But this did not address the position and relative order of the term values. On the basis of these results, Pauli obtained a general classification of every *electron* in the *atom* by the principal quantum number n and two auxiliary quantum numbers $k_1$ and $k_2$ to which he added a further quantum number $m_1$ in the presence of an external field, in agreement with experiments. In particular, his rule explained Stoner's result in a natural way and with it the period lengths 2, 8 18, 32.

---

Especially in connection with Millikan and Landé's observation that the alkali doublet can be represented by relativistic formulae and with results obtained in an earlier paper, it is suggested that this doublet and its *anomalous Zeeman effect* expresses a classically non-describable *two-valuedness* of the quantum theoretical properties of the optically active electron [German: *Leuchtelektron*], without any participation of the closed rare gas configuration of the atom core in the form of a core *angular momentum* or as the seat of the magneto-mechanical anomaly of the *atom*. We then attempt to pursue this point of view, taken as a temporary working hypothesis, as far as possible in its consequences also for *atoms* other than the alkali *atoms*, notwithstanding its difficulties from the point of view of principle. First of all it turns out that it is possible, in contrast to the usual ideas, to assign for the case of a *strong external magnetic field*, which is so strong that we can neglect the coupling between the atomic core and the optically active electrons, to those two systems, as far as the number of their *stationary states*, the values of their *quantum numbers*, and their *magnetic energy* is concerned, no other properties than those of the free atomic core of the *optically active electron* of the *alkalis*. *On the basis of these results, one is also led to a general classification of every electron in the atom by the principal quantum number n and two auxiliary quantum numbers* $k_1$ *and* $k_2$ *to which is added a further quantum number* $m_1$ *in the presence of an external field.* In conjunction with a recent paper by E. C. Stoner this classification leads to a *general quantum theoretical formulation of the completion of electron groups in atoms*.

**1. The Permanence of Quantum Numbers (Principle of Gradual Construction [German: *Aufbauprinzip*]) in Complex Structures and the Zeeman Effect.**

In a previous paper [Pauli, W. (1925). *Zeit. Phys*., 31, 373] it was emphasized that the usual ideas, according to which *the inner, completed electron shells of an atom play an essential part in the complex structure of optical spectra and their anomalous Zeeman effect in the shape of core angular momenta and as the real seat of the magneto-mechanical anomaly*, are subject to several serious difficulties. It seems therefore plausible to set against these ideas that especially the doublet structure of the alkali spectra and their *anomalous Zeeman effect* are caused by a classically indescribable *two-valuedness* of the quantum theoretical properties of the *optically active electron*.

> [The *Normal Zeeman Effect* is the splitting of spectral lines of an atomic spectrum due to the interaction between an external magnetic field and *the orbital magnetic momentum*. This effect can be observed *in the absence of electron spins*. The normal Zeeman effect can be observed as a *triplet in the observed spectrum* instead of a single spectral line in the expected spectrum. There the single spectral line has been split into three lines with equal spaces between them.
>
> The *Anomalous Zeeman Effect*, which was discovered by Thomas Preston in Dublin, Ireland, is the splitting of spectral lines of an atomic spectrum caused by the interaction between an external magnetic field and *the combined orbital and intrinsic magnetic moment*. This effect can be observed as a complex splitting of spectral lines. Sometimes the spaces between the spectral lines are wider than

expected. This happens due to the effects of *electron spin.* Since the *spin* of electrons contributes to the *angular momentum*, splitting becomes more complicated.

See Zeeman, P. (March, 1897). On the influence of magnetism on the nature of the light emitted by a substance. *Phil. Mag.*, 5, 43, 226-39; republished in *The Astrophysical Journal*, 5, 332-47; https://doi.org/_10.1086/140355; and Preston, T. (January, 1899). Radiation Phenomena in the Magnetic Field. *Nature*, 59, 1523, 224-9; https://doi.org/10.1038/059224c0; first published as Preston, T. (April, 1898). Radiation Phenomena in a Strong Magnetic Field. *Scientific Transactions of the Royal Dublin Society*, 6, 385-389; (1898). Radiation phenomena in the magnetic field. *Phil. Mag.*, 45, 275, 325-39; https://doi.org/10.1080/14786449808621140; also, both in Underwood, T. G. (2023). *Quantum Electrodynamics - annotated sources, Volume I.*]

This idea is particularly based upon the results of Millikan and Landé that the optical doublets of the alkalis are similar to the *relativity* doublets in X-ray spectra and that their magnitude is determined by a *relativistic formula.* If we now pursue this point of view, we shall assign -- as was done by Bohr and Coster for the X-ray spectra -- to the *stationary states* of the *optically active electron* involved in the emission of the alkali spectra two auxiliary *quantum numbers* $k_1$ and $k_2$ as well as the *principal quantum number* n. The first *quantum number* $k_1$ (usually simply denoted by k) has the values 1, 2, 3, ... for the s, p, d, . . . terms and changes by unity in the allowed transition processes; *it determines the magnitude of the central force interaction forces of the valence electron with the atom core.* The second *quantum number* $k_2$ is for the two terms of a doublet (e.g., $p_1$ and $p_2$) equal to $k_1 - 1$ and $k_1$, in the transition processes it changes by Â $\pm$ 1 or 0 and determines the magnitude of the *relativity correction* (which is modified according to Landé to take into account *the penetration of the optically active electron in the atom core*). If we follow Sommerfeld to define the *total angular momentum quantum number* j of an atom in general as the maximum value of the *quantum number* $m_1$ (usually simply denoted by m) which determines *the component of the angular momentum along an external field*, we must put $j = k_2 - 1$ for the alkalis. The number of *stationary states* in a *magnetic field* for given $k_1$ and $k_2$ is $2j + 1 = 2k_2$, and the number of these states for both doublet terms with given $k_1$ is altogether $2(2k_1 - 1)$.

If we now consider the case of *strong field* (*Paschen-Back effect*), we can introduce apart from $k_1$ and the just mentioned *quantum number* $m_1$, instead of $k_2$ also a *magnetic quantum number* $m_2$ which *determines directly the energy of the atom in the magnetic field*, that is, the component of the *magnetic moment* of the *valence electron* parallel to the field. For the two terms of the doublet it has, respectively, the values $m_1 + 1/2$ and $m_1 - 1/2$. Just as in the doublet structure of the alkali spectra the "anomaly of the *relativity* correction" is expressed (the magnitude of which is mainly determined by another *quantum number*, as is the magnitude of the central force *interaction energy* of the *optically active electron* and the *atom* core), so appears in the deviations of Zeeman structure from the normal Lorentz triplet the "*magnetomechanical anomaly*" which is similar to the other anomaly (the magnitude of the *magnetic moment* of the *optically active electron* is mainly determined by another *quantum number*, as is the *angular momentum*). Clearly, *the appearance of half-odd-integral (effective) quantum numbers and the thereby formally caused value* $g = 2$ *of the splitting factor of the s-term of the alkalis is closely connected with the twofoldness of the energy level*. We shall here, however, not attempt a more detailed theoretical analysis of this state of affairs and use the following considerations of the Zeeman effect of the *alkalis* as empirical data.

Without worrying about the difficulties encountered by our point of view, which we shall mention presently, *we now try to extend this formal classification of the optically active electron by four quantum numbers* $n$, $k_1$, $k_2$, $m_1$ *to atoms, more complex than the alkalis*. It now turns out that we can retain completely on the basis of this classification the *principle of permanence of quantum numbers* (*Aufbauprinzip*) also for the complex structure of the spectra and the *anomalous Zeeman effect* in contrast to the usual ideas. This principle, due to Bohr, states that *when a further electron is added to a -- possibly charged -- atom, the quantum numbers of the electrons which are already bound to the atom retain the same values as correspond to the appropriate state of the free atom core*.

Let us first of all consider the *alkaline earths*. The spectrum consists in this case of a singlet and a triplet system. The *quantum states* with a well-defined value of the quantum number $k_1$ of the *optically active electron* correspond then for the first system to altogether $1(2k_1 - 1)$ and in the last system to $3(2k_1 - 1)$ *stationary states* in an *external magnetic field*. Up to now this was interpreted as meaning that in strong fields the optically active electron in each case could take up $2k_1 - 1$ positions, while the *atom* core was able to take up in the first case one, and in the last case three positions. *The number of these positions is clearly different from the number 2 of the positions of the free atom core (alkali-like s-term) in a field*. Bohr [Bohr, N. (1923). Ann. Phys., 71, 228; especially p. 276] called this state of affairs a "*constraint*" [German: *Zwang*] which is *not analogous to the action of external fields of force*. Now, however, we can simply interpret the total $4(2k_1 - 1)$ states of the *atom* as meaning that the *atom* core always has two positions in a field, and the *optically active electron* as for the alkalis $2(2k_1 - 1)$ states.

More generally, a branching rule formulated by Heisenberg and Landé states that a stationary state of the *atom* core with N *states* in a field leads *through the addition of one more electron* to two systems of terms, corresponding to altogether $(N + 1)(2k_1 - 1)$ and

$(N - 1)(2k_1 - 1)$ *states* in a field, respectively, for a given value of the *quantum number* $k_1$ of the last *electron*. According to our interpretation, the $2N(2k_1 - 1)$ states of the complete atom in a *strong field* come about through N *states* of the atom core and $2(2k_1 - 1)$ *states* of the *optically active electron*. In the present quantum theoretical classification of the electrons the term *multiplicity* required by the branching rule is simply a consequence of the "*Aufbauprinzip*". According to the ideas presented here Bohr's *constraint* expresses itself not in a violation of the permanence of quantum numbers when the series electron is coupled to the atom core, but only in the peculiar *two-valuedness* of the quantum theoretical properties of each electron in the *stationary states* of an atom.

*We can, however, from this point of view use the "Aufbauprinzip" to calculate* not only the number of *stationary states*, but also *the energies in the case of strong fields* (at least that part which is proportional to the field) *additively from those of the free atom core and of the optically active electron*, where the latter can be taken from the alkali spectra. Because, in this case, both the total component $\mathbf{m}^{-}_1$ of the *angular momentum* of the *atom* along the field (in units [*h-bar*]) as well as the component $\mathbf{m}^{-}_2$ of the *magnetic moment* of the *atom* in the same direction (in Bohr magnetons) are equal to the sum of the quantum numbers $m_1$ and $m_2$ of the single *electrons*:

$$\mathbf{m}^{-}_1 = Sm_1, \qquad \mathbf{m}^{-}_2 = Sm_2. \qquad (1)$$

The latter can independently run through all values corresponding to the values of the *angular momentum quantum numbers* $k_1$ and $k_2$ of the *electrons* in the *stationary state* of the *atom* considered. ($\mathbf{m}^{-}_2 oh$ is here thus the part of the *energy* of the *atom* proportional to the *field strength*; o = *Larmor frequency*.)

Let us consider as an example the two *s-terms* (singlet- and triplet S-term) of the *alkaline earths*. To begin with it is sufficient to consider only the two *valence electrons*, as the contribution of the other electrons to the sums in (1) vanish when taken together. According to our general assumption we must for each of the two *valence electrons* take (independently of the other electron) the values $m_1 = - 1/2$, $m_2 = - 1$ and $m_1 = 1/2$, $m_2 = 1$ of the s-terms of the alkalis. According to (1) we then get the following values for the quantum numbers $\mathbf{m}^{-}_1$ and $\mathbf{m}^{-}_2$ of the *total atom*:

$$\mathbf{m}^{-}_1 = - 1/2 - 1/2, \quad - 1/2 + 1/2, \quad + 1/2 - 1/2, \quad + 1/2 + 1/2$$
$$\mathbf{m}^{-}_2 = - 1 - 1, \quad - 1 + 1, \quad 1 - 1, \quad 1 + 1,$$

or

$$\mathbf{m}^{-}_1 = - 1 \quad 0 \quad 1$$
$$\mathbf{m}^{-}_2 = - 2 \quad 0, 0 \quad 0\ 2$$

[Corresponding to one term with j = 0 and one with j = 1 in *weak fields*.][a]

$^{a}$ One notes that one must assign to the two cases $m_1 = -1/2$ for the first and $m_2 = 1/2$ for the second *electron*, or $m_1 = +1/2$ for the first and $m_2 = -1/2$ for the second *electron* two different terms (as far as the part of the energy independent of the field is concerned). This is perhaps a blemish of the classification given here. It will later on, however, turn out that if the inner and the outer *valence electron* are equivalent, these two terms are in fact identical.

To obtain the p-, d-, . . . terms of the *alkaline earths*, one must combine in (1) the unchanged contribution of the first *valence electron* (S-term) in an appropriate manner with the $m_1$- and $m_1$-values of the p-, d-, . . . terms of the alkalis for the second *electron*.

The rule (1) leads in general exactly to the procedure for calculating the energy values in strong field proposed recently by Landé which has been shown by this author to give correct results also in complicated cases. According to Landé this procedure leads, for instance, to the correct Zeeman terms of neon (at least in the case of *strong fields*) if one assumes$^{b}$ that in the *atom* core there is one active *electron* in a p-term (instead of in an s-term as above) and if one lets the *optically active electron* go through s-, p-, d-, f-, . . . terms.

$^{b}$ The replacement here of a seven-shell (*atom* core of neon) by one *electron* will be given a theoretical basis in the next section.

*This result now suggests that we characterize in general each electron in an atom not only by a principal quantum number n, but also by the two auxiliary quantum numbers $k_1$ and $k_2$, even when several equivalent electrons or completed electron groups are present.* Moreover, we shall allow (also in the just-mentioned cases) in our thoughts *such a strong magnetic field that we can assign to each electron, independently of the other electrons not only the quantum numbers n and $k_1$, but also the two quantum numbers $m_1$ and $m_2$* (where the last one determines the contribution of the *electron* to the *magnetic energy* of the *atom*). The connection between $k_2$ and $m_2$ for given $k_1$ and $m_1$ must be taken from the alkali spectra.

Before we apply in the next section this *quantum theoretical classification* of the *electrons* in an *atom* to the problem of the completion of the electron groups, we must discuss in more detail the difficulties encountered by the here - proposed ideas of the complex structure and the *anomalous Zeeman effect* and the limitations of the meaning of our ideas.

First of all, these ideas do not pay proper regard to the, in many respects independent, separate appearance of the different term systems (e.g., the singlet and the triplet systems of the *alkaline earths*), which also play a role in the position of the terms of these systems and in the *Landé interval rule*. Certainly, *one cannot assume two different causes for the energy differences of the triplet levels of the alkaline earths*, both the anomaly of the *relativity correction* of the optically active electron and the dependence of the interaction between the *electron* and the *atom* core on the relative orientation of these two systems.

Incompatibility with Einstein's *theory of special relativity*.

A more serious difficulty, raising a matter of principle, is however the connection of these ideas with the *correspondence principle* which is well known to be a necessary means to explain the selection rules for the *quantum numbers* $k_1$, j, and m and the polarization of the Zeeman components. It is, to be sure, not necessary according to this principle to assign in a definite *stationary state* to each *electron* an orbit uniquely determined in the sense of usual kinematics; however, *it is necessary that the totality of the stationary states of an atom corresponds to a collection (class) of orbits with a definite type of periodicity properties*. In our case, for instance, the above-mentioned selection and polarization rules require according to the *correspondence principle* a kind of motion *corresponding to a central force orbit on which is superposed a precession of the orbital plane around a definite axis of the atom to which is added in weak external magnetic fields also a precession around an axis through the nucleus in the direction of the field.* The dynamic explanation of this kind of motion of the *optically active electron*, which was based upon the assumption of deviations of the forces between the *atom* core and the *electron* from central symmetry, *seems to be incompatible with the possibility to represent the alkali doublet (and thus also the magnitude of the corresponding precession frequency) by relativistic formulae*. The situation with respect to the kind of motion in the case of *strong fields* is similar.

The difficult problem thus arises how to interpret the appearance of the kind of motion of the *optically active electron* which is required by the *correspondence principle* independently of its special dynamic interpretation which has been accepted up to now *but which can hardly be retained.* There also seems to be a close connection between this problem and the question of the magnitude of the term values of the Zeeman effect (especially of the alkali spectra).

As long as this problem remains unsolved, the ideas about the complex structure and the *anomalous Zeeman effect* suggested here can certainly not be considered to be a sufficient physical basis for the explanation of these phenomena, especially as they were in many respects better reproduced in the usually accepted point of view. *It is not impossible that in the future one will succeed in merging these two points of view*. In the present state of the problem, it seemed of interest to us to pursue as far as possible also the first point of view to see what its consequences are. This is the sense in which one must consider our discussions in the next section of the application of the tentative point of view, presented here to the problem of the completion of *electron* groups in an *atom*, notwithstanding the objections which can be made against it. *We shall here draw conclusions only about the number of possible stationary states of an atom when several equivalent electrons are present, but not about the position and relative order of the term values.*

**2. *On a General Quantum Theoretical Rule for the Possibility of the Occurrence of Equivalent Electrons in an Atom.***

*It is well known that the appearance of several equivalent electrons, that is, electrons which are fully equivalent both with respect to their quantum numbers and with respect to their binding energies, in an atom is possible only under special circumstances which are closely connected with the regularities of the complex structure of spectra.* For instance, the ground state of the *alkaline earths* in which the two *valence electrons* are equivalent corresponds to a singlet S-term, while in those *stationary states* of the *atom* which belong to the triplet system the *valence electrons* are never bound equivalently, as the lowest triplet s-term has a *principal quantum number* exceeding that of the *ground state* by unity. Let us now as second example consider the *neon spectrum*. This consists of two groups of terms with different series limits, corresponding to different states of the *atom* core. The first group, belonging to the removal of an *electron* with the quantum numbers $k_1 = 2$, $k_2 = 1$ from the *atom* core can be considered to be composed of a singlet and a triplet system, while the second group, belonging to the removal of an electron with $k_1 = k_2 = 2$ from the *atom* core, can be said to be a triplet and quintet system. The ultraviolet resonance lines of neon have not yet been observed, but there can hardly be any doubt that the *ground state* of a Ne-atom must be considered to be a p-term as far as its combination with the known *excited states* of the *atom* is concerned; in accordance with the unique definiteness and the diamagnetic behavior of the inert gas configuration there can be only one such term, namely with the value $j = 0$.[c]

[c] As already indicated, the value of j is defined here and henceforth as the maximum value of the quantum number $m_1$.

As the only p-terms with $j = 0$ are the (lowest) triplet terms of the two groups, we can thus conclude that for Ne for the value 2 of the *principal quantum number* only those two triplet terms exist and moreover are identical for both groups of terms.

In general, we can thus expect *that for those values of the quantum numbers n and $k_1$ for which already some electrons are present in the atom, certain multiplet terms of spectra are absent or coincide*. The question arises what quantum theoretical rules decide this behavior of the terms.

As is already clear from the example of the neon spectrum, this question is closely connected with the problem of the completion of *electron groups* in an *atom*, which determines the lengths 2, 8, 18, 32, . . . of the periods in the periodic table of the elements. This completion consists in that an n-quantum *electron group* neither through emission or absorption of radiation nor through other external influences is able to accept more than $2n^2$ electrons.

It is well known that Bohr in his *theory of the periodic table*, which contains a unified summary of spectroscopic and chemical data and especially a quantum theoretical basis for the occurrence of chemically similar elements such as the platinum and iron metals and the rare earths in the later period of the table, has introduced a subdivision of these *electron groups* into subgroups. By characterizing each *electron* in the *stationary states* of the *atom* by analogy with the stationary states of a central force motion by a symbol $n_k$ with k [less than or equal to] n, he obtained in general for an *electron group* with a value n of the *principal quantum number* n subgroups. In this way Bohr was led to the scheme of the structure of the inert gases given in Table 1. He has, however, emphasized himself that the equality, assumed here, of *the number of electrons in the different subgroups of a maingroup is highly hypothetical* and that for the time being no complete and satisfying theoretical explanation of the completion of the electron groups in the atom, and especially of the period lengths 2, 8, 18, 32, . . . in the *periodic table* could be given.

. . .

Recently essential progress was made in the problem of the completion of the electron groups in an atom by the considerations of E. C. Stoner[6].

[6] Stoner, E. C. (1924). *Phil. Mag.*, 48, 719.

This author suggests first of all a scheme for the atomic structure of the inert gases in which in contrast to Bohr no opening of a completed subgroup is allowed by letting other electrons of the same main group be added to it, so that the number of electrons in a closed subgroup depends only on the value of k, but not on the value of n, that is, on the existence of other subgroups in the same main group. ...

. . .

We can now make this idea of Stoner's more precise and more general, if we apply the ideas about the complex structure of the spectra and the *anomalous Zeeman effect*, discussed in the previous section, to the case where equivalent *electrons* are present in an *atom*. In that case *we arrived, on the basis of an attempt to retain the permanence of quantum numbers, at a characterization of each electron in an atom by both the principal quantum number n and the two auxiliary quantum numbers $k_1$ and $k_2$.* In *strong magnetic* *fields* also an *angular momentum quantum number* $m_1$ was added to this for each *electron* and, furthermore, one can use apart from $k_1$ and $m_1$ also a *magnetic moment quantum number* $m_2$, *instead of* $k_2$. First of all, we see that the use of the two quantum numbers $k_1$ and $k_2$ for each *electron* is in excellent agreement of Stoner's subdivision of Bohr's subgroup. Secondly, by considering the case of *strong magnetic fields* we can reduce Stoner's result, that *the number of electrons in a completed subgroup is the same as the number of the corresponding terms of the Zeeman effect of the alkali spectra*, to the following more general rule about the occurrence of equivalent electrons in an atom:

> *There can never be two or more equivalent electrons in an atom for which in strong fields the values of all quantum numbers* $n$, $k_1$, $k_2$, $m_1$ *(or, equivalently,* $n$, $k_1$, $m_1$, $m_1$*) are the same. If an electron is present in the atom for which these quantum numbers (in an external field) have definite values, this state is "occupied".*

We must bear in mind that the *principal quantum number* occurs in an essential way in this rule; of course, several (not equivalent) *electrons* may occur in an *atom* which have the same values of the *quantum numbers* $k_1$, $k_2$, $m_1$, but have different values of the *principal quantum number* $n$.

We cannot give a further justification for this rule, but it seems to be a very plausible one. It refers, as mentioned, first of all to the case of *strong fields*. However, from thermodynamic arguments (invariance of statistic weights under adiabatic transformations of the system) it follows that the number of *stationary states* of an *atom* must be the same in strong and weak fields for given values of the numbers $k_1$ and $k_2$ of the separate electrons and a value of $\mathbf{m}^{-}{}_1 = Sm_1$ (see (1)) for the whole *atom*. We can therefore also in the latter case make definite statements about the number of *stationary states* and the corresponding values of j (for a given number of equivalent electrons belonging to different values of $k_1$ and $k_2$). We can thus find the number of possibilities of realizing various *incomplete electron shells* and give an unambiguous answer to the question posed at the beginning of this section about the absence or coincidence of certain *multiplet terms in spectra* for values of the *principal quantum number* for which several equivalent *electrons* are present in an *atom*. We can, however, only say something about the number of terms and the values of their quantum numbers, but not about their magnitude and about interval relations.

We must now show that the consequences of our rule agree with experiment in the simplest cases. We must wait and see whether it will also prove itself in comparison with experiment in more complicated cases or whether it will need modifications in that case; this will become clear when complicated spectra are sorted out.

First of all, we see that Stoner's result and with it the period lengths 2, 8, 18, 32, . . . are immediately included in a natural way in our rule. Clearly, for given $k_1$ and $k_2$ there cannot be more equivalent *electrons* in an *atom* than the appropriate value of $m_1$, (that is, $2k_2$) and in the completed group there corresponds exactly one *electron* to each of these values of $m_1$. Secondly, it turns out that our rule has an immediate consequence that the triplet s-term with the same *principal quantum number* as the *ground state* is absent for the *alkaline earths*. If we investigate the possibilities for the equivalent binding of two *electrons* in s-terms (in that case we have thus $k_1 = 1$ and $k_2$ can also only have the value 1), according to our rule the cases are excluded in strong fields where both *electrons* have $m_1 = 1/2$ or both have $m_1 = -1/2$; rather, we can only have $m_1 = 1/2$ for the first *electron* and $m_1 = -1/2$ for the second *electron*, or the other way round[f] so that the *quantum number* $m_1 = Sm_1$ for the *total atom* can only have the value 0.

> [f] The second case corresponds to an interchange of the two equivalent *electrons* and gives us therefore here no new *stationary state* (compare the footnote lettered a). However, in this two-fold realizability of the *quantum state* considered is contained the fact that its statistical weight with respect to the exchangeability of the two *electrons* must be multiplied by two (compare also the discussion of statistical weights by Stoner)[6].

Therefore, also in *weak fields* (or when there is no field) only the value $j = 0$ is possible (singlet S-term).

We now investigate the case that one *electron* is removed from a closed shell, as will occur in X-ray spectra. Clearly when an *electron* is missing from one of Stoner's part-subgroups, the case is always possible that no *electron* is present with the value $m_1$; we call this the "*hole-value*" of $m_1$. The other *electrons* are then uniquely divided over the other values of $m_1$ so that for each of those values we have one *electron*. The sum of these other values of $m_1$ and thus the *quantum number* of the *total atom* is clearly in each case equal to the opposite of the *hole-value* of $m_1$. If we let it go through all possible values and take into account that an *electron* can be removed from every part-subgroup, we see that in *strong fields* the multiplicity of the *hole-values* of $m_1$ and thus also that of the values of is the same as that of the $m_1$ value of a single *electron*. Due to the invariance of statistical weights, it follows thus also for *weak fields* that the numbers of *stationary states* and of j-values of single ionized closed *electron shells* (X-ray spectra) are the same as in the alkali spectra, in accordance with experiment.

This is a special case of a general *reciprocity law*: *For each arrangement of electrons there exists a conjugate arrangement in which the hole-values of $m_1$ and the occupied values of $m_1$ are interchanged.* This interchange may refer to a single part-subgroup while the other part-subgroups are unchanged, or to a *Bohr subgroup*, or to the whole of a main group, since the different part-subgroups are completely independent of one another as far as possible arrangements are concerned. *The electron numbers of the two conjugate arrangements add up to the number of electrons in the completed state of the group (or subgroup) considered, while the j-values of the two arrangements are the same*. The latter follows from the fact that the sum of the *hole-values* of $m_1$ of an arrangement always is the opposite of the sum of the occupied $m_1$-values. Therefore, the *quantum numbers* $\bar{\mathbf{m}}_1$ of the *whole atom* are the opposite of one another for conjugate arrangements. As the j-values are defined as the upper limit of the set of $\bar{\mathbf{m}}_1$ -values, and as this set is symmetric around zero, it follows that the j-values are the same (compare the examples discussed below). Because of this *periodicity law* to some extent the relations at the end of a *period* of the *periodic table* reflect those at the beginning of a *period*. We must emphasize, however, that this for the time being refers only to the number of *stationary states* of the *shell* in question and the values of their *quantum numbers*, whereas *we can say nothing about the magnitude of their energies or about interval relations*[g].

[g] However, because of the equality of the number of $m_2$-values for conjugate arrangements it follows that also in weak fields the "g-sums" (taken over terms with the same j) of the appropriate terms are the same.

As an application of our rule, *we shall discuss now the special case of the gradual formation of the eight-shell (where of the principal quantum number considered no electrons with k = 2 are present in the ground state)*; this gives us at the same time another example of the just-derived *reciprocity rule*. The binding of the first two electrons in this shell has already been discussed and in what follows we shall assume for the sake of simplicity that no electron is missing from the $k_1$ = 1 subgroup so that it is closed (…). According to Stoner, for the following elements until the completion of the *eight-shell* (e.g., from B to Ne), the *ground state* will always be a *p-term*, in agreement with all experimental data up to now. Especially follows the alkali-like spectrum, corresponding to the binding of the third electron of the *eight-shell*, with the well-known absence of the *s-term* with the same *principal quantum number* as the *ground state*.

We can thus immediately go over to the binding of the fourth *electron* of the *eight-shell*, which appears in the not-yet analyzed arc spectrum of *carbon* and the partially already unraffled arc spectrum of *lead*. According to the *Landé-Heisenberg branching rule* (see previous section) the corresponding spectrum should have in general the same structure as the *neon* spectrum, that is, consist of a singlet-triplet group and a triplet-quintet group with different series limits, corresponding to the *$2p_1$ - and the $2p_2$ - doublet term* of the ion considered. We shall show, however, that according to our rules these spectra must differ essentially, as far as the number and *f-values* of the *p-terms* of the maximum *principal quantum number* (n = 2 for C, n = 6 for Pb) is concerned, from the Ne-spectrum (where, as we mentioned at the beginning of this section, apart from the ground state with $j = 0$ no further *p-term* exists with *principal quantum number* 2); this is in contrast to the structure of the excited states which we expect to be similar.

We must distinguish three cases, according to the number of electrons in the two part-subgroups with $k_1 = 2$, $k_2 = 1$ and with $k_1 = 2$, $k_2 = 2$ over which we must distribute two *electrons* (we have already assumed that the first two *electrons* are bound in *s-terms*, $k_1 = k_2 = 1$).

(a) Two equivalent $n_{21}$-*electrons*: Corresponding to the $p_1$-*term* of the alkalis $m_1$ can for this part-subgroup only take on the two values $m_1 = \hat{A} \pm 1/2$. It is thus closed in this case with $\mathbf{m}^-_1 = 0$ and $j = 0$.

(b) One $n_{21}$- and one $n_{22}$-*electron*: For the second part-subgroup $m_1$ can, corresponding to the $p_2$-*term* of the alkalis take on the four values $\hat{A} \pm 1/2$, $\hat{A} \pm 3/2$ and these can be combined in all possible ways with the above-mentioned values $m_1 = +1/2$ of the first *electron*, since the two *electrons* are in different part-subgroups and are thus not equivalent[h].

[h] Because of this we must count the case $m_1 = +1/2$ for the first and $m_1 = -1/2$ for the second *electron* different from the case $m_1 = -1/2$ for the first and $m_1 = +1/2$ for the second *electron*. Compare the footnote lettered a.

We have thus $\mathbf{m}^-_1 = (-3/2, -1/2, 1/2, 3/2) + (-1/2, 1/2)$

$\qquad = \hat{A} \pm (3/2 + 1/2), \hat{A} \pm (3/2 - 1/2), \hat{A} \pm (1/2 + 1/2), \hat{A} \pm (1/2 - 1/2)$

$\qquad = \hat{A} \pm 2, \hat{A} \pm 1, \hat{A} \pm 1, 0, 0.$

From this we see immediately that the terms split in two series with [absolute value] $\mathbf{m}^-_1$ [less than or equal to] 2 and with $\mathbf{m}^-_1$ [less than or equal to] 1. In the field free case these correspond clearly to two terms: one with $j = 2$, and one with $j = 1$.

(c) Two equivalent $n_{22}$-*electrons*: According to our rule the $m_1$-values of the two *electrons* must be different and we find for the possible values of $\mathbf{m}^-_1$:

$\qquad \mathbf{m}^-_1 = \hat{A} \pm (3/2 + 1/2), \hat{A} \pm (3/2 - 1/2), (3/2 - 3/2), (1/2 - 1/2)$

$\qquad = \hat{A} \pm 2, \hat{A} \pm 1, 0, 0.$

If there is no magnetic field, we find thus one term with $j = 2$ and one with $j = 0$.

*Altogether we find thus for the four-shell five different p-terms with maximum principal quantum number, of which two have $j = 2$, one $j = 1$, and two $j = 0$.*

We can say nothing about the energies or the interval relations of this group of terms. However, we can make definite statements about the Zeeman splittings of these terms to be expected.

By substituting the $m_2$-values (taken from the Zeeman terms of the *alkalis* in *strong fields*) for the separate *electrons* corresponding to the given $m_1$-values, we find from rule (1) the Zeeman splittings for the five *p-terms* of the *four-shell* in strong fields:

| | | | | | |
|---|---|---|---|---|---|
| $\bar{m}_1$ | –2 | –1 | 0 | 1 | 2 |
| $\bar{m}_2$ | –3, –2 | –2, –1, 0 | 0, 0, 0, 0, 0 | 1, 1, 2 | 2, 3 |

Using the same rule applied by Landé to higher-order multiplets, one obtains from this for the determination of the sum of the *g-values* for the two j = 2 terms (denoted by $Sg_2$) and for the *g-value* for the j = 1 term (denoted by $g_1$) the equations

$$2Sg_2 = 2 + 3 = 5, \qquad Sg_2 + g_1 = 1 + 1 + 2 = 4,$$

or

$$Sg_2 = 5/2, \qquad g_1 = 3/2.$$

The earliest test of this theoretical result for the *four-shell* is possible for *lead*. Observations certainly show four *p-terms*, while the existence of a fifth *p-term* is doubtful. So far unpublished measurements by E. Back of a few *lead* lines make it, moreover, very likely that the first four *p-terms* have *j-values* 2, 2, 1, 0, and that the *g-values* of these terms also agree with the theoretically expected ones.

Let us now return to the discussion of the gradual construction of the *eight-shell*. By means of the *reciprocity rule*, applied to the whole of the *Bohr subgroup* with k = 2, which contains in its closed state six *electrons*, we can immediately apply the results obtained for the *four-shell* to the number of possibilities to realize the *six-shell* (from electrons with $k_1 = 2$), which occurs, for instance, for O. The following cases of the *six-shell* are clearly conjugate to the cases (a), (b), and (c):

(a) Four equivalent $n_{22}$-*electrons* (two empty spaces in the $n_{21}$-group). This part-subgroup is closed; hence as before sub (a) one term with j = 0.

(b) One $n_{21}$-, three equivalent $n_{22}$-*electrons* (one empty space in the $n_{21}$-, and one empty space in the $n_{22}$-group). As before: one term with j = 2 and one term with j = 1.

(c) Two equivalent $n_{21}$- and two equivalent $n_{22}$-*electrons* (two empty spaces in the $n_{22}$-group). The first part-subgroup is closed. As before: one term with j = 2, one term with j = 0.

We must thus also here, for instance for *oxygen*, expect five *p-terms* with the smallest *principal quantum number*. So far only three such terms have been observed for O and S, with *j-values* of 2, 1, 0. We must wait and see whether two more *p-terms* of the same *principal quantum number* can be found from the observations, or whether our rule must be modified in this case.

As yet there are no observations about the *five-shell* (3 electrons with $k_1 = 2$) and we shall therefore give only the result of the discussion; according to our rule this shell gives rise to five *p-terms*, one term with j = 5/2, three terms with j = 3/2, and one term with j = 1/2. For the *seven-shell*, realised in x-ray spectra we get -- as we mentioned before -- terms similar to the *alkalis*.

We shall not discuss here further special cases, before experimental data are available, but it should be clear from the examples given that in each case our rule is able to give a unique answer to the question about the possibilities of realizing the different *shells* for a given number of equivalent *electrons*. To be sure, only in the simplest cases was it possible to verify that the results obtained in this way are in agreement with experiment.

*In general, we may note that the discussions given here are in principle based, as far as the transition from strong to weak or vanishing fields is concerned, upon the invariance of the statistical weights of quantum states*. However, on the basis of the results obtained there seem to be no reasons for a connection between the problem of the completion of *electron* groups in an *atom* and the *correspondence principle*, as Bohr suspected to be the case. It is probably necessary to improve the basic principles of quantum theory before we can successfully discuss the problem of a better foundation of the general rules, suggested here, for the occurrence of equivalent *electrons* in an *atom*.

**Pauli, W. (September, 1927). Zur Quantenmechanik des magnetischen Elektrons. (On the quantum mechanics of magnetic electrons.).**

*Zeit. Phys.*, 43, 601-23; https://doi.org/10.1007/BF01397326; translated by D. H. Delphenich: https://www.neo-classical-physics.info/uploads/3/0/6/5/3065888/pauli_-_the_magnetic_electron.pdf#:~:text=PauliEOnthequantummechanics.

Received May 3, 1927.

---

***Abstract.***

It will be shown how one can arrive at a formulation of the quantum mechanics of the *magnetic electron* by the Schrödinger method of eigenfunctions, with no use of double-valued functions, when one, on the basis of the Dirac-Jordan general theory of transformations, introduces the components of its *proper impulse moment* in a fixed direction as further independent variables in order to carry out the computations of its rotational degrees freedom, along with the position coordinates of any *electron*. In contradiction to classical mechanics, these variables can assume only the variables $+ \frac{1}{2} h/2\pi$ and $- \frac{1}{2} h/2\pi$, which is completely independent of any sort of external field. The appearance of the aforementioned new variables thus implies a simple splitting of the eigenfunctions into two position functions $\psi_\alpha$, $\psi_\beta$ for one *electron*, and more generally, for N *electrons* they split into $2^N$ functions, which are to be regarded as the "*probability amplitudes*" that in a well-defined stationary state of the system not only do the position coordinates of the *electrons* lie in a given infinitesimal interval, but also that the components of their *proper moments* in the chosen direction should have the given values, which are $+ \frac{1}{2} h/2\pi$ for $\psi_\alpha$ and $- \frac{1}{2} h/2\pi$ for $\psi_\beta$. Methods will be given for constructing as many simultaneous differential equations for the $\psi$ functions as their number suggests (thus, 2 or $2^N$, resp.) from a given Hamiltonian function. These equations are completely equivalent in their consequences to the matrix equations of Heisenberg and Jordan. Furthermore, in the case of many *electrons*, the solutions of the differential equations that satisfy the "*equivalence rule*" of Heisenberg and Dirac will be characterized by their symmetry properties under the exchange of the variable values for the two *electrons*.

## § 1. *Generalities on the nature of electronic magnetism in the Schrödinger form of quantum mechanics.*

The hypothesis that was first proposed by Goudsmit and Uhlenbeck in order to explain the complex structure of spectra and their anomalous Zeeman effect, according to which the *electron* takes on a *proper impulse moment* of magnitude $\frac{1}{2} h/2\pi$ and a *magnetic moment* of a *magneton*, was integrated into quantum mechanics by Heisenberg and Jordan[1] with the help of matrix calculations and then made quantitatively precise.

[1] Heisenberg, W. & Jordan, P. (April, 1926). Anwendung der Quantenmechanik auf das Problem der anomalen Zeemaneffekte. (Application of quantum mechanics to the problem of the anomalous Zeeman effect.) *Zeit. Phys.*, 37, 263-77, 369; https://doi.org/10.1007/BF01397100; (translation by D. H. Delphenich; https://neo-classical-physics.info/electromagnetism. html; also in Underwood, T. G. (2023). Quantum Electrodynamics – annotated sources, Volume I, pp. 369-381); examination of the quantum-mechanical behavior of the Uhlenbeck-Goudsmit electron spin hypothesis, assumes ratio of magnetic moment to mechanical angular momentum (g-factor) for the electron is 2, shows that Pauli-Dirac *non-relativistic* theory explains the *anomalous Zeeman effect* and the fine structure of the double spectra.

While the matrix method is mathematically equivalent to the *method of eigenfunctions* in many-dimensional space that was discovered by Schrödinger, one comes up against peculiar formal complications when one attempts to also treat the *forces* and *moments* that an *electron* experiences in an *external field* by the method of its *proper moment*. By the introduction of a further degree of freedom that corresponds to the orientation of the *proper impulse* of the *electron* in space, one actually expresses the empirically-established fact that this *momentum* has two possible *quantum positions* in an *external field*, so one is next led to *eigenfunctions* that are many-valued, and indeed, two-valued, in the rotational angle in question – e.g., the azimuth of the impulse around a spatially fixed axis. One has often supposed that this formally possible representation by means of two-valued *eigenfunctions* does not do justice to the true physical nature of things and has sought the solution to the problem in another direction. Thus, Darwin[2] has recently attempted to gather the facts that are summarized under the assumption of the *electron impulse* without the introduction of the top degrees of freedom for the *electron* that would correspond to new dimension in the configuration space, so he considered the amplitudes of the de Broglie waves as directed quantities – i.e., he considered the Schrödinger *eigenfunction* as vectorial.

[2] Darwin, C. G. (February, 1927). The Electron as a Vector Wave. *Nature*, 119, 2990, 282-4; 465; https://doi.org/10.1038/119282a0; also in Underwood, T. G. (2023). Quantum Electrodynamics – annotated sources, Volume I, pp. 465-70; preliminary report raises problems with Thomas's attempt to resolve difficulties with Uhlenbeck and Goudsmit theory of the spinning electron, when *relativity* transformation is applied to identify the "doublet effect" with the Zeeman effect gives value for the doublet separation twice as great as it should be, necessary to introduce factor two, this was the original difficulty of Uhlenbeck and Goudsmit that was removed by Thomas who showed that a rigid body when accelerating exhibits a sort of rotation on account of the kinematics of *relativity*, but this imports a foreign idea into mechanics, *relativity and rotation do not take at all kindly to one another*, suggests electron should be considered as a wave of two components. *wave functions* with two components should be interpreted in terms of a vector, possible to construct by a much more inductive process a system of waves of a vector character which completely reproduces the doublet spectra.

From his attempt to follow this, on first glance promising, path to its ultimate consequences, he came to complications that were again connected precisely with the number two for the positions of the *electron* in an *external field*, and which I do not believe one can surmount. On the other hand, a representation of the quantum-mechanical behavior of the *magnetic electron* using the method of *eigenfunctions*, especially in the case of atoms with many electrons, is very desirable for the fact that the variety that is realized in nature alone results for the solutions of the quantum-mechanical equations that fulfill the "*equivalence rule*" for all of the possible solutions of the present theory of Heisenberg and Dirac[3] most clearly with the help of *symmetry properties* of the *eigenfunctions* under the exchange of the variable values that belong to two *electrons*.

[3] Heisenberg, W. (1926). *Zeit. Phys*. 38, 411; 39, 499; (1927). *Zeit. Phys*., 41, 239; Dirac, P. A. M. (October, 1926). On the Theory of Quantum Mechanics. *Roy. Soc. Proc., A*, 112, 762, 661-77; https://doi.org/10.1098/rspa.1926.0133.JSTOR 94692; also in Underwood, T. G. (2023). Quantum Electrodynamics – annotated sources, Volume I, pp. 416-31; *relativistic* treatment of Schrodinger's wave theory in which the time and its *conjugate momentum* are treated from the beginning on the same footing as the other variables, applies *relativistic* formulation to system containing an atom with two electrons, finds that if the positions of the two electrons are interchanged the new state of the atom is physically indistinguishable from the original one, in order that theory only enables calculation of *observable quantities* must treat (*mn*) and (*nm*) as only one *state*, must infer that *unsymmetrical* functions of the co-ordinates (and momenta) of the two electrons cannot be represented by matrices, *symmetrical functions* such as the total *polarizations* of the atom can be considered to be represented by matrices without inconsistency, these matrices are by themselves sufficient to determine all the physical properties of the system, *theory of uniformizing variables introduced by the author can no longer apply*, allows two solutions satisfying necessary conditions, one leads to Pauli's *exclusion principle* that not more than one electron can be in any given orbit, the other leads to the Einstein-Bose statistical mechanics, accounts for the *absorption* and stimulated *emission* of radiation by an atom, elements of matrices representing total *polarization* determine *transition probabilities*, *cannot be applied to spontaneous emission*; applies to theory of ideal gas and to problem of an atomic system subjected to a perturbation from outside (e.g., an incident electromagnetic field) which can vary with time in an arbitrary manner, *with neglect of relativity mechanics* accounts for the absorption and stimulated emission of radiation and shows that the elements of the matrices representing the total polarization determine the *transition probabilities*.

We would now like to show that by a suitable use of the formulation of *quantum mechanics*, as described by Jordan and Dirac4, which makes use of general canonical transformations of the Schrödinger functions $\psi$, a *quantum-mechanical representation* of the behavior of *magnetic electrons* by the *method of eigenfunctions* is, in fact, possible, without appealing to many-valued functions.

[4] Jordan, P. (1927). *Zeit. Phys.*, 40, 809; (1926). *Gött. Nachr.*, pp. 161; cf., also London, F. (1926). *Zeit. Phys.*, 40, 193; Dirac, P. A. M. (January, 1927). The Physical Interpretation of the Quantum Dynamics. *Roy. Soc. Proc., A*, 113, 765, 621-41; https://doi.org/10.1098/rspa.1927.0012; also in Underwood, T. G. (2023). Quantum Electrodynamics – annotated sources, Volume I, pp. 443-50; *non-relativistic* matrix mechanics, Heisenberg's original matrix mechanics assumed that the elements of the diagonal matrix that represents the energy are the *energy levels* of the system, and the elements of the matrix that represents the total polarization, which are periodic functions of the time, determine the *frequencies* and *intensities* of the spectral lines in analogy to classical theory, in *Schrodinger's wave representation* physical results are based on assumption that the square of the *amplitude* of the wave function can be interpreted as a probability, enables probability of a *transition* being produced in a system by an arbitrary external perturbing force to be worked out, this paper provides a *general theory of obtaining physical results from quantum theory*, it shows all the physical information that one can hope to get from quantum dynamics and provides a general method for obtaining it, replaces special assumptions previously used, requires a theory of the more general schemes of matrix representation in which the rows and columns refer to any set of constants of integration that commute and of the laws of transformation from one such scheme to another, *does not take relativity mechanics into account*, counts time variable wherever it occurs as a parameter (a c-number), *transformation equations* that satisfy *quantum conditions* and *equations of motion*, *eigenfunctions* of Schrodinger's wave equation as *transformation functions* that enable transformation from scheme of matrix representation to scheme in which Hamiltonian is a diagonal matrix, dynamical variables represented by matrices whose rows and columns refer to the initial values of the *action variables* or to the *final values*, coefficients that enable transformation from one set of matrices to the other are those that determine the *transition probabilities*.

Namely, one achieves this by adding the components of the proper impulse of each *electron* in a fixed direction (instead of the rotational angle that is conjugate to it) as new independent variables, along with the position coordinates q of the *electron* center of mass. As we will see in what follows in § 2 in the special case of a single *electron*, in any *quantum state* (in the absence of degeneracy) the *eigenfunction* generally splits into two functions $\psi_\alpha(q_k)$ and $\psi_\beta(q_k)$, of which the square of the absolute value, when multiplied by $dq_1, \ldots, dq_f$, yields the probability that in this *state*, not only should the $q_k$ lie in the prescribed interval ($q_k$, $q_k + dq_k$), but also that the components of the *proper impulse* in the chosen fixed direction must assume the values + ½ h/2π (– ½ h/2π, resp.). It will be further shown how, by a suitable choice of linear operators for the components $s_x$, $s_y$, $s_z$ of the *proper moment* in a prescribed coordinate axis-cross, differential equations for the *eigenfunctions* of the *magnetic electron* in an *external force field* can be constructed that are equivalent to the matrix equations of Heisenberg and Jordan. This will be performed in detail in § 4 for the case of an *electron* at rest in an *external magnetic field* and for a *hydrogen atom*. It will be further investigated how the *eigenfunctions* $\psi_\alpha$, $\psi_\beta$ transform under changes of the coordinate axes (§ 3).

The differential equations for the *eigenfunctions* of the *magnetic electron* that are given in the present paper can be regarded as only provisional and approximate, since they, like the Heisenberg-Jordan matrix formulation, are not written down in a *relativistically-invariant* way, and for the *hydrogen atom* they are valid only in the approximation in which the dynamical behavior of the *proper moment* can be considered to be a secular perturbation (in the classical theory: averaged over the orbit). In particular, it thus not possible to calculate *quantum-mechanically* the corrections that are proportional to higher powers of $\alpha^2 Z^2$ ($\alpha = 2\pi e^2/hc$ = fine structure constant) in the amounts of the hydrogen *fine-structure splitting*, such as the empirically established amounts for the Röntgen spectra that are given so well by the Sommerfeld formula. These difficulties, which are still obstacles to the solution of this problem to this day, will be discussed briefly in § 4.

Thus, whether or not the formulation of the *quantum mechanics* of the *magnetic electron* that is communicated here is still completely unsatisfactory in that regard, on the other hand, it affords the advantage that in the case of many *electrons* (in contrast to the Darwin formulation), as will be shown in § 5, it gives rise to no new difficulties at all and also allows one, like Heisenberg, to easily formulate necessary *symmetry properties* of the *eigenfunction* in order for it to fulfill the "*equivalence rule*". In particular, on this basis, it already seems to me justified to communicate the method proposed at the present point in time, and one can perhaps hope that it will also prove useful in the unsolved problem of the calculation of the *hydrogen fine structure* in higher approximations.

**§ 2. …**

**(7) Elementary and composite particles with mass attract each other through the gravitational interaction or gravitational force, according to Newton's law of gravitation.**

**Gravity.**

**Underwood, T. G. (2023).** ***Gravity***: Newton's universal law of gravitation, pp. 74-7: "While Newton was able to articulate his *Law of Universal Gravitation* and verify it experimentally, he could only calculate the relative *gravitational force* in comparison to another force. It was not until Henry Cavendish's verification of the *Gravitational Constant* that the *Law of Universal Gravitation* received its final form:

$$F = GMm/r^2 = 6.674 \times 10^{-11}\ Mm/r^2\ N\ (\textit{SI units})$$

where F represents the *force* in Newtons, M and m represent the two *masses* in kilograms, and r represents the separation in meters. G represents the *Gravitational Constant*, which has a value of $6.674 \times 10^{-11}\ N\ (m/kg)^2$. Because of the magnitude of G, *gravitational force* is very small *unless large masses or short distances are involved*."

[The *kilogram* is the base unit of mass in the International System of Units (SI), having the unit symbol kg. It is a widely used measure in science, engineering and commerce worldwide, and is often simply called a kilo colloquially. It means 'one thousand grams'.

The kilogram is defined in terms of the Planck constant, the second, and the meter, both of which are based on fundamental physical constants. This allows a properly equipped metrology laboratory to calibrate a mass measurement instrument such as a Kibble balance as the primary standard to determine an exact kilogram mass.

The kilogram was originally defined in 1795 during the French Revolution as the mass of one liter of water. The current definition of a kilogram agrees with this original definition to within 30 parts per million. In 1799, the platinum *Kilogramme des Archives* replaced it as the standard of mass. In 1889, a cylinder of platinum-iridium, the International Prototype of the Kilogram (IPK), became the standard of the unit of mass for the metric system and remained so for 130 years, before the current standard was adopted in 2019.

The kilogram is defined in terms of three fundamental physical constants:

- a specific atomic transition frequency $\Delta\nu_{Cs}$, which defines the duration of the second,

- the speed of light $c$, which when combined with the second, defines the length of the meter,
- and the Planck constant $h$, which when combined with the meter and second, defines the mass of the kilogram.

The formal definition according to the General Conference on Weights and Measures (CGPM) is: The *kilogram*, symbol kg, is the SI unit of mass. It is defined by taking the fixed numerical value of the Planck constant $h$ to be $6.62607015 \times 10^{-34}$ when expressed in the unit J·s, which is equal to $kg \cdot m^2 \cdot s^{-1}$, where the meter and the second are defined in terms of $c$ and $\Delta\nu_{Cs}$.

Defined in term of those units, the kg is formulated as:

$$1\ kg = (299792458)^2/(6.62607015 \times 10^{-34})(9192631770) h\Delta\nu_{Cs}/c^2$$
$$= 917097121160018/621541050725904751 0^{42} h\Delta\nu_{Cs}/c^2$$
$$\approx (1.475521399735270 \times 10^{40}) h\Delta\nu_{Cs}/c^2\ .$$

This definition is generally consistent with previous definitions: the *mass* remains within 30 ppm of the *mass* of one liter of water.

The *newton* (N) is the unit of force in the International System of Units (SI). It is defined as 1 kg . m . $s^{-2}$, the force which gives a mass of 1 kilogram an acceleration of 1 meter per second per second.]

**Underwood, T. G. (2023). *Gravity*:** Newton's calculation of Kepler's laws, pp. 71-3: "Newton used his mathematical description of gravity to derive Kepler's laws of planetary motion (which Kepler had first obtained empirically), to account for tides, the trajectories of comets, the precession of the equinoxes and other phenomena, eradicating doubt about the Solar System's heliocentricity. He demonstrated that the motion of objects on Earth and celestial bodies could be accounted for by the same principles. Newton's inference that the Earth is an oblate spheroid was later confirmed by the geodetic measurements of Maupertuis, La Condamine, and others, convincing most European scientists of the superiority of Newtonian mechanics over earlier systems."

In his *Principia*, Newton computed the *acceleration* of a planet moving according to Kepler's first and second laws.

1. The *direction* of the *acceleration* is towards the Sun.

2. The *magnitude* of the *acceleration* is inversely proportional to the square of the planet's distance from the Sun (the *inverse square law*).

Newton defined the *force* acting on a planet to be the product of its *mass* and the *acceleration*. So:

1. Every planet is attracted towards the Sun.

2. The force acting on a planet is directly proportional to the *mass* of the planet and is inversely proportional to the square of its *distance* from the Sun.

The Sun plays an unsymmetrical part, which is unjustified. So, Newton assumed, in his *law of universal gravitation*:

1. All bodies in the Solar System attract one another.

2. The force between two bodies is in direct proportion to the product of their *masses* and in inverse proportion to the square of the *distance* between them.

As the planets have small *masses* compared to that of the Sun, the orbits conform approximately to Kepler's laws. Newton's model improves upon Kepler's model, and fits actual observations more accurately.

By *Newton's second law*, the *gravitational force* that acts on the planet is:

$$F = m_{planet}\, d^2\mathbf{r}/dt^2 = -\, m_{planet}\, \alpha\, r^{-2}\, \underline{\mathbf{r}}$$

where $m_{planet}$ is the *mass* of the planet and $\alpha$ has the same value for all planets in the Solar System. According to *Newton's third law*, the Sun is attracted to the planet by a *force* of the same magnitude. Since the *force* is proportional to the *mass* of the planet, under the symmetric consideration, it should also be proportional to the *mass* of the Sun, $m_{Sun}$. So

$$\alpha = G\, m_{Sun}$$

where G is the *gravitational constant*.

The *acceleration* of Solar System body number i is, according to Newton's laws:

$$d^2\mathbf{r}/dt^2 = G \sum_{j \neq I} m_j\, r_{ij}^{-2}\, \underline{\mathbf{r}}_{ij}$$

where $m_j$ is the *mass* of body j, $r_{ij}$ is the *distance* between body i and body j, $\underline{\mathbf{r}}_{ij}$ is the unit vector from body i towards body j, and the vector summation is over all bodies in the Solar System, besides i itself.

In the special case where there are only two bodies in the Solar System, Earth and Sun, the *acceleration* becomes

$$d^2\mathbf{r}/dt^2{}_{Earth} = G\, m_{Moon,Earth}\, r^{-2}\, \underline{\mathbf{r}}_{Moon,Earth}$$

which is the *acceleration* of the Kepler motion. So, the Earth moves around the Sun according to Kepler's laws.

Using *Newton's law of gravitation, Kepler's third law* ($a^3/T^2$ = const.) can be found in the case of a circular orbit *by setting the centripetal force equal to the gravitational force*:

$$mr\omega^2 = GmM/r^2.$$

Then, expressing the angular velocity ω in terms of the orbital period T and then rearranging, results in *Kepler's third law*:

$$mr\,(2\pi/T)^2 = GmM/r^2 \rightarrow T^2 = (4\pi^2/GM)\, r^3 \rightarrow T^2 \propto r^3.$$

A more detailed derivation can be done with general elliptical orbits, instead of circles, as well as orbiting the center of *mass*, instead of just the large *mass*. This results in replacing a circular radius, r, with the semi-major axis, *a*, of the elliptical relative motion of one *mass* relative to the other, as well as replacing the large *mass* M with M + m . However, with planet *masses* being so much smaller than the Sun, this correction is often ignored. The full corresponding formula is:

$$a^3/T^2 = G(M + m)/4\pi^2 \approx GM/4\pi^2 \approx 7.496 \text{ x } 10^{-6}\ AU^3t^{-2} = \text{const.}$$

where M is the *mass* of the Sun (= $1.9885 \times 10^{30}$ kg), m is the *mass* of the planet (in kg), G is the *gravitational constant* (= $6.674 \times 10^{-11}\ N \cdot m^2 . kg^{-2}$), T is the *orbital period* (in days), *a* is the elliptical semi-major axis (in meters), and AU is the astronomical unit (= 149,597,870,700 m), the average *distance* from earth to the sun.

**Newton, I. (July, 1687).** ***Philosophiæ Naturalis Principia Mathematica.*** **(The Mathematical Principles of Natural Philosophy.) Book I: The Motion of Bodies.**

1st Edition, London; 2nd Edition, Cambridge, 1713; 3rd Edition, London, 1726. (In Latin); translation below of 3rd Edition by A. Motte, (1729). London.); https://en.wikisource.org/wiki/The_Mathematical_Principles_of_Natural_Philosophy_(1729); also at https://ia601604.us.archive.org/1/items/newtonspmathema00newtrich/newtonspmathema00newtrich_bw.pdf.; also in Underwood, T. G. (2023). *Gravity*, Part I, pp. 50-3.

*Philosophiæ Naturalis Principia Mathematica* (Mathematical Principles of Natural Philosophy) is a work in three books written in Latin, first published July 5, 1687, with encouragement and financial help from Edmond Halley. After annotating and correcting his personal copy of the first edition, Newton published two further editions, during 1713 with errors of the 1687 corrected, and an improved version in 1726. The *Principia* includes *Newton's three laws of motion*, laying the foundation for classical mechanics; *Newton's law of universal gravitation*; and a derivation of *Johannes Kepler's laws of planetary motion* (which Kepler had first obtained empirically). In Book I: The Motion of Bodies, Newton addresses the motion of bodies attracted to each other by centripetal forces.

---

CONTENTS

Section VII: Concerning the rectilinear ascent and descent of bodies
Section VIII: Of the invention of orbits wherein bodies will revolve, being acted upon by any sort of centripetal force
Section IX: Of the motion of bodies in movable orbits; and of the motion of the apsides
Section X: Of the motion of bodies in given superficies; and of the reciprocal motion of funependulous bodies
Section XI: Of the motions of bodies tending to each other with centripetal forces
Section XII: Of the attractive forces of sphaerical bodies
Section XIII: Of the attractive forces of bodies which are not of a sphaerical figure
Section XIV: Of the motion of very small bodies when agitated by centripetal forces tending to the several parts of any very great body

---

...

***Section II: Of the invention of centripetal forces.***

Proposition IX Theorem XXIII. *If two bodies S and P, attracting each other with forces reciprocally proportional to the squares of their distance, revolve about their common center of gravity; I say that the principal axis of the ellipse which either of the bodies as P describes by this motion about the other S, will be to the principal axis of the ellipse, which the same body P may describe in the same periodical time about the other body S quiescent, as the sum of the two bodies S+P to the first of two mean proportionals between that sum and the other body S.*

...

***Section XI: Of the motions of bodies tending to each other with centripetal forces***

Proposition LVII. Theorem XX. *Two bodies attracting each other mutually, describe similar figures about their common center of gravity, and about each other mutually.*

For the distances of the bodies from their common center of gravity are reciprocally as the bodies; and therefore, in a given ratio to each other; and thence by composition of ratio's, in a given ratio to the whole distance between the bodies. Now these distances revolve about their common term with an equable angular motion, because lying in the same right line they never change their inclination to each other mutually. But right lines that are in a given ratio to each other, and revolve about their terms with an equal angular motion, describe upon planes, which either rest with those terms, or move with any motion not angular, figures entirely similar round those terms. Therefore, the figures described by the revolution of these distances are similar. *Q. E. D.*"

> [Newton specifically considers forces of attraction between bodies which are equal to the reciprocal to the squares of the distance between them, which applies to the force between two electrically charged bodies, under Coulomb's law of electrical forces, as well as to gravitational forces. In Proposition LXXIV he argues that the

if the force between particles is a reciprocal of the squares of the distance between them, then the force between a sphere and a particle external to the sphere is a reciprocal of the distance between the external particle and the center of the sphere.]

...

**Section XII: Of the attractive forces of sphaerical bodies.**

Proposition LXXIV. Theorem XXXIV. *The same things supposed (if to the several points of a given sphere there tend equal centripetal forces decreasing in a duplicate ratio of the distances from the points), I say, that a corpuscle situate without the sphere is attracted with a force reciprocally proportional to the square of its distance from the center.*

For suppose the sphere to be divided into innumerable concentric spherical superficies, and the attractions of the corpuscle arising from the several superficies will be reciprocally proportional to the square of the distance of the corpuscle from the center of the sphere (by prop. 7 1.) And by composition, the sum of those attractions, that is, the attraction of the corpuscle towards the entire sphere, will be in the fame ratio. *Q. E. D.*

Cor. 1. Hence the attractions of homogeneous spheres at equal distances from the centers will be as the spheres themselves. For (by prop. 72) if the distances be proportional to the diameters of the spheres, the forces will be as the diameters. Let the greater distance be diminished in that ratio; and the distances now being equal, the attraction will be increased in the duplicate of that ratio; and therefore, will be to the other attraction in the triplicate of that ratio; that is, in the ratio of the spheres.

Cor. 2. At any distances whatever, the attractions are as the spheres applied to the squares of the distances.

Cor. 3. If a corpuscle placed without a homogeneous sphere is attracted by a force reciprocally proportional to the square of its distance from the center, and the sphere consists of attractive particles; the force of every particle will decrease in a duplicate ratio of the distance from each particle.

Proposition LXXV. Theorem XXXV. *If to the several points of a given sphere there tend equal centripetal forces decreasing in a duplicate ratio of the distances from the points; I say that another similar sphere will be attracted to it with a force reciprocally proportional to the square of the distance of the centers.*

For the attraction of every particle is reciprocally as the square of its distance from the center of the attracting sphere (by prop. 7.4.) and is therefore the same as if that whole attracting force issued from one single corpuscle placed in the center of this sphere. But this attraction is as great, as on the other hand the attraction of the same corpuscle would be, if that were itself attracted by the several particles of the attracted sphere with the same force with which they are attracted by it. But that attraction of the corpuscle would be (by prop. 7.4.) reciprocally proportional to the square of its distance from the center of the sphere; therefore, the attraction of the sphere, equal thereto, is also in the same ratio. *Q E.D.*

Cor. 1. The attractions of spheres towards other homogeneous spheres, are as the attracting spheres applied to the squares of the distances of their centers from the centers of those which they attract.

Cor. 2. The case is the same when the attracted sphere does also attract. For the several points of the one attract the several points of the other with the same force with which they themselves are attracted by the others again; and therefore since in all attractions (by law 3.) the attracted and attracting point are both equally acted on, the force will be doubled by their mutual attractions, the proportions remaining.

> [In Book III of *Principia* (below), Newton notes *that the centripetal force which arises between planets is the same as the gravitational force attracting matter to the Earth* and focusses on gravitational attraction. Newton used the Latin word gravitas (weight) for the effect that would become known as *gravity.*
> "Book III. Proposition V. Theorem V. Scholium. *The force which retains the celestial bodies in their orbits has been hitherto called centripetal force; but it being now made plain that it can be no other than a gravitating force, we shall hereafter call it gravity.*
> For the cause of that centripetal force which retains the moon in its orbit will extend itself to all the planets."]

**Newton, I. (July, 1687). *Philosophiæ Naturalis Principia Mathematica.* Book III: Of the System of the World.**

1st Edition, London; 2nd Edition, Cambridge, 1713; 3rd Edition, London, 1726. (In Latin); translation below of 3rd Edition by A. Motte, (1729). London.); https://en.wikisource.org/wiki/The_Mathematical_Principles_of_Natural_Philosophy_(1729); also at https://ia601604.us.archive.org/1/items/newtonspmathema00newtrich/newtonspmathema00newtrich_bw.pdf.; also in Underwood, T. G. (2023). *Gravity*, Part I, pp. 54-70.

In Book III, Newton notes that the *centripetal force* which arises between planets is the same as the *gravitational force* attracting matter to the Earth and focusses on gravitational attraction. He then proposes that "all bodies gravitate towards; every Planet and that the Weights of bodies towards any the same Planet, at equal distances from the center of the Planet, are proportional to the quantities of matter which they severally contain; and that there is a power of gravity tending to all bodies, proportional to the several quantities of matter which they contain; and that the force of gravity towards the several equal particles of any body, is reciprocally as the square of the distance of places from the particles".

In Proposition VI Newton provides his definition of *gravitational mass*, and in Proposition VII, together with its corollary 2, Newton restates his *universal law of gravitation*,

$$F = Gm_1m_2/r^2,$$

where F is the force, $m_1$ and $m_2$ are the masses of the objects interacting, r is the distance between the centers of the masses and G is the gravitational constant. *Newton's law of gravitation* states that every point mass in the universe attracts every other point mass with a force that is directly proportional to the product of their masses, and inversely proportional to the square of the distance between them.

---

**BOOK III: OF THE SYSTEM OF THE WORLD**

***Rules of Reasoning in Philosophy*.**

Rule I. *We are to admit no more causes of natural things than such as are both true and sufficient to explain their appearances.*

To this purpose the philosophers say that Nature does nothing in vain, and more is in vain when less will serve; for Nature is pleased with simplicity, and affects not the pomp of superfluous causes.

Rule II. *Therefore, to the same natural effects we must, as far as possible, assign the same causes.*

As to respiration in a man and in a beast; the descent of stones in Europe and in America; the light of our culinary fire and of the sun; the reflection of light in the earth, and in the planets.

Rule III. *The qualities of bodies, which admit neither intension nor remission of degrees, and which are found to belong to all bodies within the reach of our experiments, are to be esteemed the universal qualities of all bodies whatsoever.*

For since the qualities of bodies are only known to us by experiments, we are to hold for universal all such as universally agree with experiments; and such as are not liable to diminution can never be quite taken away. We are certainly not to relinquish the evidence of experiments for the sake of dreams and vain fictions of our own devising; nor are we to recede from the analogy of Nature, which uses to be simple, and always consonant to itself.

…

…

Rule IV. *In experimental philosophy we are to look upon propositions collected by general induction from, phenomena as accurately or very nearly true, notwithstanding any contrary hypotheses that may be imagined, till such time as other phenomena occur, by which they may either be made more accurate, or liable to exceptions.*

This rule we must follow, that the argument of induction may not be evaded by hypotheses

***Phenomena, or Appearances.***

Phenomenon I. *That the circumjovial planets, by radii drawn to Jupiter's center, describe areas proportional to the times of description; and that their periodic times, the fixed stars being at rest, are in the sesquiplicate proportion of their distances from, its center.* This we know from astronomical observations. For the orbits of these planets differ but insensibly from circles concentric to Jupiter; and their motions in those circles are found to be uniform. And all astronomers agree that *their periodic times are in the sesquiplicate proportion of the semi-diameters of their orbits*; and so it manifestly appears from the following table.

> [The *sesquiplicate* ratio of given terms is the ratio between the square roots of the cubes of those terms.]

…

Phenomenon II. *That the circumsaturnal planets, by radii drawn, to Saturn's center, describe areas proportional to the times of description; and that their periodic times, the fixed stars being at rest, are in the sesqniplicata proportion of their distances from its center.*

For, as Cassini from his own observations has determined, their distances from Saturn's center and their periodic times are as follows.

…

Phenomenon III. *That the five primary planets, Mercury, Venus, Mars, Jupiter, and Saturn, with their several orbits, encompass the sun.*

That Mercury and Venus revolve about the sun, is evident from their moon-like appearances. When they shine out with a full face, they are, in respect of us, beyond or above the sun; when they appear half full, they are about the same height on one side or other of the sun; when horned, they are below or between us and the sun; and they are sometimes, when directly under, seen like spots traversing the sun's disk. That Mars surrounds the sun, is as plain from its full face when near its conjunction with the sun. and from the gibbous figure which it shews in its quadratures. And the same thing is demonstrable of Jupiter and Saturn, from their appearing full in all situations; for the shadows of their satellites that appear sometimes upon their disks make it plain that the light they shine with is not their own, but borrowed from the sun.

Phenomenon IV. *That the fixed Stars being at rest, the periodic times of the five primary Planets, and (whether of the Sun about the Earth, or) of the Earth about the Sun, are in the sesquiplicate proportion of their mean distances from the Sun.*

This proportion, first observ'd by Kepler, is now receiv'd by all astronomers. For the periodic times are the same, and the dimensions of the orbits are the same, whether the Sun revolves about the Earth, or the Earth about the Sun. And as to the measures of the periodic times, all astronomers are agreed about them. But for the dimensions of the orbits, Kepler and Bullialdus, above all others, have determin'd them from observations with the greatest accuracy: and the mean distances corresponding to the periodic times, differ but insensibly from those which they have assign'd, and for the most part fall in between them; as we may see from the following table. The periodic times, with respect to the fixed Stars, of the Planets and Earth revolving about the Sun, in days and decimal parts of a day. 10759,275. 4332,514. 686,9785. 365,2565. 224,6176. 87,9692. The mean distances of the Planets and of the Earth from the Sun. According to Kepler 951000. 519650. 152350. To Bullialdus 954198. 522520. 152350. To the periodic Times 954006. 520096. 152369. According to Kepler 100000. 72400. 38806. To Bullialdus 100000. 72398. 38585. To the periodic times 100000. 72333. 38710. As to Mercury and Venus, there can be no doubt about their distances from the Sun; for they are determin'd by the elongations of those Planets from the Sun. And for the distances of the superior Planets, all dispute is cut off by the eclipses of the satellites of Jupiter. For, by those eclipses, the position of the shadow, which Jupiter projects, is determin'd; whence we have the heliocentric longitude of Jupiter. And from its heliocentric and geocentric longitudes compar'd together, we determine its distance.

Phænomenon V. *Then the primary Planets, by radii drawn to the Earth, describe areas no wise proportional to the times: But that the areas, which they describe by radii drawn to the Sun, are proportional to the times of description.*

For to the Earth, they appear sometimes direct, sometimes stationary, nay and sometimes retrograde. But from the Sun they are always seen direct, and to proceed with a motion nearly uniform, that is to say, a little swifter in the perihelion and a little slower in the aphelion distances, so as to maintain an equality in the description of the areas. This is a noted proposition among astronomers, and particularly demonstrable in Jupiter, from the eclipses of his satellites; by the help of which eclipses, as we have said, the heliocentric longitudes of that Planet, and its distances from the Sun. are determined.

Phænomenon VI. *That the Moon by a radius drawn to the Earth's center, describes an area proportional to the time of description.*

This we gather from the apparent motion of the Moon, compar'd with its apparent diameter. It is true that the motion of the Moon is a little disturbed by the action of the Sun. But in laying down these phænomena, I neglect those small and inconsiderable errors.

***Propositions.***

Proposition I. Theorem I. *That the forces by which the circumjovial planets are continually drawn off from rectilinear motions, and retained in their proper orbits, tend to Jupiter's center; and are reciprocally as the squares of the distances of the places of those planets from that center.*

The former part of this Proposition appears from Phaen. I, and Prop. II or III, Book I: the latter from Phaen. I, and Cor. 6, Prop. IV, of the same Book. The same thing we are to understand of the planets which encompass Saturn, by Phaen. II.

Proposition II. Theorem II. *That the forces by which the primary planets are continually drawn off from rectilinear motions, and retained in their proper orbits, tend to the sun; and are reciprocally as the squares of the distances of the places of those planets from the sun's center.*

The former part of the Proposition is manifest from Phaen. V, and Prop. II, Book I; the latter from Phaen. IV, and Cor. 6, Prop. IV, of the same Book. But this part of the Proposition is, with great accuracy, demonstrable from the quiescence of the aphelion points; for a very small aberration from the reciprocal duplicate proportion would (by Cor. 1, Prop. XLV, Book I) produce a motion of the apsides sensible enough in every single revolution, and in many of them enormously great.

Proposition III. Theorem III. *That the force by which the moon is retained in its orbit tends to the earth; and is reciprocally as the square of the distance of its place from the earths center.*

The former part of the Proposition is evident from Phaen. VI, and Prop. II or III, Book I; the latter from the very slow motion of the moon's apogee; which in every single revolution amounting but to $3^{\circ}\ 3'$ *in consequentia*, may be neglected. For (by Cor. 1. Prop. XLV, Book I) it appears, that, if the distance of the moon from the earth's center is to the semi-diameter of the earth as D to 1, the force, from which such a motion will result, is reciprocally as $D^2$ 4/243, i.e., reciprocally as the power of D, whose exponent is 2 4/243; that is to say, in the proportion of the distance something greater than reciprocally duplicate, but which comes 59 ¾ times nearer to the duplicate than to the triplicate proportion. But in regard that this motion is owing to the action of the sun (as we shall afterwards shew), it is here to be neglected. The action of the sun, attracting the moon from the earth, is nearly as the moon's distance from the earth; and therefore (by what we have shewed in Cor. 2, Prop. XLV. Book I) is to the centripetal force of the moon as 2 to 357.45, or nearly so; that is, as 1 to 178 29/40. And if we neglect so inconsiderable a force of the sun, the remaining force, by which the moon is retained in its orb, will be reciprocally as $D^2$. This will yet more fully appear from comparing this force with the force of gravity, as is done in the next Proposition.

Corollary. If we augment the mean centripetal force by which the moon is retained in its orb, first in the proportion of 177 29/40 to 178 29/40, and then in the duplicate proportion of the semi-diameter of the earth to the mean distance of the centers of the moon and earth, we shall have the centripetal force of the moon at the surface of the earth; supposing this force, in descending to the earth's surface, continually to increase in the reciprocal duplicate proportion of the height.

Proposition IV. Theorem IV. *That the moon gravitates towards the earth, and by the force of gravity is continually drawn off from a rectilinear motion, and retained in its orbit.*

The mean distance of the moon from the earth in the syzygies in semidiameters of the earth, is, according to *Ptolemy* and most astronomers, 59: according to *Vendelin* and *Huygens*, 60; to *Copernicus*, 60 1/3 to *Street*, 60 2/3; and to *Tycho*, 56 1/2. But *Tycho*, and all that follow his tables of refraction, making the refractions of the sun and moon (altogether against the nature of light) to exceed the refractions of the fixed stars, and that by four or five minutes *near the horizon*, did thereby increase the moon's horizontal parallax by a like number of minutes, that is, by a twelfth or fifteenth part of the whole parallax. Correct this error, and the distance will become about 60 ½ semi-diameters of the earth, near to what others have assigned. Let us assume the mean distance of 60 diameters in the syzygies; and suppose one revolution of the moon, in respect of the fixed stars, to be completed in 27d.

7 h. 43', as astronomers have determined; and the circumference of the earth to amount to 123,249,600 *Paris* feet, as the French have found by mensuration. And now if we imagine the moon, deprived of all motion, to be let go, so as to descend towards the earth with the impulse of all that force by which (by Cor. Prop. III) it is retained in its orb, it will in the space of one minute of time, describe in its fall 15 1/12 *Paris* feet. This we gather by a calculus, founded either upon Prop. XXXVI, Book I, or (which comes to the same thing; upon Cor. 9, Prop. IV, of the same Book. For the versed sine of that arc, which the moon, in the space of one minute of time, would by its mean motion describe at the distance of 60 semi-diameters of the earth, is nearly 15 1/12 *Paris* feet, or more accurately 15 feet, 1 inch, and 1 line 4/9. Wherefore, since that force, in approaching to the earth, increases in the reciprocal duplicate proportion of the distance, and, upon that account, at the surface of the earth, is 60 x 60 times greater than at the moon, a body in our regions, falling with that force, ought in the space of one minute of time, to describe 60 x 60 x 15 1/12 *Paris* feet; and, in the space of one second of time, to describe 15 1/12 of those feet; or more accurately 15 feet, 1 inch, and 1 line 4/9. And with this very force we actually find that bodies here upon earth do really descend: for a pendulum oscillating seconds in the latitude of *Paris* will be 3 *Paris* feet, and 8 lines ½ in length, as Mr. *Huygens* has observed. And the space which a heavy body describes by falling in one second of time is to half the length of this pendulum in the duplicate ratio of the circumference of a circle to its diameter (as Mr. *Huygens* has also shewn), and is therefore 15 *Paris* feet, 1 inch, 1 line 7/9. And therefore, the force by which the moon is retained in its orbit becomes, at the very surface of the earth, equal to the force of gravity which we observe in heavy bodies there. And therefore (by Rule I and II) *the force by which the moon is retained in its orbit is that very same force which we commonly call gravity*; for, were gravity another force different from that, then bodies descending to the earth with the joint impulse of both forces would fall with a double velocity, and in the space of one second of time would describe 30 1/6 *Paris* feet; altogether against experience.

This calculus is founded on the hypothesis of the earth's standing still; for if both earth and moon move about the sun, and at the same time about their common center of gravity, the distance of the centers of the moon and earth from one another will be 60 ½ semi-diameters of the earth; as may be found by a computation from Prop. LX, Book I.

Scholium. The demonstration of this Proposition may be more diffusely explained after the following manner. Suppose several moons to revolve about the earth, as in the system of Jupiter or Saturn: the periodic times of these moons (by the argument of induction) would observe the same law which *Kepler* found to obtain among the planets; and therefore, their centripetal forces would be reciprocally as the squares of the distances from the center of the earth, by Prop. I, of this Book. Now if the lowest of these were very small, and were so near the earth as almost to touch the tops of the highest mountains, the centripetal force thereof, retaining it in its orb, would be very nearly equal to the weights of any terrestrial bodies that should be found upon the tops of those mountains, as may be known by the foregoing computation. Therefore, if the same little moon should be deserted by its centrifugal force that carries it through its orb, and so be disabled from going onward therein, it would descend to the earth; and that with the same velocity as heavy bodies do actually fall with upon the tops of those very mountains; because of the equality of the forces that oblige them both to descend. And if the force by which that lowest moon would descend were different from gravity, and if that moon were to gravitate towards the earth, as we find terrestrial bodies do upon the tops of mountains, it would then descend with twice the velocity, as being impelled by both these forces conspiring together. Therefore, since both these forces, that is, the gravity of heavy bodies, and the centripetal forces of the moons, respect the center of the earth, and are similar and equal between themselves, they will (by Rule I and II) have one and the same cause. And therefore, *the force which retains the moon in its orbit is that very force which we commonly call gravity*; because otherwise this little moon at the top of a mountain must either be without gravity, or fall twice as swiftly as heavy bodies are wont to do.

Proposition V. Theorem V. *That the circumjovial planets gravitate towards Jupiter; the circumsaturnal towards Saturn; the circumsolar towards the sun; and by the forces of their gravity are drawn off from rectilinear motions, and retained in curvilinear orbits.*

For the revolutions of the circumjovial planets about Jupiter, of the circumsaturnal about Saturn, and of Mercury and Venus, and the other circumsolar planets, about the sun, are appearances of the same sort with the revolution of the moon about the earth; and therefore, by Rule II, must be owing to the same sort of causes; especially since it has been demonstrated, that the forces upon which those revolutions depend tend to the centers of Jupiter, of Saturn, and of the sun; and that those forces, in receding from Jupiter, from Saturn, and from the sun, decrease in the same proportion, and according to the same law, as the force of gravity does in receding from the earth.

Corollary 1. There is, therefore, a power of gravity tending to all the planets; for, doubtless, Venus, Mercury, and the rest, are bodies of the same sort with Jupiter and Saturn. And since all attraction (by Law III) is mutual, Jupiter will therefore gravitate towards all his own satellites, Saturn towards his, the earth towards the moon, and the sun towards all the primary planets.

Corollary 2. The force of gravity which tends to any one planet is reciprocally as the square of the distance of places from that planet's center.

Corollary 3. All the planets do mutually gravitate towards one another, by Cor. 1 and 2. And hence it is that Jupiter and Saturn, when near their conjunction; by their mutual attractions sensibly disturb each other's motions. So, the sun disturbs the motions of the moon; and both sun and moon disturb our sea, as we shall hereafter explain.

Scholium. *The force which retains the celestial bodies in their orbits has been hitherto called centripetal force; but it being now made plain that it can be no other than a gravitating force, we shall hereafter call it gravity*. For the cause of that centripetal force which retains the moon in its orbit will extend itself to all the planets, by Rule I, II, and IV.

Proposition VI. Theorem VI. *That all bodies gravitate towards; every Planet and that the Weights of bodies towards any the same Planet, at equal distances from the center of the Planet, are proportional to the quantities of matter which they severally contain.*

It has been, now of a long time, observed by others, that all sorts of heavy bodies, (allowance being made for the inequality of retardation, which they suffer from a small power of resistance in the air) descend to the Earth from equal heights in equal times: and that equality of times we may distinguish to a great accuracy, by the help of pendulums. I tried the thing in gold, silver, lead, glass, sand, common salt, wood, water, and wheat. I provided two wooden boxes, round and equal. I filled the one with wood, and suspended an equal weight of gold (as exactly as I could) in the center of oscillation of the other. The Boxes hanging by equal threads of 11 feet, made a couple of pendulums perfectly equal in weight and figure, and equally receiving the resistance of the air. And placing the one by the other, I observed them to play together forwards and backwards, for a long time, with equal vibrations. And therefore, the quantity of matter in the gold (by Cor. 1. and 6. Prop. XXIV, Book II.) was to the quantity of matter in the wood, as the action of the motive force (or vis matrix) upon all the gold, to the action of the same upon all the wood; that is, as the weight of the one to the weight of the other. And the like happened in the other bodies. By these experiments, in bodies of the same weight, I could manifestly have discovered a difference of matter less than the thousandth part of the whole, had any such been. But without all doubt, the nature of gravity towards the Planets, is the same as towards the Earth. For, should we imagine our terrestrial bodies removed to the orb of the Moon, and there, together with the Moon, deprived of all motion, to be let go, so as to fall together towards the Earth: it is certain, from what we have demonstrated before, that, in equal times, they would describe equal spaces with the Moon, and of consequence are to the Moon, in quantity of matter, as their weights to its weight. Moreover, since the satellites of Jupiter perform their revolutions in times which observe the sesquiplicate proportion of their distances from Jupiter's center, their accelerative gravities towards Jupiter will be reciprocally as the squares of their distances from Jupiter's center; that is, equal, at equal distances. And therefore, these satellites, if supposed to fall *towards* Jupiter from equal heights, would describe equal spaces in equal times, in like manner as heavy bodies do on our Earth. And by the same argument, if the circumsolar Planets were supposed to be let sail at equal distances from the Sun, they would, in their descent towards the Sun, describe equal spaces in equal times. But forces, which equally accelerate unequal bodies, must be as those bodies; that is to say, the weights of the Planets *towards* the Sun must be as their quantities of matter. Further, that the weights of Jupiter and of his satellites towards the Sun are proportional to the several quantities of their matter, appears from the exceeding regular motions of the satellites, (by Cor. 3. Prop. LXV, Book I.) For if some of those bodies were more strongly attracted to the Sun in proportion to their quantity of matter, than others; the motions of the satellites would be disturbed by that inequality of attraction (by Cor. 2. Prop. LXV, Book I.) If, at equal distances from the Sun, any satellite in proportion to the quantity of its matter, did gravitate towards the Sun, with a force greater than Jupiter in proportion to his, according to any given proportion, suppose of d to e; then the distance between the centers of the Sun and of the satellite's orbit would be always greater than the distance between the centers of the Sun and of Jupiter, nearly in the subduplicate of that proportion; as by some computations I have found. And if the satellite did gravitate towards the Sun with a force, lesser in the proportion of e to d, the distance of the center of the satellite's orb from the Sun, would be less than the distance of the center of Jupiter from the Sun, in the subduplicate of the same proportion. Therefore if, at equal distances from the Sun, the accelerative gravity of any satellite towards the Sun were greater or less than the accelerative gravity of Jupiter towards the Sun, but by one 1/1000 part of the whole gravity; the distance of the center of the satellite's orbit from the Sun would be greater or less than the distance of Jupiter from the Sun, by one 1/2000 part of the whole distance; that is, by a fifth part of the distance of the utmost satellite from the center of Jupiter; an eccentricity of the orbit, which would be very sensible. But the orbits of the satellites are concentric to Jupiter, and therefore the accelerative gravities of Jupiter, and of all its satellites towards the Sun, are equal among themselves. And by the same argument, the weights of Saturn and of his satellites towards the Sun, at equal distances from the Sun, are as their several quantities of matter: and the weights of the Moon and of the Earth towards the Sun, are either none, or accurately proportional to the masses of matter which they contain. But some they are by Cor. 1. and 3, Prop. V.

But further, the weights of all the parts of every Planet towards any other Planet, are one to another as the matter in the several parts. For if some parts did gravitate more, others less, than for the quantity of their matter; then the whole Planet, according to the sort of parts with which it most abounds, would gravitate more or less, than in proportion to the quantity of matter in the whole. Nor is it of any moment, whether these parts are external or internal. For, if, for example, we would imagine the terrestrial bodies with us to be raised up to the orb of the Moon, to be there compared with its body: If the weights of such bodies were to the weights of the external parts of the Moon, as the quantities of matter in the one and in the other respectively; but to the weights of the internal parts, in a greater or less proportion, then likewise the weights of those bodies would be to the weight of the whole Moon, in a greater or less proportion; against what we have shewed above.

Corollary 1. Hence the weights of bodies do not depend upon their forms and textures. For if the weights could be altered with the forms, they would be greater or less, according to the variety of forms, in equal matter; altogether against experience.

Corollary 2. Universally, all bodies about the Earth, gravitate towards the Earth; and the weights of all, ar equal distances from the Earth's center, are as the quantities of matter which they severally contain. This is the quality of all bodies, within the reach of our experiments; and therefore, (by Rule III) to be affirmed of all bodies whatsoever. Is the *æther*, or any other body, were either altogether void of gravity, or were to gravitate less in proportion to its quantity of matter; then, because (according to *Aristotle*, *Des Cartes*, and others) there is no difference betwixt that and other bodies, but in mere form of matter, by a successive change from form to form, it might be changed at last into a body of the same condition with those which gravitate most in proportion to their quantity of matter; and, on the other hand, the heaviest bodies, acquiring the first form of that body, might by degrees, quite lose their gravity. And therefore, the weights would depend upon the forms of bodies, and with those forms might be changed, contrary to what was proved in the preceding Corollary.

Corollary 3. All spaces are not equally Full. For if all spaces were equally full, then the specific gravity of the fluid which fills the region of the air, on account of the extreme density of the matter, would fall nothing short of the specific gravity of quick-silver, or gold, or any other the most dense body; and therefore, neither gold, nor any other body, could descend in air. For bodies do not descend in fluids, unless they are specifically heavier than the fluids. And if the quantity of matter in a given space, can, by any rarefaction, be diminished, what should hinder a diminution to infinity?

Corollary 4. If all the solid particles of all bodies are of the same density, nor can be rarified without pores a void space or vacuum must be granted. By bodies of the same density, I mean those, whose *vires inertia* are in the proportion of their bulks.

Corollary 5. *The power of gravity is of a different nature from the power of magnetism.* For the magnetic attraction is not as the matter attracted. Some bodies are attracted more by the magnet, others less; most bodies not at all. The power of magnetism, in one and the same body, may be increased and diminished; and is sometimes far stronger, for the quantity of matter, than the power of gravity; and in receding from the magnet, decreases not in the duplicate, but almost in the triplicate proportion of the distance, as nearly as I could, judge from some rude observations.

Proposition VII. Theorem VII. *That there is a power of gravity tending to all bodies, proportional to the several quantities of matter which they contain.*

That all the Planets mutually gravitate one towards another, we have prov'd before; as well as that the force of gravity towards every one of them, consider'd apart, is reciprocally as the square of the distance of places from the center of the planet. And thence (by Prop. LXIX, Book I, and its Corollaries) it follows, that the gravity tending towards all the Planets, is proportional to the matter which they contain. Moreover, since all the parts of any planet A gravitate towards any other planet B; and the gravity of every part is to the gravity of the whole, as the matter of the part to the matter of the whole; and (by Law III) to every action corresponds an equal re-action: therefore the planet B will, on the other hand, gravitate towards all the parts of the planet A; and its gravity towards any one part will be to the gravity towards the whole, as the matter of the part to the matter of the whole. Q.E.D.

Corollary 1. Therefore, the force of gravity towards any whole planet, arises from, and is compounded of, the forces of gravity towards all its parts. Magnetic and electric attractions afford us examples of this. For all attraction towards the whole arises from the attractions towards the several parts. The thing may be easily understood in gravity, if we consider a greater planet, as form'd of a number of lesser planets, meeting together in one globe. For hence it would appear that the force of the whole must arise from the forces of the component parts. If it is objected, that, according to this law, all bodies with us must mutually gravitate one towards another, whereas no such gravitation anywhere appears: I answer, that since the gravitation towards these bodies is to the gravitation towards the whole Earth, as these bodies are to the whole Earth, the gravitation, towards them must be far less than to fall under the observation of our senses.

Corollary 2. The force of gravity towards the several equal particles of any body, is reciprocally as the square of the distance of places from the particles; as appears from Cor. 3, Prop. LXXIV, Book I.

Proposition VIII. Theorem VIII. *In two spheres mutually gravitating each towards the other, if the matter in places on all sides round about and equidistant from the centers, is similar; the weight of either sphere towards the other, will be reciprocally as the square of the distance between their centers.*

After I had found that the force of gravity towards a whole planet did arise from, and was compounded of the forces of gravity towards all its parts; and towards every one part, was in the reciprocal proportion of the squares of the distances from the part: I was yet in doubt, whether that reciprocal duplicate proportion did accurately hold, or but nearly so, in the total force compounded of so many partial ones. For it might be that the proportion which accurately enough took place in greater distances, should be wide of the truth near the surface of the planet, where the distances of the particles are unequal, and their situation dissimilar. But by the help of Prop. LXXV and LXXVI, Book I, and their Corollaries, I was at last satisfy'd of the truth of the proposition, as it now lies before us.

Corollary 1. Hence, we may find and compare together the weights of bodies towards different planets. For the weights of bodies revolving in circles about planets, are (by Cor. 2, Prop. IV, Book I.) as the diameters of the circles directly, and the squares of their periodic times reciprocally; and their weights at the surfaces of the planets, or at any other distances from their centers, are (by this proposition) greater or less, in the reciprocal duplicate proportion of the distances. Thus from the periodic times of Venus, revolving about the Sun, in 224 d. 16 ¾ h. of the utmost circumjovial satellite revolving about Jupiter, in 16 d. 16 9/13 h; of the Hugenian satellite about Saturn in 15 d. 22 2/3 h; and of the Moon about the Earth in 27 d. 7h. 43'; compared with the mean distance of Venus from the Sun, and with the greatest heliocentric elongations of the outmost circumjovial satellite from Jupiter's center, 8' 16" of the Hugenian satellite from the center of Saturn, 3' 4", and of the Moon from the Earth, 10' 33"; by computation I found, that the weight of equal bodies, at equal distances from the centers of the Sun, of Jupiter, of Saturn, and of the Earth, towards the Sun, Jupiter, Saturn, and the Earth, were one to another, as 1, 1/1067, 1/3021, and 1/169282 respectively. Then because as the distances are increased or diminished, the weights are diminished or increased in a duplicate ratio; the weights of equal bodies towards the Sun, Jupiter, Saturn, and the Earth, at the distances 10000, 997, 791 and 109 from their centers, that is, at their very superficies, will be as 10000, 943, 529 and 435 respectively. How much the weights of bodies are at the superficies of the Moon, will be'shewn hereafter.

Corollary 2. Hence likewise we discover the quantity of matter in the several Planets. For their quantities of matter are as the forces of gravity at equal distances from their centers, that is, in the Sun, Jupiter, Saturn, and the Earth, as 1, 1/1067, 1/3021, and 1/169282 respectively. If the parallax of the Sun be taken greater or less than 10", 30"', the quantity of matter in the Earth must be augmented or diminished in the triplicate of that proportion.

Corollary 3. Hence also we find the densities of the Planets. For (by Prop. LXXII, Book I) the weights of equal and similar bodies towards similar spheres, are, at the surfaces of those spheres, as the diameters of the spheres. And therefore, the densities of dissimilar spheres are as those weights applied to the diameters of the spheres. But the true diameters of the Sun, Jupiter, Saturn, and the Earth, were one to another as 10000, 997, 791 and 109; and the weights towards the same, as 10000, 943, 529, and 435 respectively; and therefore, their densities are as 100, 94 ½, 67 and 400. The density of the Earth, which comes out by this computation, does not depend upon the parallax of the Sun, but is determined by the parallax of the Moon, and therefore is here truly defin'd. The Sun therefore is a little denser than Jupiter, and Jupiter than Saturn, and the Earth four times denser than the Sun; for the Sun, by its great heat, is kept in a sort of a rarefy'd state. The Moon is denser than the Earth, as shall appear afterwards.

Corollary 4. The smaller the Planets are, they are, *ceteris paribus*, of so much the greater density. For so the powers of gravity on their several surfaces, come nearer to equality. They are likewise, ceteris paribus, of the greater density, as they are nearer to the Sun. So, Jupiter is more dense than Saturn, and the Earth than Jupiter. For the Planets were to be placed at different distances from the Sun, that according to their degrees of density, they might enjoy a greater or less proportion of the Sun's heat. Our water, if it were remov'd as far as the orb of Saturn, would be converted into ice, and the orb of Mercury would quickly fly away in vapor. For the light of the Sun, to which its heat is proportional, is seven times denser in the orb of the Mercury than with us: and by the thermometer I have found, that a sevenfold heat of our summer-sun will make water boil. Nor are we to doubt, that the matter of Mercury is adapted to its heat, and is therefore more dense than the matter of our Earth; since, in a denser matter, the operations of nature require a stronger heat.

Proposition IX. Theorem IX. *That the force of gravity, consider'd downwards from the surface of the planets, decreases nearly in the proportion of the distances from their centers.*

If the matter of the planet were of a uniform density, this proposition would be accurately true, (by Prop. LXXIII, Book I). The error therefore can be no greater than what may arise from the inequality of the density.
…

General Scholium.

The hypotheses of Vortices is press'd with many difficulties. That every Planet by a radius drawn to the Sun may describe areas proportional to the times of description, the periodic times of the several parts of the Vortices should observe the duplicate proportion of their distances from the Sun. But that the periodic times of the Planets may obtain the sesquiplicate proportion of their distances from the Sun, the periodic times of the parts of the Vortex ought to be in sesquiplicate proportion of their distances. That the smaller Vortices may maintain their lesser revolutions about Saturn, Jupiter, and other Planets, and swim quietly and undisturb'd in the greater Vortex of the Sun, the periodic times of the parts of the Sun's Vortex should be equal. But the rotation of the Sun and Planets about their axes, which ought to correspond with the motions of their Vortices, recede far from all these proportions. The motions of the Comets are exceedingly regular, are govern'd by the same laws with the motions of the Planets, and can by no means be accounted for by the hypotheses of Vortices. For Comets are carry'd with very eccentric motions through all parts of the heavens indifferently, with a freedom that is incompatible with the notion of a Vortex. Bodies, projected in our air, suffer no resistance but from the air. Withdraw the air, as is done in Mr. Boyle's vacuum, and the resistance ceases. For in this void a bit of fine down and a piece of solid gold descend with equal velocity. And the parity of reason must take place in the celestial spaces above the Earth's atmosphere; in which spaces, where there is no air to resist their motions, all bodies will move with the greatest freedom; and the Planets and Comets will constantly pursue their revolutions in orbits given in kind and position, according to the laws above explain'd. But though these bodies may indeed persevere in their orbits by the mere laws of gravity, yet they could by no means have at first deriv'd the regular position of the orbits themselves from those laws. The six primary Planets are revolv'd about the Sun, in circles concentric with the Sun, and with motions directed towards the same parts and almost in the same plan. Ten Moons are revolv'd about the Earth, Jupiter and Saturn, in circles concentric with them, with the same direction of motion, and nearly in the planes of the orbits of those Planets. But it is not to be conceived that mere mechanical causes could give birth to so many regular motions: since the Comets range over all parts of the heavens, in very eccentric orbits. For by that kind of motion they pass easily through the orbits of the Planets, and with great rapidity; and in their aphelions, where they move the slowest, and are detain'd the longest, they recede to the greatest distances from each other, and thence suffer the least disturbance from their mutual attractions. …

…

Hitherto we have explain'd the phænomena of the heavens and of our sea, by the power of Gravity, *but have not yet assign'd the cause of this power*. This is certain, that it must proceed from a cause that penetrates to the very centers of the Sun and Planets, without suffering the least diminution of its force; that operates, not according to the quantity of surfaces of the particles upon which it acts, (as mechanical causes use to do,) but *according to the quantity of the solid matter which they contain*, and propagates its virtue on all sides, to immense distances, decreasing always in the duplicate proportion of the distances. Gravitation towards the Sun, *is made up out of the gravitations towards the several particles of which the body of the Sun is compos'd*; and in receding from the Sun, decreases accurately in the duplicate proportion of the distances, as far as the orb of Saturn, as evidently appears from the quiescence of the aphelions of the Planets; nay, and even to the remotest aphelions of the Comets, if those aphelions are also quiescent. *But hitherto I have not been able to discover the cause of those properties of gravity from phænomena, and I framed no hypotheses*. For whatever is not deduc'd from the phænomena, is to be called a hypothesis; and *hypotheses*, whether metaphysical or physical, whether of occult qualities or mechanical, *have no place in experimental philosophy*. In this philosophy particular propositions are inferr'd from the phænomena, and afterwards render'd general by induction. Thus, it was that the impenetrability, the mobility, and the impulsive force of bodies, and the laws of motion and of gravitation, were discovered. And to us it is enough, that gravity does really exist, and act according to the laws which we have explained, and abundantly serves to account for all the motions of the celestial bodies, and of our sea. And now we might add something concerning a certain most subtle Spirit, which pervades and lies hid in all gross bodies; by the force and action of which Spirit, the particles of bodies mutually attract one another at near distances, and cohere, if contiguous; and electric bodies operate to greater distances, as well repelling as attracting the neighboring corpuscles; and light is emitted, reflected, refracted, inflected, and heats bodies; and all sensation is excited, and the members of animal bodies move at the command of the will, namely, by the vibrations of this Spirit, mutually propagated along the solid filaments of the nerves, from the outward organs of sense to the brain, and from the brain into the muscles. But these are things that cannot be explain'd in few words, nor are we furnish'd with that sufficiency of experiments which is required to an accurate determination and demonstration of the laws by which this electric and elastic spirit operates.

## Quantum entanglement between matter.

In addition to *quantum entanglement* between the *spin states* of elementary and composite particles which creates the *weak interaction* or attractive *weak force*, it is possible that there may also be *quantum entanglement* between *matter* creating the *gravitational interaction* or attractive *gravitational force*.

**Underwood, T. G. (2023).** ***Gravity***, Application of quantum theory to gravity - quantum entanglement, pp. 199-201: "… *Quantum entanglement* is the phenomenon that occurs when a duet of particles are generated, interact, or share spatial proximity *in such a way that the quantum state of each particle of the group cannot be described independently of the state of the others, including when the particles are separated by a large distance*. The topic of quantum entanglement is at the heart of the disparity between classical and quantum physics: *entanglement* is a primary feature of *quantum mechanics* not present in classical mechanics.

Measurements of physical properties such as position, momentum, spin, and polarization performed on *entangled* particles can, in some cases, be found to be perfectly correlated. For example, if a pair of *entangled* particles is generated such that their *total spin* is known to be zero, and one particle is found to have clockwise *spin* on a first axis, then the *spin* of the other particle, measured on the same axis, is found to be anticlockwise. However, this behavior gives rise to seemingly paradoxical effects: any measurement of a particle's properties results in an apparent and irreversible wave function collapse of that particle and changes the original quantum state. With *entangled* particles, such measurements affect the entangled system as a whole.

Such phenomena were the subject of a 1935 paper by Albert Einstein, Boris Podolsky, and Nathan Rosen,[1]

> [1] Einstein, A., Podolsky, B., Rosen, N. (1935). Can Quantum-Mechanical Description of Physical Reality Be Considered Complete? *Phys. Rev.*, 47, 10: 777–80. doi:10.1103/PhysRev.47.777.

…

Einstein and others considered such behavior impossible, as it violated the local realism view of causality (Einstein referring to it as "spooky action at a distance") and argued that the accepted formulation of quantum mechanics must therefore be incomplete.

Later, however, the counterintuitive predictions of *quantum mechanics* were verified in tests where the polarization or *spin* of entangled particles were measured at separate locations, statistically violating *Bell's inequality*.

[In 1964, John Stewart Bell, a physicist from Northern Ireland, published a paper "*On the Einstein–Podolsky–Rosen paradox*" which pointed out that under restricted conditions, local hidden-variable models can reproduce the predictions of quantum mechanics. He then demonstrates that this cannot hold true in general. Bell deduced that if measurements are performed independently on the two separated particles of an *entangled pai*r, then the assumption that the outcomes depend upon hidden variables within each half implies a mathematical constraint on how the outcomes on the two measurements are correlated. Such a constraint would later be named a *Bell inequality*. Bell then showed that *quantum physics* predicts correlations that violate this *inequality*.]

...

According to *some* interpretations of *quantum mechanics*, the effect of one measurement occurs instantly. Other interpretations, which do not recognize wavefunction collapse, dispute that there is any "effect" at all. However, all interpretations agree that *entanglement* produces correlation between the measurements, and that the mutual information between the *entangled* particles can be exploited, but that any transmission of information at faster-than-light speeds is impossible.

... Schrödinger shortly thereafter published his seminal paper defining and discussing the notion of "*entanglement*."[2]

[2] Schrödinger, E. (1935). Discussion of probability relations between separated systems. *Mathematical Proceedings of the Cambridge Philosophical Society*. 31, 4, 555–63. doi:10.1017/S0305004100013554.

In the paper, he recognized the importance of the concept, and stated: "I would not call [*entanglement*] one but rather the characteristic trait of *quantum mechanics*, the one that enforces its entire departure from classical lines of thought."

Like Einstein, Schrödinger was dissatisfied with the concept of *entanglement*, because it seemed to violate the speed limit on the transmission of information implicit in the *theory of relativity*. ...

## Introduction of a quantum theory of gravity.

Unburdened by Einstein's theories of relativity, a *quantum theory of gravity* can easily be developed by following Dirac's procedure for *non-relativistic quantum electrodynamics* in Dirac, P. A. M. (March, 1927); see pages 248-66 above. In the quotation from Dirac (1927) below, "*gravitation*" and "*graviton*" have been inserted in square brackets after Dirac's "*radiation*' and "*light-quantum*" to illustrate this.

**Dirac, P. A. M. (March, 1927). *The quantum theory of the emission and absorption of radiation*** (applied to gravity).

"The underlying ideas of the theory are very simple. Consider *an atom interacting with a field of radiation* [*gravitation*], which we may suppose for definiteness to be confined in an enclosure so as to have only a discrete set of degrees of freedom. Resolving the *radiation* [*gravitation*] into its Fourier components, we can consider the *energy* and *phase* of each of the components to be dynamical variables describing the *radiation field* [*gravitation field*]. Thus, if $E_r$ is the *energy* of a component labelled r and $\vartheta_r$ is the corresponding *phase* (defined as the time since the wave was in a standard phase), we can suppose each $E_r$ and $\vartheta_r$ to form a pair of *canonically conjugate* variables. In the absence of any *interaction* between the field and the atom, the whole system of field plus atom will be describable by the Hamiltonian

$$H = \Sigma_r E_r + H_0 \qquad (1)$$

equal to the total *energy*, $H_0$ being the Hamiltonian for the atom alone, since the variables $E_r$, $\vartheta_r$ obviously satisfy their *canonical equations of motion*

$$E^{\cdot}_r = \delta H/\delta\vartheta_r = 0, \qquad \vartheta^{\cdot}_r = \delta H/\delta E_r = 1.$$

*When there is interaction between the field and the atom, it could be taken into account on the classical theory by the addition of an interaction term to the Hamiltonian* (1), which would be a function of the variables of the *atom* and of the variables $E_r$, $\vartheta_r$ that describe the *field*. This *interaction term* would give the effect of the *radiation* [*gravitation*] on the atom, and also the reaction of the atom on the *radiation field* [*gravitation field*].

In order that an analogous method may be used on the *quantum theory, it is necessary to assume that the variables $E_r$, $\vartheta_r$ are q-numbers* satisfying the standard *quantum conditions* $\vartheta_r E_r - E_r \vartheta_r = ih$, etc., where h is $(2\pi)^{-1}$ times the usual Planck's constant, like the other dynamical variables of the problem. *This assumption immediately gives light-quantum* [*graviton*] *properties to the radiation* [*gravitation*]*.

* Similar assumptions have been used by Born and Jordan [Born, M. & Jordan, P. (December, 1925). Zur Quantenmechanik. (On Quantum Mechanics.) *Zeit. Phys.*, 34, 858-88] p. 886, for the purpose of taking over the classical formula for the emission of radiation

by a dipole into the quantum theory, and by Born, Heisenberg and Jordan [Born, M., Heisenberg, W. & Jordan, P. (August, 1926). Zur Quantenmechanik II. (On Quantum Mechanics II.) *Zeit. Phys*., 35, 557-615] p. 606, for calculating the energy fluctuations in a field of *black-body radiation*.

For if $\nu_r$ is the *frequency* of the component r, $2\pi\nu_r\vartheta_r$ is an *angle variable*, so that its *canonical conjugate* $E_r/2\pi\nu_r$ can only assume a discrete set of values differing by multiples of h, which means that $E_r$ can change only by integral multiples of the quantum $(2\pi h)\ \nu_r$. If we now add an *interaction term* (taken over from the classical theory) to the Hamiltonian (1), the problem can be solved according to the rules of *quantum mechanics*, and we would expect to obtain the correct results for the action of the radiation [*gravitation*] and the atom on one another. It will be shown that we actually get the correct laws for the *emission* and *absorption* of radiation [*gravitation*], and the correct values for Einstein's A's and B's.

[*Einstein coefficients* are mathematical quantities which are a measure of the probability of absorption or emission of light by an atom or molecule. The Einstein A coefficients are related to the rate of *spontaneous emission* of light, and the Einstein B coefficients are related to the *absorption* and *stimulated emission* of light.]

In the author's previous theory[#],

[#] Dirac, P. A. M. (October, 1926). On the Theory of Quantum Mechanics. *Roy. Soc. Proc., A*, 112, 762, 661-77, § 5[; *relativistic* treatment of Schrodinger's wave theory in which the time t and its *conjugate momentum* –W are treated from the beginning on the same footing as the other variables, sets $x_4 = ict$ (so that $x_1^2 + x_2^2 + x_3^2 + x_4^2 = 0$ and $x_1^2 + x_2^2 + x_3^2 = c^2t^2$) and $p_4 = iW/c$ where – W is the *momentum* conjugate to t, substitutes $(t - x_1/c)$ for t as *uniformizing variable* in order that its contribution to the exchange of energy with the radiation field may vanish, applies to system containing an atom with two electrons, finds that if the positions of the two electrons are interchanged the new state of the atom is physically indistinguishable from the original one, in order that theory only enables calculation of *observable quantities* must treat (*mn*) and (*nm*) as only one *state*, must infer that *unsymmetrical* functions of the co-ordinates (and momenta) of the two electrons cannot be represented by matrices, *symmetrical functions* such as the total *polarization* of the atom can be considered to be represented by matrices without inconsistency, these matrices are by themselves sufficient to determine all the physical properties of the system, theory of uniformizing variables introduced by the author can no longer apply, allows two solutions satisfying necessary conditions, one leads to Pauli's exclusion principle that not more than one electron can be in any given orbit, the other leads to the Einstein-Bose statistical mechanics, accounts for the *absorption* and stimulated *emission* of radiation by an atom, elements of matrices representing total *polarization* determine *transition probabilities*, *cannot be applied to spontaneous emission*; applies to theory of ideal gas and to problem of an atomic system subjected to a perturbation from outside (e.g., an incident electromagnetic field) which can vary with time in an arbitrary manner, *with neglect of relativity mechanics* accounts for the absorption and stimulated emission of radiation and shows that the elements of the matrices representing the total polarization determine the *transition probabilities*].

*where the energies and phases of the components of radiation were c-numbers*, only the B's could be obtained, and *the reaction of the atom on the radiation could not be taken into account*.

It will also be shown that the Hamiltonian which describes the *interaction of the atom and the electromagnetic waves* [*gravitation*] can be made identical with the Hamiltonian for the problem of the *interaction of the atom with an assembly of particles moving with the velocity of light and satisfying the Einstein-Bose statistics, by a suitable choice of the interaction energy for the particles*. The number of particles having any specified *direction of motion* and *energy*, which can be used as a dynamical variable in the Hamiltonian for the particles, is equal to the number of quanta of *energy* in the corresponding wave in the Hamiltonian for the waves. *There is thus a complete harmony between the wave and light-quantum* [*graviton*] *descriptions of the interaction*. We shall actually build up the theory from the light-quantum [*graviton*] point of view, and show that the Hamiltonian transforms naturally into a form which resembles that for the waves.

The mathematical development of the theory has been made possible by the author's *general transformation theory* of the quantum matrices[$].

[$] Dirac, P. A. M. (January, 1927). The Physical Interpretation of the Quantum Dynamics. *Roy. Soc. Proc., A*, 113, 765, 621-41[; *non-relativistic* matrix mechanics, Heisenberg's original matrix mechanics assumed that the elements of the diagonal matrix that represents the energy are the *energy levels* of the system, and the elements of the matrix that represents the total polarization, which are periodic functions of the time, determine the *frequencies* and *intensities* of the spectral lines in analogy to classical theory, in *Schrodinger's wave representation* physical results are based on assumption that the square of the *amplitude* of the wave function can be interpreted as a probability, enables probability of a *transition* being produced in a system by an arbitrary external perturbing force to be worked out, this paper provides a *general theory of obtaining physical results from quantum theory*, it shows all the physical information that one can hope to get from quantum dynamics and provides a general method for obtaining it, replaces special assumptions previously used, requires a theory of the more general schemes of matrix representation in which the rows and columns refer to any set of constants of integration that commute and of the laws of transformation from one such scheme to another, *does not take relativity mechanics into account*, counts time variable wherever it occurs as a parameter (a c-number), *transformation equations* that satisfy *quantum conditions* and *equations of motion*, *eigenfunctions* of Schrodinger's wave equation as *transformation functions* that enable transformation from scheme of matrix representation to scheme in which Hamiltonian is a diagonal matrix, dynamical variables represented by matrices whose rows and columns refer to the initial values of the *action variables* or to the *final values*, coefficients that enable transformation from one set of matrices to the other are those that determine the *transition probabilities*].

An essentially equivalent theory has been obtained independently by Jordan [Jordan, P. (November, 1927; received October 18, 1926). Über eine neue Begründung der Quantenmechanik. (On a new justification for quantum mechanics.) *Zeit. Phys*., 40, 809-38; https://doi.org/10.1007/BF01390903]. See also, London, F. (1926). Winkelvariable und Kanonische Transformationen in der Undulationsmechanik. *Zeit. Phys*., 40, 193-210.

Owing to the fact that we count the time as a *c-number*, we are allowed to use the notion of the value of any dynamical variable at any instant of time. This value is a *q-number*, capable of being represented by a generalized "matrix" according to many different matrix schemes, some of which may have continuous ranges of rows and columns, and may require the matrix elements to involve certain kinds of infinities (of the type given by the S functions). A matrix scheme can be found in which any desired set of constants of integration of the dynamical system that commute are represented by diagonal matrices, or in which a set of variables that commute are represented by matrices that are diagonal at a specified time*.

* One can have a matrix scheme in which a set of variables that commute are at all times represented by diagonal matrices if one will sacrifice the condition that the matrices must satisfy the *equations of motion*. The transformation function from such a scheme to one in which the *equations of motion* are satisfied will involve the time explicitly. See p. 628 in Dirac (January, 1927). The Physical Interpretation of the Quantum Dynamics [*loc. cit.*]

The values of the diagonal elements of a diagonal matrix representing any *q-number* are the *characteristic values* of that *q-number*. A Cartesian *co-ordinate* or *momentum* will in general have all *characteristic values* from $-\infty$ to $+\infty$, while an *action variable* has only a discrete set of *characteristic values*. (We shall make it a rule to use unprimed letters to denote the *dynamical variables* or *q-numbers*, and the same letters primed or multiply primed to denote their *characteristic values*. *Transformation functions* or *eigenfunctions* are functions of the *characteristic values* and not of the *q-numbers* themselves, so they should always be written in terms of primed variables.)

If $f(\xi, \eta)$ is any function of the *canonical variables* $\xi_k$, $\eta_k$, the matrix representing $f$ at any time t in the matrix scheme in which the $\xi_k$ at time t are diagonal matrices may be written down without any trouble, since the matrices representing the $\xi_k$ and $\eta_k$ themselves at time t are known, namely,

$$\xi_k\,(\xi'\,\xi'') = \xi'_k\,\delta(\xi'\,\xi''),$$
$$\eta_k\,(\xi'\,\xi'') = -\,ih\,\delta(\xi'_1 - \xi''_1)\ \ldots\ \delta(\xi'_{k-1} - \xi''_{k-1})\delta(\xi'_k - \xi''_k)\ \delta(\xi'_{k+1} - \xi''_{k+1})\ \ldots \quad (2)$$

Thus if the Hamiltonian H is given as a function of the $\xi_k$ and $\eta_k$ we can at once write down the matrix $H(\xi' \xi'')$. We can then obtain the *transformation function*, $(\xi' /\alpha')$ say, which transforms to a matrix scheme $(\alpha)$ in which the Hamiltonian is a diagonal matrix, as $(\xi' /\alpha')$ must satisfy the integral equation

$$\int H(\xi' \xi'')\, d\xi''(\xi''/\alpha') = W(\alpha') \,.\, (\xi'/\alpha'), \qquad (3)$$

of which the *characteristic values* $W(\alpha')$ are the *energy levels*. This equation is just Schrodinger's *wave equation* for the *eigenfunctions* $(\xi' /\alpha')$ which becomes an ordinary differential equation when H is a simple algebraic function of the $\xi_k$ and $\eta_k$ on account of the special equations (2) for the matrices representing $\xi_k$ and $\eta_k$. Equation (3) may be written in the more general form

$$\int H(\xi' \xi'')\, d\xi''(\xi''/\alpha') = ih\, \partial(\xi' /\alpha')/\partial t, \qquad (3')$$

in which it can be applied to systems for which the Hamiltonian involves the time explicitly.

One may have a dynamical system specified by a Hamiltonian H which cannot be expressed as an algebraic function of any set of *canonical variables*, but which can all the same be represented by a matrix $H(\xi' \xi'')$. Such a problem can still be solved by the present method, since one can still use equation (3) to obtain the *energy* levels and *eigenfunctions*. We shall find that the Hamiltonian which describes the interaction of a *light-quantum* [*graviton*] and an atomic system is of this more general type, so that the interaction can be treated mathematically, although one cannot talk about an interaction *potential energy* in the usual sense.

*It should be observed that there is a difference between a light-wave and the de Broglie or Schrodinger wave associated with the light-quanta. Firstly, the light-wave is always real, while the de Broglie wave associated with a light-quantum moving in a definite direction must be taken to involve an imaginary exponential.* A more important difference is that their *intensities* are to be interpreted in different ways. The number of *light-quanta* [*gravitons*] per unit volume associated with a monochromatic *light-wave* [*gravitation*] equals the *energy* per unit volume of the wave divided by the *energy* $(2\pi h)\nu$ of a single *light-quantum* [*graviton*]. On the other hand, a monochromatic de Broglie wave of *amplitude a* (multiplied into the imaginary exponential factor) must be interpreted as representing $a^2$ *light-quanta* per unit volume for all *frequencies*. This is a special case of the general rule for interpreting the matrix analysis*,

* *Loc. cit.* Dirac, P. A. M. (January, 1927). The Physical Interpretation of the Quantum Dynamics, §§ 6, 7.

according to which, if ($\xi'$ /$\alpha'$) or $\psi_{\alpha'}$ ($\xi_k'$) is the *eigenfunction* in the variables $\xi_k$ of the state $\alpha'$ of an atomic system (or simple particle), $|\psi_{\alpha'} (\xi_k')|^2$ is the probability of each $\xi_k$ having the value $\xi_k'$, [or $|\psi_{\alpha'} (\xi_k')|^2 d\xi_1' d\xi_2'$ ... is the probability of each $\xi_k$ lying between the values $\xi_k'$ and $\xi_k' + d\xi_k'$, when the $\xi_k$ have continuous ranges of characteristic values] on the assumption that all phases of the system are equally probable. *The wave whose intensity is to be interpreted in the first of these two ways appears in the theory only when one is dealing with an assembly of the associated particles satisfying the Einstein-Bose statistics* [such as the *graviton*]. *There is thus no such wave associated with electrons.*

## § 2. *The Perturbation of an Assembly of Independent Systems.*

We shall now consider *the transitions produced in an atomic system by an arbitrary perturbation*. The method we shall adopt will be that previously given by the author, [#]

[#] *Loc. cit.* Dirac, P. A. M. (October, 1926). On the Theory of Quantum Mechanics.

which leads in a simple way to equations which determine the *probability* of the system being in any *stationary state* of the unperturbed system at any time[$].

[$] The theory has recently been extended by Born [(Born, M. (192[7]). Das Adiabatenprinzip in der Quantenmechanik. *Zeit. Phys.*, 40. 167-192; https://doi.org/10.1007/bf01400360] so as to take into account the adiabatic changes in the stationary states that may be produced by the perturbation as well as the transitions. This extension is not used in the present paper.

This, of course, gives immediately the probable number of systems in that *state* at that time for an assembly of the systems that are independent of one another and are all perturbed in the same way. The object of the present section is to show that the equations for the rates of change of these probable numbers can be put in the Hamiltonian form in a simple manner, which will enable further developments in the theory to be made.

Let $H_0$ be the Hamiltonian for the unperturbed system and V the perturbing *energy*, which can be an arbitrary function of the dynamical variables and may or may not involve the time explicitly, so that the Hamiltonian for the perturbed system is $H = H_0 + V$. The *eigenfunctions* for the perturbed system must satisfy the *wave equation*

$$ih\ \delta\psi/\delta t = (H_0 + V)\ \psi,$$

where $(H_0 + V)$ is an operator. If $\psi = \Sigma_r\ a_r\psi_r$ is the solution of this equation that satisfies the proper initial conditions, where the $\psi_r$'s are the *eigenfunctions* for the unperturbed system, each associated with one *stationary state* labelled by the suffix *r*, and the $a_r$'s are functions of the time only, then $|a_r|^2$ is the *probability* of the system being in the state at any time. The $a_r$'s must be normalized initially, and will then always remain normalized. The theory will apply directly to an assembly of N similar independent systems if we multiply each of these $a_r$'s by $N^{1/2}$ so as to make $\Sigma_r |a_r|^2 = N$. We shall now have that $|a_r|^2$ is the probable number of systems in the *state r*.

The equation that determines the rate of change of the $a_r$'s is*

$$ih\, \dot{a}_r = \Sigma_s V_{rs}\, a_s, \qquad (4)$$

where the $V_{rs}$'s are the elements of the matrix representing V.

* *Loc. cit.* Dirac, P. A. M. (October, 1926). On the Theory of Quantum Mechanics, equation (25).

The *conjugate* imaginary equation is

$$ih\, \dot{a}_r^* = \Sigma_s V_{rs}^* a_s^* = \Sigma_s a_s^* V_{sr}^*, \qquad (4')$$

If we regard $a_r$ and ih $a_r^*$ as *canonical conjugates*, equations (4) and (4') take the Hamiltonian form with the Hamiltonian function $F_1 = \Sigma_{rs}\, a_r^* V_{rs} a_s$, namely,

$$da_r/dt = 1/ih\, \partial F_1/\partial a_r^*, \qquad ih\, da_r^*/dt = -\, \partial F_1/\partial a_r.$$

We can transform to the *canonical variables* $N_r$, $\phi_r$ by the *contact transformation*

$$a_r = N_r^{\frac{1}{2}}\, e^{-i\phi/h}, \qquad a_r^* = N_r^{\frac{1}{2}}\, e^{i\phi/h}.$$

This transformation makes the new variables $N_r$ and $\phi_r$ real, $N_r$ being equal to $a_r a_r^* = |a_r|^2$, the probable number of systems in the state r, and $\phi_r/h$ being the *phase* of the *eigenfunction* that represents them. The Hamiltonian $F_1$ now becomes

$$F_1 = \Sigma_{rs} V_{rs}\, N_r^{\frac{1}{2}}\, N_s^{\frac{1}{2}}\, e^{i(\phi r - \phi s)/h},$$

and the equations that determine the rate at which transitions occur have the *canonical* form

$$\dot{N}_r = \partial F_1/\partial \phi_r, \qquad \dot{\phi}_r = \partial F_1/\partial N_r.$$

A more convenient way of putting the *transition equations* in the Hamiltonian form may be obtained with the help of the quantities

$$b_r = a_r\, e^{-iWr\, t/h}, \qquad b_r^* = a_r^*\, e^{iWr\, t/h},$$

$W_r$ being the *energy* of the *state* r. We have $|b_r|^2$ equal to $|a_r|^2$, the probable number of systems in the *state* r. For $\dot{b}_r$ we find

$$ih\, \dot{b}_r = W_r b_r + ih\, \dot{a}_r\, e^{-iWr\, t/h}$$
$$= W_r b_r + \Sigma_s V_{rs} b_s\, e^{i(Wr - Ws)\, t/h}$$

with the help of (4). If we put $V_{rs} = \upsilon_{rs}\, e^{i(Wr - Ws)\, t/h}$, so that $\upsilon_{rs}$ is a constant when V does not involve the time explicitly, this reduces to

$$ih\, b^{\cdot}_r = W_r b_r + \Sigma_s\, \upsilon_{rs}\, b_s = \Sigma_s\, H_{rs} b_s, \tag{5}$$

where $H_{rs} = W_r\, \delta_{rs} + \upsilon_{rs}$, which is a matrix element of the total Hamiltonian $H = H_0 + V$ with the time factor $e^{i(Wr- Ws)\, t/h}$ removed, so that $H_{rs}$ is a constant when H does not involve the time explicitly. Equation (5) is of the same form as equation (4), and may be put in the Hamiltonian form in the same way.

It should be noticed that equation (5) is obtained directly if one writes down the Schrodinger equation in a set of variables that specify the *stationary states* of the unperturbed system. If these variables are $\xi_h$, and if $H(\xi'\xi'')$ denotes a matrix element of the total Hamiltonian H in the ($\xi$) scheme, this Schrodinger equation would be

$$ih\, \partial\psi(\xi')/\partial t = \Sigma_{\xi''}\, H(\xi'\xi'')\, \psi(\xi''), \tag{6}$$

like equation (3'). This differs from the previous equation (5) only in the notation, a single suffix r being there used to denote a *stationary state* instead of a set of numerical values $\xi'_k$ for the variables $\xi'_k$, and $b_r$ being used instead of $\psi(\xi')$. Equation (6), and therefore also equation (5), can still be used when the Hamiltonian is of the more general type which cannot be expressed as an algebraic function of a set of *canonical variables*, but can still be represented by a matrix $H(\xi'\xi'')$ or $H_{rs}$.

We now take $b_r$ and $ihb_r^*$ to be *canonically conjugate variables* instead of $a_r$ and $iha_r^*$. The equation (5) and its *conjugate imaginary equation* will now take the Hamiltonian form with the Hamiltonian function

$$F = \Sigma_{rs}\, b_r^* H_{rs} b_s. \tag{7}$$

Proceeding as before, we make the *contact transformation*

$$b_r = N_r^{\,½}\, e^{-i\theta/h}, \qquad b_r^* = N_r^{\,½}\, e^{i\theta/h}, \tag{8}$$

to the new *canonical variables* $N_r$, $\theta_r$, where $N_r$ is, as before, the probable number of systems in the *state* r, and $\theta_r$ is a new *phase*. The Hamiltonian F will now become

$$F = \Sigma_{rs}\, H_{rs}\, N_r^{½}\, N_s^{½}\, e^{i(\theta r- \theta s)/h},$$

and the equations for the rates of change of $N_r$ and $\theta_r$ will take the *canonical* form

$$N^{\cdot}_r = \partial F/\partial\theta_r, \qquad \theta^{\cdot}_r = \partial F/\partial N_r.$$

The Hamiltonian may be written

$$F = \Sigma_r\, W_r N_r + \Sigma_{rs}\, \upsilon_{rs}\, N_r^{½}\, N_s^{½}\, e^{i(\theta r- \theta s)/h}. \tag{9}$$

The first term $\Sigma_r\, W_r N_r$ is the *total proper energy* of the assembly, and the second may be regarded as the additional *energy* due to the perturbation. If the perturbation is zero, the phases $\theta_r$ would increase linearly with the time, while the previous *phases* $\phi_r$ would in this case be constants.

**§3. *The Perturbation of an Assembly satisfying the Einstein-Bose Statistics.***

According to the preceding section we can describe the effect of a perturbation on an assembly of independent systems by means of *canonical variables* and Hamiltonian *equations of motion*. The development of the theory which naturally suggests itself is to make these *canonical variables q-numbers* satisfying the usual *quantum conditions* instead of *c-numbers*, so that their Hamiltonian *equations of motion* become true *quantum equations*. The Hamiltonian function will now provide a Schrodinger *wave equation*, which must be solved and interpreted in the usual manner. The interpretation will give not merely the probable number of systems in any *state*, but the *probability* of any given distribution of the systems among the various *states*, this *probability* being, in fact, equal to the square of the modulus of the normalized solution of the *wave equation* that satisfies the appropriate initial conditions. We could, of course, calculate directly from elementary considerations the *probability* of any given distribution *when the systems are independent*, as we know the probability of each system being in any particular *state*. We shall find that the *probability* calculated directly in this way *does not agree with that obtained from the wave equation except in the special case when there is only one system in the assembly*. In the general case it will be shown that *the wave equation leads to the correct value for the probability of any given distribution when the systems obey the Einstein-Bose statistics instead of being independent.*

We assume the variables $b_r$, $ihb_r^*$ of §2 to be *canonical* q-numbers satisfying the *quantum conditions*

$$b_r \,.\, ihb_r^* - ihb_r^* .\, b_r = ih,$$
$$b_r b_r^* - b_r^* b_r = 1,$$

and $$b_r b_s - b_s b_r = 0, \qquad b_r^* b_s^* - b_s^* b_r^* = 0,$$
$$b_r b_s^* - b_s^* b_r = 0 \qquad (s \neq r).$$

The *transformation equations* (8) must now be written in the *quantum form*

$$b_r = (N_r + 1)^{1/2} e^{-i\theta/h} = e^{-i\theta/h} N_r^{1/2}$$
$$b_r^* = N_r^{1/2} e^{i\theta/h} = e^{i\theta/h} (N_r + 1)^{1/2}, \qquad (10)$$

in order that the $N_r$, $\theta_r$ may also be *canonical variables*. These equations show that the $N_r$ can have only integral *characteristic values* not less than zero, #

# See § 8 of the author's paper, p. 281. [Dirac, P. A. M. (May, 1926). The elimination of the nodes in quantum mechanics. *Roy. Soc. Proc., A*, 111, 757, 281–305 [; the laws of classical mechanics must be generalized when applied to atomic systems, *the commutative*

*law of multiplication* as applied to dynamical variables is replaced by certain *quantum conditions* which are just sufficient to enable one to evaluate xy – yx when x and y are given, it follows that the dynamical variables cannot be ordinary numbers expressible in the decimal notation (which numbers will be called *c-numbers*), but may be considered to be numbers of a special kind (which will be called *q-numbers*), whose nature cannot be exactly specified, but which can be used in the algebraic solution of a dynamical problem in a manner closely analogous to the way the corresponding classical variables are used, the object of this paper is to simplify the *non-relativistic* quantum treatment by the introduction of *quantum variables*, in the classical treatment of the dynamical problem of a number of particles or electrons moving in a central field of force and disturbing one another one always begins by making the initial simplification known as the *elimination of the nodes*, this consists in obtaining a *contact transformation* from the Cartesian co-ordinates and momenta of the electrons to a set of canonical variables of which all except three are independent of the orientation of the system as a whole while these three determine the orientation, introduces *action variables and their canonical conjugate angle variables*, *transformation equations*, substitutes set of *c-numbers* for *action variables* to fix *stationary state* and obtain physical results, applies to *anomalous Zeeman effect*, showed that *non-relativistic* theory gave the correct g-formula for *energy* of stationary states and Kronig's results for the relative intensities of the lines of a multiplet and their components in a weak magnetic field].

which provides us with a justification for the assumption that the variables are *q-numbers* in the way we have chosen. The numbers of systems in the different *states* are now ordinary quantum numbers.

The Hamiltonian (7)

$$[F = \Sigma_{rs}\, b_r{}^* H_{rs} b_s. \qquad (7)]$$

now becomes

$$F = \Sigma_{rs}\, b_r{}^* H_{rs} b_s = \Sigma_{rs}\, N_r{}^{1/2}\, e^{i\theta r/h}\, H_{rs}(N_s + 1)^{1/2}\, e^{-i\theta s/h}$$
$$= \Sigma_{rs}\, H_{rs} N_r{}^{1/2}\, (N_s + 1 - \delta_{rs})^{1/2}\, e^{i(\theta r-\theta s)/h} \qquad (11)$$

in which the $H_{rs}$ are still *c-numbers*. We may write this F in the form corresponding to (9)

$$[F = \Sigma_r\, W_r N_r + \Sigma_{rs}\, \upsilon_{rs}\, N_r{}^{1/2}\, N_s{}^{1/2}\, e^{i(\theta r-\theta s)/h}. \qquad (9)]$$

$$F = \Sigma_r\, W_r N_r + \Sigma_{rs}\, \upsilon_{rs}\, N_r{}^{1/2}\, (N_s + 1 - \delta_{rs})^{1/2}\, e^{i(\theta r-\theta s)/h} \qquad (11')$$

in which it is again composed of a *proper energy* term $\Sigma_r W_r N_r$ and an *interaction energy* term.

…

*Consider first the case when there is only one system in the assembly*. The probability of its being in the *state* q is determined by the *eigenfunction* $\psi(N_1', N_2', ...)$ in which all the N's are put equal to zero except $N'_q$, which is put equal to unity. This *eigenfunction* we shall denote by $\psi\{q\}$. When it is substituted in the left-hand side of (13), all the terms in the summation on the right-hand side vanish except those for which r = q, and we are left with

$$ih\, \partial/\partial t\, \psi\{q\} = \Sigma_r H_{qs}\, \psi\{s\},$$

which is the same equation as (5)

$$[ih\, b^{\cdot}_r = W_r b_r + \Sigma_s \upsilon_{rs}\, b_s = \Sigma_s H_{rs} b_s, \qquad (5)]$$

with {q} playing the part of $b_q$. This establishes the fact that the present theory is equivalent to that of the preceding section when there is only one system in the assembly.

*Now take the general case of an arbitrary number of systems in the assembly, and assume that they obey the Einstein-Bose statistical mechanics.* This requires that, in the ordinary treatment of the problem, only those *eigenfunctions* that are *symmetrical* between all the systems must be taken into account, these *eigenfunctions* being by themselves sufficient to give a complete quantum solution of the problem[#].

[#] *Loc. cit.* Dirac, P. A. M. (October, 1926). On the Theory of Quantum Mechanics, § 3.

We shall now obtain the equation for the rate of change of one of these symmetrical *eigenfunctions*, and show that it is identical with equation (13). ...

... We have thus established that the Hamiltonian (11)

$$[F = \Sigma_{rs}\, b_r^* H_{rs} b_s = \Sigma_{rs}\, N_r^{1/2}\, e^{i\theta r/h}\, H_{rs}(N_s + 1)^{1/2}\, e^{-i\theta s/h}$$
$$= \Sigma_{rs}\, H_{rs} N_r^{1/2}\, (N_s + 1 - \delta_{rs})^{1/2}\, e^{i(\theta r - \theta s)/h} \qquad (11)]$$

describes the effect of a perturbation on an assembly satisfying the *Einstein-Bose statistics*.

## § 4. *The Reaction of the Assembly on the Perturbing System.*

Up to the present we have considered only perturbations that can be represented by a perturbing *energy* V added to the Hamiltonian of the perturbed system, V being a function only of the dynamical variables of that system and perhaps of the time. The theory may readily be extended to the case when the perturbation consists of interaction with a perturbing dynamical system, *the reaction of the perturbed system on the perturbing system being taken into account*. (The distinction between the perturbing system and the perturbed system is, of course, not real, but it will be kept up for convenience.)

We now consider a perturbing system, described, say, by the canonical variables $J_k$, $w_k$, the J's being its first integrals when it is alone, interacting with an assembly of perturbed systems with no mutual interaction, that satisfy the *Einstein-Bose statistics*. The total Hamiltonian will be of the form

$$H_T = H_P\,(J) + \Sigma_n\, H\,(n),$$

where $H_P$ is the Hamiltonian of the perturbing system (a function of the J's only) and H(n) is equal to the *proper energy* $H_0(n)$ plus the *perturbation energy* V(n) of the nth system of the assembly. H(n) is a function only of the variables of the nth system of the assembly and of the J's and ϖ's, and does not involve the time explicitly.

The Schrodinger equation corresponding to equation (14) is now …

…

This is the Schrodinger equation corresponding to the Hamiltonian function

$$F = H_P(J) + \Sigma_{rs} H_{rs}N_r^{1/2} (N_s + 1 - \delta_{rs})^{1/2} e^{i(\theta r - \theta s)/h}, \qquad (19)$$

in which $H_{rs}$ is now a function of the J's and ϖ's, being such that when represented by a matrix in the (J) scheme its (J' J") element is H ($J'_r$; $J''_s$). (It should be noticed that $H_{rs}$ still commutes with the N's and θ's.)

Thus, *the interaction of a perturbing system and an assembly satisfying the Einstein-Bose statistics can be described by a Hamiltonian of the form (19)*

$$[F = H_P(J) + \Sigma_{rs} H_{rs}N_r^{1/2} (N_s + 1 - \delta_{rs})^{1/2} e^{i(\theta r - \theta s)/h}. \qquad (19)]$$

We can put it into the form corresponding to (11')

$$[F = \Sigma_r W_rN_r + \Sigma_{rs} \upsilon_{rs} N_r^{1/2} (N_s + 1 - \delta_{rs})^{1/2} e^{i(\theta r - \theta s)/h} \qquad (11')]$$

by observing that the matrix element H($J'_r$; $J''_s$) is composed of the sum of two parts, a part that comes from the *proper energy* $H_0$, which equals $W_r$ when $J''_k = J'_k$ and s = r and vanishes otherwise, and a part that comes from the *interaction energy* V, which may be denoted by υ($J'_r$; $J''_s$). Thus, we shall have

$$H_{rs} = W_r \delta_{rs} + \upsilon_{rs},$$

where $\upsilon_{rs}$ is that function of the J's and ϖ's which is represented by the matrix whose (J' J") element is υ($J'_r$; $J''_s$), and so (19) becomes

$$F = H_P(J) + \Sigma_r W_rN_r + \Sigma_{rs} \upsilon_{rs}N_r^{1/2} (N_s + 1 - \delta_{rs})^{1/2} e^{i(\theta r - \theta s)/h}, \qquad (20)$$

*The Hamiltonian is thus the sum of the proper energy of the perturbing system $H_P(J)$, the proper energy of the perturbed systems $\Sigma_r W_rN_r$ and the perturbation energy $\Sigma_{rs} \upsilon_{rs}N_r^{1/2} (N_s + 1 - \delta_{rs})^{1/2} e^{i(\theta r - \theta s)/h}$.*

## § 5. *Theory of Transitions in a System from One State to Others of the Same Energy.*

Before applying the results of the preceding sections to *light-quanta* [*gravitons*] we shall consider the solution of the problem presented by a Hamiltonian of the type (19)

$$[F = H_P(J) + \Sigma_{rs} H_{rs}N_r^{1/2} (N_s + 1 - \delta_{rs})^{1/2} e^{i(\theta r - \theta s)/h}. \qquad (19)]$$

The essential feature of the problem is that it refers to a dynamical system which can, under the influence of a perturbation *energy* which does not involve the time explicitly, make transitions from one state to others of the same *energy*. The problem of collisions between an atomic system and an *electron*, which has been treated by Born*, is a special case of this type.

* Born, M. (November, 1926) Quantenmechanik der Stoßvorgänge. (Quantum mechanics of collision processes.) *Zeit. Phys.*, 38, 803-27; https://doi.org/10.1007/BF01397184.

Born's method is to find a *periodic* solution of the *wave equation* which consists, in so far as it involves the *co-ordinates* of the colliding *electron*, of plane waves, representing the incident *electron*, approaching the atomic system, which are scattered or diffracted in all directions. The square of the *amplitude* of the waves scattered in any direction with any *frequency* is then assumed by Born to be the *probability* of the *electron* being scattered in that direction with the corresponding *energy*.

*This method does not appear to be capable of extension in any simple manner to the general problem of systems that make transitions from one state to others of the same energy*. Also, there is at present no very direct and certain way of interpreting a periodic solution of a *wave equation* to apply to a non-periodic physical phenomenon such as a collision. (The more definite method that will now be given shows that Born's assumption is not quite right, it being necessary to multiply the square of the *amplitude* by a certain factor.)

*An alternative method of solving a collision problem is to find a solution of the wave equation which consists initially simply of plane waves moving over the whole of space in the necessary direction with the necessary frequency to represent the incident electron*. In course of time waves moving in other directions must appear in order that the *wave equation* may remain satisfied. The *probability* of the *electron* being scattered in any direction with any *energy* will then be determined by the rate of growth of the corresponding harmonic component of these waves. The way the mathematics is to be interpreted is by this method quite definite, being the same as that of the beginning of §2.

We shall apply this method to the general problem of a system which makes transitions from one *state* to others of the same *energy* under the action of a perturbation. Let $H_0$ be the Hamiltonian of the unperturbed system and V the *perturbing energy*, which must not involve the time explicitly. If we take the case of a continuous range of *stationary states*, specified by the first integrals, $\alpha_k$ say, of the unperturbed motion, then, following the method of § 2, we obtain

$$ih\, \dot{a}(\alpha') = \int V(\alpha'\alpha'')\, d\alpha''.\ a(\alpha''), \qquad (21)$$

corresponding to equation (4)

$$[ih\, \dot{a}_r = \Sigma_s V_{rs}\, a_s. \qquad (4)]$$

…

This result differs by the factor $(2\pi h)^2/2mE'$. $P'/P^0$ from Born's* [where $P^0$ refers to the *initial momentum*, and E' and P' refer, respectively, to the *resultant energy and momentum* of the scattered *electron*].

* In a more recent paper [Born, M. (1926). *Nachr. Gesell. d. Wiss., Gottingen*, 146] Born has obtained a result in agreement with that of the present paper for *non-relativity* mechanics, by using an interpretation of the analysis based on the conservation theorems. I am indebted to Prof. N. Bohr for seeing an advance copy of this work.

The necessity for the factor $P'/P^0$ in (26) could have been predicted from the *principle of detailed balancing*, as the factor $| v (p'; p^0) |^2$ is symmetrical between the direct and reverse processes#.

# See Klein, O., Rosseland, S. (March, 1921).Über Zusammenstöße zwischen Atomen und freien Elektronen. (About collisions between atoms and free electrons.) *Zeit. Phys.*, 4, 46-51; https://doi.org/10.1007/BF01328041, eq. (4).

### § 6. *Application to Light-Quanta*$ [*Gravitons*].

[$ Dirac, P. A. M. (May, 1927). The quantum theory of dispersion. *Roy. Soc. Proc., A*, 114, 769, 710-28, page 711, ff: "In Dirac (March, 1927)., § 6, *it was in error assumed that* $V_{mn}$ *caused transitions from state m to state n*, and consequently the information there obtained about an absorption (or emission) process in terms of the number of *light-quanta* existing *before the process* should really apply to an emission (or absorption) process in terms of the number of *light-quanta* in existence *after the process*. This change, of course, does not affect the results (namely the proof of Einstein's laws) which can depend on $|V_{mn}|^2 = |V_{nm}|^2$."]

We shall now apply the theory of §4 to the case when the systems of the assembly are *light-quanta* [*gravitons*], the theory being applicable to this case since *light-quanta* [*gravitons*] obey the *Einstein-Bose statistics* and have no mutual interaction [???]. A *light-quantum* [*graviton*] is in a *stationary state* when it is moving with constant *momentum* in a straight line. Thus, a *stationary state r* is fixed by the three components of *momentum* of the *light-quantum* [*graviton*] and a variable that specifies its state of *polarization*. We shall work on the assumption that there are a finite number of these *stationary states*, lying very close to one another, as it would be inconvenient to use continuous ranges. The interaction of the *light-quanta* [*gravitons*] with an atomic system will be described by a Hamiltonian of the form (20)

$$[F = H_P(J) + \Sigma_r W_r N_r + \Sigma_{rs} \upsilon_{rs} N_r^{1/2} (N_s + 1 - \delta_{rs})^{1/2} e^{i(\theta r - \theta s)/h}, \qquad (20)]$$

in which $H_P(J)$ is the Hamiltonian for the *atomic system alone*, and the coefficients $\upsilon_{rs}$ are for the present unknown. *We shall show that this form for the Hamiltonian, with the* $\upsilon_{rs}$ *arbitrary, leads to Einstein's laws for the emission and absorption of radiation.*

*The light-quantum [graviton] has the peculiarity that it apparently ceases to exist when it is in one of its stationary states, namely, the zero state, in which its momentum, and therefore also its energy, are zero.* When a *light-quantum [graviton]* is absorbed, it can be considered to jump into this *zero state*, and when one is emitted, it can be considered to jump from the *zero state* to one in which it is physically in evidence, so that it appears to have been created. Since there is no limit to the number of *light-quanta [gravitons]* that may be created in this way, we must suppose that there are an infinite number of *light-quanta [gravitons]* in the *zero state, so that the $N_0$ of the Hamiltonian (20) is infinite.* We must now have $\theta_0$, the variable *canonically conjugate* to $N_0$, a constant, since

$$\theta^{\cdot}_0 = \partial F/\partial N_0 = W_0 + \text{terms involving } N_0^{-1/2} \text{ or } (N_0 + 1 - \delta_{rs})^{-1/2}$$

and $W_0$ is zero. *In order that the Hamiltonian (20) may remain finite it is necessary for the coefficients $\upsilon_{r0}$, $\upsilon_{0r}$ to be infinitely small.* We shall suppose that they are infinitely small in such a way as to make $\upsilon_{r0}N_0^{1/2}$ and $\upsilon_{0r}N_0^{1/2}$ finite, in order that the transition probability coefficients may be finite. Thus, we put

$$\upsilon_{r0}\,(N_0 + 1)^{1/2}\, e^{-i\theta_0/h} = \upsilon_r, \qquad \upsilon_{r0}\, N_0^{1/2}\, e^{i\theta_0/h} = \upsilon_r^*,$$

where $\upsilon_r$ and $\upsilon_r^*$ are finite and *conjugate imaginaries*. We may consider the $\upsilon_r$ and $\upsilon_r^*$ to be functions only of the J's and ϖ's of the *atomic system*, since their factors $(N_0 + 1)^{1/2}\, e^{-i\theta_0/h}$ and $N_0^{1/2}\, e^{i\theta_0/h}$ are practically constants, the rate of change of $N_0$ being very small compared with $N_0$. The Hamiltonian (20)

$$[F = H_P(J) + \Sigma_r\, W_rN_r + \Sigma_{rs}\, \upsilon_{rs}N_r^{1/2}\,(N_s + 1 - \delta_{rs})^{1/2}\, e^{i(\theta r-\theta s)/h}, \qquad (20)]$$

now becomes

$$F = H_P(J) + \Sigma_r\, W_rN_r + \Sigma_{r\neq 0}\, [\upsilon_rN_r^{1/2}\, e^{i\theta r/h} + \upsilon_r^*(N_r + 1)^{1/2}e^{-i\theta r/h}] + \Sigma_{r\neq 0}\, \Sigma_{s\neq 0}\, \upsilon_{rs}N_r^{1/2}\,(N_s + 1 - \delta_{rs})^{1/2}\, e^{i(\theta r-\theta s)/h} \qquad (27)$$

The *probability of a transition* in which a *light-quantum [graviton]* in the *state* r is absorbed is proportional to the square of the modulus of that matrix element of the Hamiltonian which refers to this transition. This matrix element must come from the term $\upsilon_rN_r^{1/2}\, e^{i\theta r/h}$ in the Hamiltonian, and must therefore be proportional to $N'^{1/2}_r$ where $N'_r$ is the number of *light-quanta [gravitons]* in *state* r *before the process*.

The *probability of the absorption* process is thus proportional to $N'_r$. In the same way the probability of a *light-quantum [graviton]* in *state* r being *emitted* is proportional to ($N'_r$ + 1), and the *probability* of a *light-quantum [graviton]* in *state* r being *scattered* into *state* s is proportional to $N'_r$ ($N'_r$ + 1). *Radiative [gravitational] processes of the more general type considered by Einstein and Ehrenfest*[$] *in which more than one light-quantum [graviton] take part simultaneously, are not allowed on the present theory.*

[$] Einstein, A. & Ehrenfest, P. (December, 1923). Zur Quantentheorie des

Strahlungsgleichgewichts. (On the quantum theory of radiation equilibrium.) *Zeit. Phys.*, 19, 301-6; https://doi.org/10.1007/BF01327565.

To establish a connection between the number of *light-quanta* [*gravitons*] per *stationary state* and the *intensity* of the *radiation* [*gravitation*], we consider an enclosure of finite volume, A say, containing the *radiation* [*gravitation*]. The number of *stationary states* for *light-quanta* [*gravitons*] of a given type of *polarization* whose *frequency* lies in the range $\nu_r$, to $\nu_r + d\nu_r$ and whose direction of motion lies in the solid angle about the direction of motion for *state* r will now be $A\nu_r^2 d\nu_r d\varpi_r/c^3$. The *energy* of the *light-quanta* [*gravitons*] in these *stationary states* is thus $N'_r \,.\, 2\pi h\nu_r \,.\, A\nu_r^2 d\nu_r d\varpi_r/c^3$. This must equal $Ac^{-1}I_r d\nu_r d\varpi_r$, where $I_r$ is the *intensity* per unit *frequency* range of the *radiation* [*gravitation*] about the *state* r. Hence

$$I_r = N'_r\,(2\pi h)\,\nu_r^3/c^2, \qquad (28)$$

so that $N'_r$ is proportional to $I_r$ and $(N'_r + 1)$ is proportional to $I_r + (2\pi h)\,\nu_r^3/c^2$.

We thus obtain that the *probability* of an *absorption* process is proportional to $I_r$, the incident *intensity* per unit *frequency* range, and that of an *emission* process is proportional to $I_r + (2\pi h)\nu_r^3/c^2$, which are just *Einstein's laws**.

* The ratio of stimulated to *spontaneous emission* in the present theory is just twice its value in Einstein's. This is because in the present theory either polarized component of the incident radiation can stimulate only radiation polarized in the same way, while in Einstein's the two polarized components are treated together. This remark applies also to the *scattering process*.

In the same way the *probability* of a process in which a *light-quantum* [*graviton*] is *scattered* from a state *r* to a *state s* is proportional to $I_r\,[I_s + (2\pi h)\nu_r^3/c^2)]$, which is *Pauli's law for the scattering of radiation by an electron.*#

\# Pauli, W. (December, 1923). Über das thermische Gleichgewicht zwischen Strahlung und freien Elektronen. (About the thermal equilibrium between radiation and free electrons.) *Zeit. Phys.*, 18, 272-86; https://doi.org/10.1007/BF01327708.

## §7. *The Probability Coefficients for Emission and Absorption.*

We shall now consider the *interaction of an atom and radiation* [*gravitation*] from the *wave* point of view. We resolve the *radiation* [*gravitation*] into its *Fourier components*, and suppose that their number is very large but finite. Let each component be labelled by a suffix *r*, and suppose there are $\sigma_r$ components associated with the *radiation* [*gravitation*] of a definite type of *polarization* per unit solid angle per unit *frequency* range about the component *r*. Each component *r* can be described by a vector potential $\kappa_r$ chosen so as to make the scalar potential zero. The perturbation term to be added to the Hamiltonian will now be, according to the classical theory *with neglect of relativity mechanics*, $c^{-1}\,\Sigma_r\,\kappa_r\,\dot{X}_r$, where $X_r$ is the component of the total *polarization* of the atom in the direction of $\kappa_r$, which is the direction of the *electric vector* of the component *r*.

We can, as explained in § 1, suppose the field to be described by the *canonical variables* $N_r$, $\vartheta_r$, of which $N_r$ is the number of *quanta* of *energy* of the component *r*, and $\vartheta_r$ is its *canonically conjugate phase*, equal to $2\pi h\nu_r$ times the $\vartheta_r$ of § 1. …

…

The Hamiltonian for the whole system of *atom plus radiation* [*gravitation*] would now be, according to the classical theory,

$$F = H_P(J) + \Sigma_r (2\pi h\nu_r) N_r + 2^{c-1} \Sigma_r (h\nu_r/c\sigma_r)^{1/2} \dot{X}_r N_r^{1/2} \cos \theta_r/h, \qquad (29)$$

*where $H_P(J)$ is the Hamiltonian for the atom alone*. On the *quantum theory* we must make the variables $N_r$ and $\theta_r$ canonical q-numbers like the variables $J_k$, $\varpi_k$ that describe the *atom*. We must now replace the $N_r^{1/2} \cos \theta_r/h$ in (29) by the real q-number

$$\tfrac{1}{2} \{N_r^{1/2} e^{i\theta r/h} + e^{-i\theta r/h} N_r^{1/2}\} = \tfrac{1}{2} \{N_r^{1/2} e^{i\theta r/h} + (N_r + 1)^{1/2} e^{-i\theta r/h}\}$$

so that the Hamiltonian (29) becomes

$$F = H_P(J) + \Sigma_r (2\pi h\nu_r) N_r + h^{1/2}c^{-3/2} \Sigma_r (\nu_r/\sigma_r)^{1/2} \dot{X}_r \{N_r^{1/2} e^{i\theta r/h} + (N_r + 1)^{1/2} e^{-i\theta r/h}] \qquad (30)$$

This is of the form (27)

$$[F = H_P(J) + \Sigma_r W_r N_r + \Sigma_{r\neq 0} [\upsilon_r N_r^{1/2} e^{i\theta r/h} + \upsilon_r^*(N_r + 1)^{1/2} e^{-i\theta r/h}]$$
$$+ \Sigma_{r\neq 0} \Sigma_{s\neq 0} \upsilon_{rs} N_r^{1/2} (N_s + 1 - \delta_{rs})^{1/2} e^{i(\theta r-\theta s)/h}, \qquad (27)]$$

with $\upsilon_r^* = \upsilon_r^* = h^{1/2}c^{-3/2}(\nu_r/\sigma_r)^{1/2} \dot{X}_r$,

$\upsilon_{rs} = 0 \qquad (r, s \neq 0). \qquad (31)$

*The wave point of view is thus consistent with the light-quantum* [*graviton*] *point of view and gives values for the unknown interaction coefficient $\upsilon_{rs}$ in the light-quantum* [*graviton*] *theory*. These values are not such as would enable one to express the *interaction energy* as an algebraic function of *canonical variables*. Since the *wave* theory gives $\upsilon_{rs} = 0$ for r, s ≠ 0, *it would seem to show that there are no direct scattering processes, but this may be due to an incompleteness in the present wave theory*.

We shall now show that the Hamiltonian (30)

$$[F = H_P(J) + \Sigma_r (2\pi h\nu_r) N_r + h^{1/2}c^{-3/2} \Sigma_r (\nu_r/\sigma_r)^{1/2} \dot{X}_r \{N_r^{1/2} e^{i\theta r/h} + (N_r + 1)^{1/2} e^{-i\theta r/h}\} \qquad (30)]$$

leads to the correct expressions for *Einstein's A's and B's*. We must first modify slightly the analysis of § 5 so as to apply to the case when the system has a large number of discrete *stationary states* instead of a continuous range. …

…

The present theory, since it gives a proper account of *spontaneous emission*, must presumably give the effect of *radiation reaction* on the emitting system, and enable one to calculate the natural breadths of *spectral lines*, *if one can overcome the mathematical difficulties involved in the general solution of the wave problem corresponding to the Hamiltonian (30)*. Also, the theory enables one to understand how it comes about that there is no violation of the law of the *conservation of energy* when, say, a *photo-electron* is emitted from an atom under the action of extremely weak incident *radiation*. The *energy of interaction* of the atom and the *radiation* is a *q-number* that does not commute with the first integrals of the *motion* of the atom alone or with the *intensity* of the *radiation*. Thus, one cannot specify this energy by a *c-number* at the same time that one specifies the *stationary state* of the atom and the *intensity* of the *radiation* by *c-numbers*. *In particular, one cannot say that the interaction energy tends to zero as the intensity of the incident radiation tends to zero*. There is thus always an unspecifiable amount of *interaction energy* which can supply the *energy* for the *photo-electron*.

***Summary.***

The problem is treated of *an assembly of similar systems satisfying the Einstein-Bose statistical mechanics, which interact with another different system*, a Hamiltonian function being obtained to describe the motion. *The theory is applied to the interaction of an assembly of light-quanta* [*graviton*] *with an ordinary atom, and it is shown that it gives Einstein's laws for the emission and absorption of radiation* [*gravitation*]. The interaction of an atom with *electromagnetic* [*gravitational*] *waves* is then considered, and it is shown that if one takes the *energies* and *phases* of the waves to be *q-numbers* satisfying the proper *quantum conditions* instead of *c-numbers*, the Hamiltonian function takes the same form as in the *light-quantum* [*graviton*] treatment. *The theory leads to the correct expressions for Einstein's A's and B's*."

## (8) Elementary and composite particles with the same electric charge attract each other, and elementary and composite particles with opposite electric charge are repulsed, through the electromagnetic interaction or electromagnetic force, according to Coulomb's law.

**Coulomb's Law; Ampère's Force Law for Magnetism.**

**Underwood, T. G. (2023).** ***Electricity & Magnetism.*** Part II: Coulomb's Law (1785), pp. 52-4: "Coulomb's inverse-square law, or simply *Coulomb's Law*, is an experimental law that calculates *the amount of force between two electrically charged particles at rest*. This electric force is conventionally called the *electrostatic force* or *Coulomb force*. Although the law was known earlier, it was first published by Coulomb in 1785. *Coulomb's Law was essential to the development of the theory of electromagnetism and maybe even its starting point*, as it allowed meaningful discussions of the amount of *electric charge* in a particle.

> *Coulomb's Law states that the magnitude, or absolute value, of the attractive or repulsive electrostatic force between two point-charges is directly proportional to the product of the magnitudes of their charges and inversely proportional to the squared distance between them.*

Coulomb discovered that *bodies with like electrical charges repel*: "It follows therefore from these three tests, that *the repulsive force* that the two balls – [that were] electrified with the same kind of electricity – exert on each other, *follows the inverse proportion of the square of the distance*". He also showed that *oppositely charged bodies attract according to an inverse-square law.*

*Coulomb's Law* states that the magnitude of the *electrostatic force* F between two point-charges is given by

$$| F | = k_e | q_1 || q_2 |/r^2$$

where $k_e$ is the *Coulomb constant*, $q_1$ and $q_2$ are the *quantities of each charge*, and r is the *distance between the charges.*

The *Coulomb constant* is a proportionality factor that appears in Coulomb's law and related formulas. Denoted $k_e$, it is also called the *electric force constant* or *electrostatic constant* hence the subscript 'e'. The *Coulomb constant* is given by $k_e = 1/4\pi\varepsilon_0$. The constant $\varepsilon_0$ is the *vacuum electric permittivity* (also known as the *electric constant*).

Since the 2019 redefinition of the SI base units, the *Coulomb constant*, as calculated from CODATA 2018 recommended values, is

$k_e = 8.9875517923(14) \times 109\ N \cdot m^2 \cdot C^{-2}$.]

[Ancient cultures around the Mediterranean knew that certain objects, such as rods of amber, could be rubbed with cat's fur to attract light objects like feathers and pieces of paper. Thales of Miletus made the first recorded description of static electricity around 600 BC, when he noticed that friction could make a piece of amber attract small objects.

In 1600, English scientist William Gilbert made a careful study of *electricity* and magnetism, *distinguishing the lodestone effect from static electricity produced by rubbing amber*. He coined the Neo-Latin word *electricus* ("of amber" or "like amber", from ἤλεκτρον [*elektron*], the Greek word for "amber") to refer to the property of attracting small objects after being rubbed. This association gave rise to the English words "electric" and "electricity", which made their first appearance in print in Thomas Browne's *Pseudodoxia Epidemica* of 1646.

Early investigators of the 18th century who suspected that the electrical force diminished with distance as the force of gravity did (i.e., as the inverse square of the distance) included Daniel Bernoulli and Alessandro Volta, both of whom measured the force between plates of a capacitor, and Franz Aepinus who supposed the inverse-square law in 1758. Based on experiments with electrically charged spheres, Joseph Priestley of England was among the first to propose that electrical force followed an inverse-square law, similar to Newton's law of universal gravitation. However, he did not generalize or elaborate on this. In 1767, he conjectured that the force between charges varied as the inverse square of the distance. In 1769, Scottish physicist John Robison announced that, according to his measurements, the force of repulsion between two spheres with charges of the same sign varied as $x^{-2.06}$. In the early 1770s, the dependence of the force between charged bodies upon both distance and charge had already been discovered, but not published, by Henry Cavendish of England. In his notes, Cavendish wrote, "We may therefore conclude that the electric attraction and repulsion must be inversely as some power of the distance between that of the 2 + 1/50th and that of the 2 − 1/50th, and there is no reason to think that it differs at all from the inverse duplicate ratio". Finally, in 1785, the Coulomb published his first three reports of *electricity* and *magnetism* where he stated his law.]

Being an *inverse-square law*, the law is similar to Isaac Newton's inverse-square law of universal gravitation, but gravitational forces always make things attract, while electrostatic forces make charges attract or repel.

> [*Coulomb's Law* can be used to derive *Gauss's Law* (below), and vice versa. In the case of a single point charge at rest, the two laws are equivalent, expressing the same physical law in different ways. The law has been tested extensively, and observations have upheld the law on the scale from $10^{-16}$ m to $10^{8}$ m.]

The *scalar form* of *Coulomb's Law* gives the magnitude of the vector of the *electrostatic force* F between two point-charges $q_1$ and $q_2$, but not its direction. The magnitude of the force is

$$| F | = | q_1 q_2 | / 4\pi\varepsilon_0 r^2$$

where $\varepsilon_0$ is the *electric constant*, and r is the *distance between the charges*. If the product $q_1q_2$ is positive, *the force between the two charges is repulsive; if the product is negative, the force between them is attractive.*

The *vector form* of *Coulomb's Law* states that the *electrostatic force* $\mathbf{F}_1$ experienced by a charge, $q_1$ at position $\mathbf{r}_1$, in the vicinity of another charge, $q_2$ at position $\mathbf{r}_2$, in a vacuum is equal to

$$\mathbf{F}_1 = q_1 q_2 / 4\pi\varepsilon_0 \, \mathbf{r}\hat{}_{12} / | \mathbf{r}_{12} |^3.$$

where $\mathbf{r}_{12} = \mathbf{r}_1 - \mathbf{r}_2$ is the *displacement vector between the charges*, $\mathbf{r}\hat{}_{12} \equiv \mathbf{r}_{12}/| \mathbf{r}_{12} |$ a *unit vector* pointing from $q_2$ to $q_1$, and $\varepsilon_0$ the *electric constant*. Here, $\mathbf{r}\hat{}_{12}$ is used for the vector notation.

The force is along the straight line joining the two charges. *If the charges have the same sign, the electrostatic force between them makes them repel; if they have different signs, the force between them makes them attract.*

There are three conditions to be fulfilled for the validity of Coulomb's inverse square law:

1. The charges must have a spherically symmetric distribution (e.g. be point charges, or a charged metal sphere).
2. The charges must not overlap (e.g. they must be distinct point charges).
3. The charges must be stationary with respect to a nonaccelerating frame of reference.

The last of these is known as the *electrostatic approximation.*

*Coulomb's Law holds even within atoms, correctly describing the force between the positively charged atomic nucleus and each of the negatively charged electrons*. This simple law also correctly accounts for the forces that bind atoms together to form molecules and for the forces that bind atoms and molecules together to form solids and liquids. Generally, as the distance between ions increases, the force of attraction, and binding energy, approach zero and ionic bonding is less favorable. As the magnitude of opposing charges increases, energy increases and ionic bonding is more favorable."

**Underwood, T. G. (2023).** ***Electricity & Magnetism.*** Part II: Ampère's Force Law (June, 1822), pp. 124-5: "*Ampère's Force Law* was first described in Ampère, A.-M. (1822). *Memoire sur la Determination de la formule qui represente l'action mutuelle de deux portions infiniment petites de conducteurs voltaiques*. (Memoir on the Determination of the Formula which Represents the Mutual Action of Two Infinitely Small Portions of Voltaic Conductors.) read at the Royal Academy of Sciences, in the session of June 10, 1822.

*Ampere's Force Law is a relationship between the magnetic field of a closed path and the current around this path.* It states that *there is an attractive or repulsive force between two parallel wires carrying an electric current which is proportional to their lengths and to the intensities of their currents.*

It can be viewed as an alternative version of the *Biot-Savart Law* and can be applied to various physical situations. This law is particularly useful when calculating the current distributions with considerable symmetry, or determining the resulting magnetic field from a symmetric current setup. This is a similar concept to Gauss' Law, which calculates *electric flux*. The physical origin of the force is *that each wire generates a magnetic field, following the Biot–Savart Law, and the other wire experiences a magnetic force as a consequence, following the Lorentz Force Law.*

The best-known and simplest example of *Ampère's Force Law*, states that the *magnetic force* between two straight parallel conductors is given by

$$F_m = 2k_A\, I_1 I_2 L/r,$$

where $k_A$ is the *magnetic force constant* from the *Biot–Savart law*, $F_m$ is the *total force* on either wire (the longer is approximated as infinitely long relative to the shorter), L is the length of the shorter wire, r is the distance between the two wires, and $I_1$, $I_2$ are the *direct currents* carried by the wires.

This is a good approximation if one wire is sufficiently longer than the other, so that it can be approximated as infinitely long, and if the distance between the wires is small compared to their lengths (so that the one infinite-wire approximation holds), but large compared to their diameters (so that they may also be approximated as infinitely thin lines). The value of $k_A$ depends upon the system of units chosen, and the value of $k_A$ decides how large the unit of current will be.

In the SI system, by definition,

$$k_A = \mu_0/4\pi$$

with $\mu_0$ the *magnetic constant*, in SI units

$$\mu_0 = 1.25663706212(19) \times 10^{-6}\ \text{H/m}.$$

The general formulation of the magnetic force for arbitrary geometries is based on iterated line integrals and combines the *Biot–Savart Law* and *Lorentz Force* in one equation. To determine the force between wires in a material medium, the *magnetic constant* is replaced by the actual *permeability* of the medium.

In this form, it is immediately obvious that the force on wire 1 due to wire 2 is equal and opposite the force on wire 2 due to wire 1, in accordance with *Newton's third law of motion.*

The form of *Ampère's Force Law* commonly given was derived by James Clerk Maxwell in 1873 and is one of several expressions consistent with the original experiments of André-Marie Ampère and Carl Friedrich Gauss. The x-component of the force between two linear currents I and I', as depicted in the diagram below, was given by Ampère in 1825 and Gauss in 1833 as follows:

$$dF_x = kII'ds' \int ds \frac{\cos(xds)\cos(rds') - \cos(rx)\cos(dsds')}{r^2}.$$

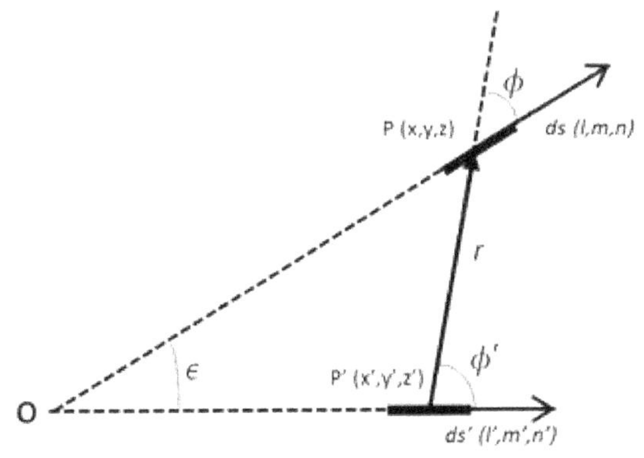


Diagram of original Ampère experiment."

## Coulomb, C-A. (1785). Premier mémoire sur l'électricité et le magnétisme.

*Histoire de l'Académie Royale des Sciences avec les mémoires de mathématiques et de physique, partie "Mémoires"*, pages 569–577; http://www.ampere.cnrs.fr/i-corpuspic/tab/Sources/coulomb/Coulomb_El_1785.pdf; translation by L. L. Bucciarelli, Emeritus Professor of Engineering and Technology Studies (MIT), MIT, 2000 (revised and notes added by Christine Blondel and Bertrand Wolff, 2012); also in Underwood, T. G. (2024). *Electricity & Magnetism.* Part II: pp. 55-61.

**Construction et usage d'une balance électrique, fondée sur la propriété qu'ont les fils de métal, d'avoir une force de réaction de torsion proportionnelle à l'angle de torsion. (Construction and use of an electric Balance based on the property that filaments of metal produce a reactive force in torsion proportional to the angle of twist.)**

---

***Experimental determination of the law according to which the elements of bodies, electrified with the same kind of electricity, are mutually repelled.***

Coulomb described the construction of a torsion balance and used this to demonstrate what he described as the *fundamental law of electricity*, now known as *Coulomb's Law. The law states that the magnitude, or absolute value, of the attractive or repulsive electrostatic force between two point-charges is directly proportional to the product of the magnitudes of their charges and inversely proportional to the squared distance between them.*

---

In a memoir presented to the Academy, in 1784, I have determined from experiments *the laws governing the torsional resistance of a filament of metal* and I have found that this force is proportional to the angle of torsion, to the fourth power of the diameter of the suspended filament and inversely proportional to its length - all multiplied by a constant coefficient which depends on the nature of the metal and is easily determined by experiment.

I have shown in the same Memoir that *by means of this force of torsion, it was possible to precisely measure extremely small forces* as, for example, one ten thousandths of a grain[1].

[1] 1 grain = 0,053 g

In the same Memoir I described a first application of this theory, seeking to evaluate the constant force attributed to adhesion in the formula for the surface friction of a solid body moving through a fluid.

*Today, I set before the eyes of the Academy, an electric balance constructed according to the same principles*. It measures with the greatest precision the state and the electric force of a body, however weak the degree of electricity.

***Construction of the Balance (Pl. XIII)***

While practice has taught me that, in order to execute several electric experiments in a convenient way, it is necessary to correct some defaults in the first balance of this kind that I put to use, still, as this has been up until now the only one which I have employed, I am going to give its description, noting how its form and dimensions can and ought to be changed according to the nature of the experiments one plans to carry out. The first figure presents the balance in perspective. Here are the details.

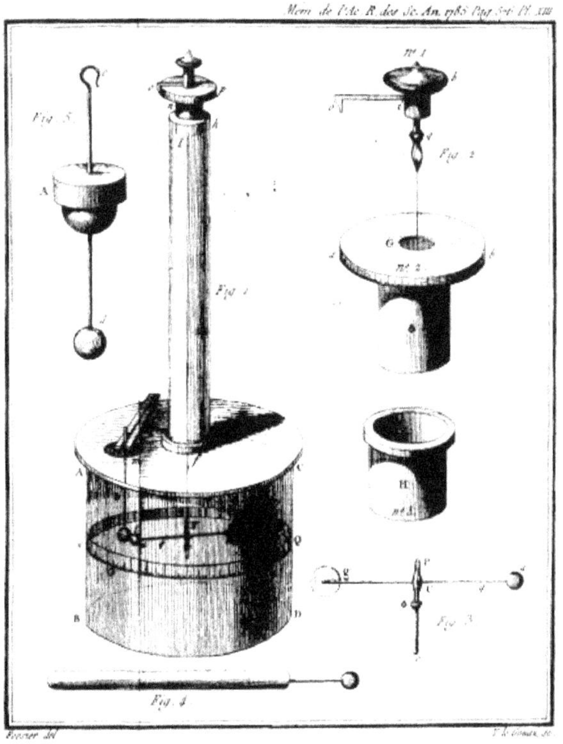

On a cylinder of glass ABCD, of 12 [pouces[2]]

[2] 1 pouce = 2,7 cm

in diameter and 12 [pouces] high, we place a plate of glass of 13 [pouces] in diameter, which covers the whole vessel of glass. This cover is pierced with two holes of approximately 20 [lignes[3]]

[3] 1 ligne = 0, 23 cm

in diameter, one in the middle, at f, on which is elevated a tube of glass of 24 [pouces] in height. This tube is cemented over the hole f, with the cement used in electric apparel: at the highest extremity of the tube at h, is placed a torsion micrometer the details of which are shown in *figure 2*. The top, n° 1, bears a knob b, the pointer io, and the clasp of suspension, q. This piece goes into the hole G of part n° 2. Part n° 2 is formed of a circle *a*b divided on its edge into 360 degrees, and of a copper tube Φ which fits into the tube H, n° 3, sealed at the interior at its highest extremity of the tube or of the glass stem fh of figure 1. The clasp q (*figure 2*, n° 1) has approximately the shape of the extremity of a solid mechanical pencil clamp [porte-crayon], which can be tightened by means of the annulus q. The clasp of this pencil clamp holds the end of a filament of very thin silver. The other end of the filament of silver is fixed (*fig. 3*) at P, by the clasp of a cylinder Po of copper or of iron, whose diameter is but a [ligne], and whose end P is split and forms a clasp which is tightened by means of the collar Φ. This small cylinder has a hole in C, in order to allow the needle *a*g to slide through (*fig. 1*). It is necessary that the weight of this small cylinder be of sufficient magnitude in order to put the filament of silver in tension without breaking it. The needle that one sees (*fig. 1*) at *a*g, suspended horizontally at approximately the midpoint of the height of the big vase which encloses it, is formed, either of a filament of silk plastered with sealing wax, or of a straw likewise covered with Spanish wax, and finished off from q to *a*, a distance of 18 [lignes], by a cylindrical filament of shellac. At the end *a* of this needle, is a small ball of pith of two to three [lignes] diameter. At g, is a small vertical plane of paper coated with turpentine which serves as a counter-weight to the ball a and which dampens the oscillations.

We have said that the cover AC is pierced with a second hole at m. It is in this second hole that one introduces a small cylinder mΦt, of which the lower part Φt is shellac. At t is a ball likewise made of pith. Around the vase, at the height of the needle, one scribes a circle zQ divided into 360 degrees: for simplicity, I have made use of a band of paper divided into 360 degrees, which I glue around the vase, at the height of the needle.

To begin to operate with this instrument, in placing the cover (atop the vase), I position the hole m approximately at the first division or at the point O of the circle zOQ traced on the vase. I place the pointer oi of the micrometer on the point O, at the first division of this micrometer. I then turn the micrometer within the vertical tube fh until, keeping in view the vertical filament which suspends the needle and the center of the ball, the needle *a*g is directed towards the first division of the circle zOQ. I then introduce through the hole m the other ball t suspended by the filament mΦ t such that it touches the ball *a* and that, in keeping in view the center of the filament of suspension and the ball t, one encounters the first division o of the circle zOQ. The balance is now in a state ready for all operations; *we go on to give as an example, the means by which we are able to determine the fundamental law according to which electrified bodies repel themselves.*

***The fundamental law of Electricity.***

*The repulsive force of two small globes electrified with the same kind of electricity, is inversely proportional to the square of the distance between the centers of the two globes.*

***The Experiment***

One *electrifies*, *fig*. 4, a small conductor, which is nothing but a pin with a large head, which is isolated by forcing its point into the end of a rod of Spanish wax. One introduces this pin into the hole m, bringing it in contact with the ball t, [which is] in contact with the ball *a*. In retracting the pin *the two balls find themselves electrified with the same kind of electricity and they repel themselves mutually* to a distance that one measures in viewing, (keeping in line) the filament of suspension and the center of the ball *a*, the division corresponding to the circle zOQ. Turning then the pointer of the micrometer in the direction pno, one torques the filament of suspension *l*p, and one produces a force proportional to the angle of torsion, which tends to make the ball *a* approach the ball t. One observes, by this means, the distance to which different angles of torsion bring the ball *a* back towards the ball t and in comparing the forces of torsion with the corresponding distances of the two balls, one determines the law of repulsion.

I will only present here, some tests which are easy to repeat and which will immediately reveal the law of repulsion.

*First Test*. Having electrified the two balls with the head of the pin, with the pointer of the micrometer positioned at o, the ball *a* of the needle is displaced from the ball t by 36 degrees.

*Second Test*. Having torqued the filament of suspension by means of the knob o of the micrometer by 126 degrees, the two balls approach each other and stop at 18 degrees distance the one from the other.

*Third Test*. Having torqued the filament of suspension by 567 degrees, the two balls approach until 8 and one-half degrees.

***Explication and results of this experiment.***

Before the balls are electrified, yet touching, the center of the ball *a*, fixed to the needle, is at a distance equal to the diameter of the balls from the point where the torsion of the filament of suspension is null. It is necessary to be warned that the filament of silver *l*P, 28 [pouces] long, which forms the suspension is so fine that the weight of one pied[4]

[4] 1 pied = 32,5 cm

of length is but 1/16 of a grain. In calculating the force required to twist this filament, in acting at the point *a* elongated some four [pouces] from the filament *l*P or from the center of suspension, I have found, using the formulas derived in a *Memoire on the laws of the force of torsion of filaments of metal*, printed in the volume of the *Academie* for 1784, that to torque this filament 360 degrees, requires at point *a*, in acting with the lever *a*P of four [pouces] of length, a force of 1/340 of a grain. Thus, as the forces of torsion are, as proved in the *Memoire*, proportional to the angles of torsion, the least repulsive force between the two balls, displaces them sensibly one from the other.

We find in our *first test*, where the pointer of the micrometer is at the point o, that the balls are displaced 36 degrees, which produces in the same time a force of torsion of 36° = 1/3400 of a grain; in the *second test*, the distance of the balls is 18 degrees, but as one has torqued the micrometer 126 degrees, it results that at the distance of 18 degrees, the repulsive force is 144 degrees: thus at half of the first distance, *the repulsion of the balls is quadrupled.*

In the *third test*, where we have twisted the filament of suspension 567 degrees, the two balls find themselves no further apart than 8- and one-half degrees. The total torsion, is consequently, 576 degrees, *quadruple the one of the second test*, and it is only off by one half a degree that the distance of the two balls in this third test was reduced to half of the distance it was in the second. *It results thus from these three tests, that the repulsive action of the two balls electrified with the same kind of electricity exert on upon the other was the inverse ratio of the square of the distances.*

***First Remark***

In repeating the preceding tests, we will observe that in making use of a filament of silver, as thin as the one we have employed (which only requires a torsional force of approximately 24 thousands of a grain to twist it through an angle of 5 degrees) that however calm be the air, and whatever precautions that one can take, we could not answer for the natural position of the needle when the torsion is zero, to within 2 or 3 degrees. Thus, in order to have a first test to compare with the following ones, it requires, after having electrified the two balls, to torque the filament of suspension some 30 to 40 degrees, this which will give a force of torsion strong enough so that the 2 or 3 degrees of uncertainty in the initial position of the needle, when the torsion is zero, does not produce any sensible error in the results. It is necessary furthermore to be warned that the filament of silver, which I used in this test, is so fine that it breaks with the least disturbance. I have found in the following that it would be more useful to employ in these tests a filament of suspension of nearly double the diameter, although its flexibility to torsion be 14 to 15 times smaller than that of the first. It is necessary to take care, before making use of this filament of silver, over the course of two or three days, to tension it by a weight which is approximately half this that might break it. It is necessary still yet to warn the reader, that in using this last filament of silver, never to torque it beyond 300 degrees because in exceeding this degree of torsion it begins to strain-harden and reacts, as we have proven in the *Memoir* already cited, printed in 1784, with a force smaller than the one corresponding to the torsion angle.

### *Second Remark*

*The electricity of the two balls diminishes somewhat over the duration of the experiment.* I noticed that, the day where I have made the preceding tests, the electrified balls, finding themselves repulsed to 30 degrees one from the other, under an angle of torsion of 50 degrees, they come back toward each other about one degree in three minutes. But as I have only used two minutes to make the three preceding tests, one can, in these tests, neglect the error which results from the loss of electricity. If one desires greater precision, as when the air is humid, and the electricity dissipates rapidly, one ought, by a preliminary observation, determine the law of diminution of the electric action of the two balls in each minute, and then, on the basis of this preliminary observation, use it to correct the results of tests that one wishes to make that day.

### *Third Remark*

The distance of the two balls, when they are displaced one from the other by their reciprocal repulsive action, is not precisely measured by the angle they make, but by the cord of the arc which joins their centers. In the same way that the lever at whose extremities acts the action, is not measured by the mean of the length of the needle, or by the radius, but by the sine of the half of the angle formed by the distance of the two balls. These two quantities, of which one is smaller than the arc and diminishes consequently the distance measured by this arc, while the other reduces the lever arm, compensating themselves in some way; and in the tests of this sort with which we are concerned, one can without sensible error, hold to the evaluation that we have given, if the distance of the two balls does not exceed 25 to 30 degrees; in the other cases, one must make the rigorous calculation.

### *Fourth Remark*

As experience has shown, in a well closed chamber, one can determine with the first filament of silver, to within 2 or 3 degrees, the position of the needle, when the torsion is null, this which gives, after the calculation of the forces of torsion, proportional to the angle of torsion, a force more or less of a 40 thousandth of a grain, the weakest degree of electricity will be measurable easily with this balance. For this operation, one makes pass, *fig 5*, across a cork of Spanish wax, a small filament of copper cd, terminating at c by a crochet and in d, by a small ball of gilded pith, and we put the cork A in the trough m of the balance, *fig 1*, in such a way that the center of the ball d, viewed by the string of suspension, retakes to point o of the circle zOQ. In approaching then an electrified body of the crochet o, however weak be the electricity of this body, the ball *a* separates from the ball d, giving the sign of the electricity and the distance of the two balls as a measure of the force, according to the principle of the inverse ratio of the square of the distances.

But I ought to warn you that, since these first tests, I have had different small electrometers made according to the same principles of the force of torsion, using a filament of silk for the suspension, such as it leaves the cocoon, or a thread of the sheep of Angora. One of these electrometers which has almost the same shape as the electric balance, described in this *Memoir*, is much smaller. It is only 5 to 6 [pouces] in diameter, a stem of one pouce; the needle is a small filament of shellac of 12 [lignes] of length, terminated at *a* by a small very light circle of tinsel.

The needle and the tinsel weigh a little more than a quarter of a grain; the filament of suspension, such as it leaves from the cocoon, is 4 [pouces] long, having a flexibility such that in acting with a lever arm of one pouce, it only requires a sixtieth thousand of a grain to twist it an entire circle or 360 degrees. In presenting in this electrometer at the crochet C of figure 5, an ordinary rod of Spanish wax, *electrified by friction*, at a 3 feet distance from this crochet, the needle is chased to more than 90 degrees. We will describe in more detail in the following this electrometer, when we will determine the nature and the degree of electricity of different bodies which through rubbing each other, take on a very weak degree of electricity.

## Ampère, A-M. (1822). Memoire sur la Determination de la formule qui represente l'action mutuelle de deux portions infiniment petites de conducteurs voltaıques. (Memoir on the Determination of the Formula which Represents the Mutual Action of Two Infinitely Small Portions of Voltaic Conductors.)

*Annales de Chimie et de Physique*, 20, 398–419 (in French); also in Ampère, A-M. (1822), *Recueil d'observations électro-dynamiques: contenant divers mémoires, notices, extraits de lettres ou d'ouvrages périodiques sur les sciences, relatifs a l'action mutuelle de deux courans électriques, à celle qui existe entre un courant électrique et un aimant ou le globe terrestre, et à celle de deux aimans l'un sur l'autre* (in French), Chez Crochard, Paris; http://www.ampere. cnrs.fr/bibliographies/pdf/1822-P097.pdf, pp. 293-318; translation by T. G. Underwood; also in Mémoires sur l'Électromagnétisme et l'Électrodynamique. *Nature*, 109, 677–678 (May 27, 1922); https://doi.org/10.1038/109677d0; also in Underwood, T. G. (2024). *Electricity & Magnetism.* Part II: pp. 126-8.

Read at the Royal Academy of Sciences, in the session of June 10, 1822.

In this paper Ampère derived his force law, $F_m = 2k_A I_1 I_2 L/r$ where $k_A$ is the *magnetic force constant* from the *Biot–Savart law*, $F_m$ is the *total force* on either wire (the longer is approximated as infinitely long relative to the shorter), L is the length of the shorter wire, r is the distance between the two wires, and $I_1$, $I_2$ are the *direct currents* carried by the wires.

---

***Abstract.***

THE two memoirs given in this volume have been taken from Ampère's wonderful "*Recueil d'observations electrodynamiques*" published in 1822. Ørsted had described a few years previously the action of an electric current on a compass needle, and in the first memoir under notice, the mutual action of two electric currents on one another is described. The author then describes the apparatus he made and the experiments he carried out. Finally, he formulates the laws which we use to-day. *In the second memoir the formula for the mutual action between two infinitely small elements of conductors carrying currents is proved.* Ampère's researches paved the way for much of Faraday's work, and Clerk Maxwell makes full use of his results in his treatise. …

---

**Second Memoir**

***On the Determination of the Formula which Represents the Mutual Action of Two Infinitely Small Portions of Voltaic Conductors.***

[p. 293.] When a new kind of action hitherto unknown is discovered, the first object of the physicist must be to determine the principal phenomena which result from it, and the circumstances under which they occur; it then remains to find the means of applying the calculation by representing by formulas the value of the forces exerted on each other by the particles of the bodies in which this kind of action is manifested. As soon as I had recognized that two voltaic conductors acted on each other, sometimes by attracting, sometimes by repelling each other, and that I had distinguished and described the actions they exercise in the different situations in which they may find themselves in relation to each other, in order to be able to deduce, by the known methods of integration, the action which takes place between two given portions of conductors of form and situation.

The impossibility of subjecting infinitely small portions of the voltaic circuit directly to experiment made necessary observations made on conductive filaments of finite magnitude, and this must consequently satisfy two conditions, that the observations be capable of great precision, and that they be suitable for determining the value of the mutual action of two infinitely small portions. These can be satisfied in two ways: one consists in measuring with the greatest accuracy the values of the mutual action of two portions of a finite quantity, by placing them successively in relation to each other, at different distances and in different positions, for it is evident that here the action does not depend only on the distance; it is then necessary to make an assumption about the value of the mutual action of two infinitely small portions, to conclude that the action that must result from it for the conductors of finite magnitude on which we have operated, and to modify the hypothesis until the results of the calculation agree with those of the observation. It is this procedure that I had first proposed to follow, as I explained in detail in a Memoir read to the Academy of Sciences on October 9, 1820[1]; and although it leads us to the truth only by the indirect way of hypotheses, it is none the less valuable, since it is often the only one that can be employed in researches of this kind.

[1] This Memoir has not been published separately, but the principal results have been inserted in that which I published in 1820 in volume xv of *the Annals of Chemistry and Physics.*

One of the members of this Academy, whose works have embraced all parts of physics, has perfectly described this in the *Notice on the magnetization imparted to metals in motion*, which he read to us on April 2, 1821, calling him to a work in a way of divination which is the end of almost all physical research[2].

[2] See the *Journal des Savants*, April, 1821, page 255.

But there is another way of attaining the same end more directly, and that is that which I have since followed, and which has led me to the result I desired; it is necessary to ascertain by experiment that the moving parts of the conductors are, in certain cases, exactly in equilibrium between equal forces or equal moments of rotation, whatever the shape of the moving part, and to seek directly, with the help of calculation, what must be the value of the mutual action of two infinitely small portions, so that the balance is indeed independent of the shape of the moving part.

This is how I determined this value by combining two experiments of this kind; one which I described in a Memoir read at the Academy on December 26, 1820, and in this collection, pag. 216 and following; the other of which I have just ascertained with all possible accuracy.

…

[p. 309.] … that the intensities of action of the two small portions of conductors which I have named g and h in the note in the *Journal de Physique*, will be represented here because their lengths are ds and ds' by ids and i' ds' and that their mutual action will be represented by

$$\rho \, ii' \, ds ds'/r^n$$

the exponent n being equal to 2, if this action is, all other things being equal, in inverse proportion to the square of the distance, as I have admitted from my first work on electrodynamic phenomena, basing myself in truth rather on analogy than on direct proofs.

…

## 9. The spin of elementary and composite particles creates an attractive force – the weak interaction or weak force - through exchange interaction resulting from entanglement between two quantum spin states.

**Quantum entanglement between spin states.**

*Quantum entanglement* is the phenomenon of a group of particles being generated, interacting, or sharing spatial proximity in such a way that the quantum state of each particle of the group cannot be described independently of the state of the others, including when the particles are separated by a large distance. The topic of *quantum entanglement* is at the heart of the disparity between classical and quantum physics: *entanglement* is a primary feature of *quantum mechanics* not present in *classical mechanics*.

One of the most common forms of *quantum entanglement* is between the two *quantum spin states* of a particle. For example, if a pair of *entangled* particles is generated such that their *total spin* is known to be zero, and *one particle is found to have clockwise spin on a first axis, then the spin of the other particle, measured on the same axis, is found to be anticlockwise*. However, this behavior gives rise to seemingly paradoxical effects: *any measurement of a particle's properties results in an apparent and irreversible wave function collapse of that particle and changes the original quantum state*.

## Heitler, W. & London, F. (June, 1927). Wechselwirkung neutraler Atome und homöopolare Bindung nach der Quantenmechanik. (Interaction of neutral atoms and homeopolar bonding according to quantum mechanics.)

*Zeit. Phys.*, 44, 455–72. https://doi.org/10.1007/BF01397394; also at http://quantum-chemistry-history.com/Heitler_London_Dat/WechselWirk1927/WechselWirk1927.htm (in German); translation by T. G. Underwood; also in Underwood, T. G. (2024). *Quantum Entanglement*, Part II, pp. 158-72.

Received 30 June 1927.

Zürich, Physikal. Institut der Universität.

[1] Presented at the conference of the German Physical Society Freiburg i. Br., June 12, 1927.

With 2 illustrations.

Heitler and London examined the interaction between *neutral atoms* though non-polar bonds, in what is known as valance bonds; and applied quantum mechanics to calculate the *interaction energy* of the atoms when they move closer together. They found that two neutral atoms could interact with each other in two ways; *the problem was twofold degenerate, corresponding to the two ways of assigning the electrons to the neutral atoms* (known as *quantum entanglement*). Examination of the different cases of two H atoms and two He atoms showed that by applying the *Pauli principle*, the selected eigenfunctions of the system should change or maintain their sign respectively, when two electrons were swapped, if the two electrons compared had the same or different *spin*. They found that in the case of He there was only one solution, which yielded about the right size of the He gas kinetic-radius, *due to the fact that 2 He atoms (and the same applies to all noble gases) cannot differ in their spin* – in contrast to hydrogen (and all atoms with unfinished shells) – so that 2 He atoms have only one possible mode of behaving. They questioned whether the *exchange phenomenon*, which was so decisive for the interaction of atoms, was also noticeable in other branches of physics.

---

***Abstract***

The interplay of forces between *neutral atoms* shows a characteristic quantum mechanical ambiguity. This ambiguity is therefore capable of encompassing the various modes of relation which experience furnishes: in the case of hydrogen, for example, the possibility of a homopolar bond or elastic reflection, whereas in the case of noble gases only the latter

— and this already as an effect of a first approximation of approximately the right magnitude. In the selection and discussion of the different modes of behavior, the Pauli principle also proves itself here, when applied to systems of several atoms.

---

The interaction between *neutral atoms* has given rise to a theoretic treatment which has so far caused considerable difficulties. Whilst the attractions of ions make a simple picture which has been known for a long time, the conditions in the case of neutral atoms, in particular the possibility of a *non-polar bond*, are quite difficult to understand, if one did not want to resort to very artificial explanations [2].

[2] For literature on this subject, see e.g. *Handb. d. Phys.*, XXII, 1926, article Herzfeld.

The development of *quantum mechanics has provided new points of view for the treatment of these problems: first of all, in the new "models" the charge distribution is completely different* [3] *than in the Bohr models* (i.e. decay like $e^{-r}$), which would already result in a completely different interplay of forces between "neutral" atoms.

[3] As is well known, *the correspondence between classical and quantum mechanical quantities relates only to the electric moments,* i.e., to the focal points of the charge, not to the charge distribution itself, which difference does not come into play in the usual spectroscopic questions.

However, a characteristic quantum mechanical *beating phenomenon*, which is closely related to the resonance oscillations introduced by Heisenberg, is crucial for the understanding of the possible behaviors between *neutral atoms*. We will study these ratios using the example of two H-atoms (§ 1) and two He-atoms (§ 3). To anticipate the result (§ 2): for the *interaction energy*, one obtains a solution: one which is attractive at medium distances of the atoms, and one at small distances and which are necessary for a *homopolar* molecule formation (already in a first approximation, in which perturbations caused by polarization are to be avoided). *However, this solution is not permitted in the ground state due to the Pauli prohibition* (§ 4) for He. A second solution, which is the only one that can be considered for He, provides repulsion everywhere (van der Waals' b-forces).

## § 1. *Interaction of two hydrogen atoms.*

We set ourselves the task of determining the change in energy, which *two neutral hydrogen atoms* in the ground state experience when we approach them to the distance R (measured by the distance of the nuclei) from each other). Depending on whether this additional energy decreases or increases as the atoms gradually approach, we infer attraction or repulsion.[1]

[1] It is not superfluous to point out that the *connection between wave-mechanical frequency and mechanical energy assumed here is quite hypothetical*, for we know that it is precisely the Lorentz force approach, which governs the action of the field on matter, that is *replaced in quantum mechanics by something quite different, namely, the wave equation.*

**1.** We designate the two nuclei, the distance of which is once and for all fixed and equal to R, by *a* and b, the two electrons by numbers 1 and 2, and finally the distances of the electrons from the nuclei and from each other by $r_{a1}, \ldots, r_{12}$.

The *wave equation of our problems - of the 2-hydrogen atom problems* - then reads (with $\Delta_{12} = \Delta_1 + \Delta 2 = \partial^2/\partial x_1^2 + \partial^2/\partial y_1^2 + \partial^2/\partial z_1^2 + \partial^2/\partial x_2^2 + \partial^2/\partial y_2^2 + \partial^2/\partial z_2^2$):

$$(\chi) \equiv \Delta_{12}\,\chi + 8\pi^2 m/h^2\,\{E - (\varepsilon^2/R + \varepsilon^2/r_{12} - \varepsilon^2/r_{a1} - \varepsilon^2/r_{a2} - \varepsilon^2/r_{b1} - \varepsilon^2/r_{b2})\}\,\chi = 0 \qquad (1)$$

We are interested in those solutions $\chi$ *which correspond to the perturbations of two neutral H-atoms in the ground state*, and accordingly we shall approximate them from the well-known eigenfunctions of the hydrogen ground state.

If the electron 1 is located at the nucleus *a*, then the well-known hydrogen eigenfunction is to be assigned to it

$$\psi(1) = 1/\sqrt{\pi}\,(1/a_0)^{3/2}\,e^{-r_{a1}/a_0} \qquad (2)$$

where $a_0$ means Bohr's orbital radius of the $1_1$-orbit. The argument 1 denotes the coordinates of the electron 1 in q-space, we will write it in the future as an index, also $\psi_1$, (a measurement with the eigenvalue index is not to be feared, since we are always looking at atoms in their ground state). Accordingly,

$$\varphi_1 = 1/\sqrt{\pi}\,(1/a_0)^{3/2}\,e^{-r_{b1}/a_0} \qquad (2')$$

means that the first electron is located at nucleus b. The two eigenfunctions (2) and (2') are quite different, although they can be brought to coincide by a simple translation; because they take place in different areas of space[1].

[1] Accordingly, it should always be borne in mind in the following that the other conclusions remain unchanged, even if $\psi$ and $\varphi$ are different in their form, whereby our considerations are at once reduced to an essentially universal concept.

Corresponding eigenfunctions exist for electron 2:

$$\psi_2 = 1/\sqrt{\pi}\,(1/a_0)^{3/2}\,e^{-r_{a2}/a_0} \qquad \varphi_2 = 1/\sqrt{\pi}\,(1/a_0)^{3/2}\,e^{-r_{b2}/a_0} \qquad (2a)$$

which indicates that electron 2 is located at nucleus $a$ or b. The eigenfunctions (2a) differ from the eigenfunctions (2) and (2') only in q-space.

**2.** As unperturbed eigenfunctions we shall have to choose those which say that an electron is in one nucleus and the other electron is in another.[2]

[2] The possibility of ionization is excluded here for the time being; to what extent this is justified, we will only show later (§ 5).

If one thinks of these two as yet uncoupled systems as one system, it is well known that the product of these two eigenfunctions must be regarded as the common eigenfunction. However, depending on the distribution of the two electron ions on the two nuclei, this is possible in two ways. First of all, you have:

$$\psi_1\varphi_2 \quad \text{(1 is at } a\text{, 2 is at b).} \tag{3a}$$

With the same right, however, one also obtains:

$$\psi_2\varphi_1 \quad \text{(2 is at } a\text{, 1 is at b).} \tag{3b}$$

Both possibilities belong to the same energy of the entire system (double hydrogen energy). It is a case of twofold degeneration: all pairs of orthogonal linear combinations of $\psi_1\varphi_2$ and $\psi_2\varphi_1$:

$$\begin{aligned} \alpha &= a\, \psi_1\varphi_2 + \mathrm{b}\, \psi_2\varphi_1 \\ \beta &= \mathrm{c}\, \psi_1\varphi_2 + \mathrm{d}\, \psi_2\varphi_1 \end{aligned} \tag{4}$$

with the conditions of normalization and orthogonality

$$\left.\begin{aligned} &a^2 + \mathrm{b}^2 + 2a\mathrm{bS} = 1 \\ &\mathrm{c}^2 + \mathrm{d}^2 + 2\mathrm{cdS} = 1 \\ &a\mathrm{c} + \mathrm{bd} + (a\mathrm{d} + \mathrm{bc})\mathrm{S} = 0 \end{aligned}\right\} \tag{5}$$

(where $\mathrm{S} = \int \psi_1\varphi_1\psi_2\varphi_2\, d\tau_1\, d\tau_2$)

are to be regarded as intrinsically perturbed eigenfunctions of these problems.

**3.** The exact eigenfunctions of the differential equation (1) are derived from linear combinations (4), the eigenfunctions of "zero approximation" and differ from the latter only by a small perturbation The coefficients $a$, b, c, d in (4) for these eigenfunctions of zero approximation can be determined according to known rules. Although equation (1) lacks the character of a perturbation problem, because no term in the potential function can be separated as a "perturbation potential", the method for determining the said coefficients and the perturbation energies is exactly the same as if there were a perturbation potential. In order to make the calculations more transparent, we want to choose from the outset the correct linear combinations that are standardized according to (Ö) (and verify their correctness afterwards);

$$\left.\begin{array}{l}\alpha = 1/\sqrt{(2 + 2S)}\,(\psi_1\varphi_2 + \psi_2\varphi_1)\\ \beta = 1/\sqrt{(2 - 2S)}\,(\psi_1\varphi_2 - \psi_2\varphi_1).\end{array}\right\} \quad (4a)$$

For the perturbed eigenfunctions, we will now have to start:

$$\left.\begin{array}{l}\chi_\alpha = \alpha + \upsilon_\alpha\\ \chi_\beta = \beta + \upsilon_\beta.\end{array}\right\} \quad (6)$$

With this approach, we enter into equation (1) and take into account that the functions $\psi_1$, $\psi_2$, $\varphi_1$ and $\varphi_2$ satisfy the four equations,

$$\left.\begin{array}{l}\Delta_1\psi_1 + 8\pi^2 m/h^2\,(E_0 + \varepsilon^2/r_{a1})\,\psi_1 = 0,\\ \Delta_2\psi_2 + 8\pi^2 m/h^2\,(E_0 + \varepsilon^2/r_{a2})\,\psi_2 = 0,\\ \Delta_1\varphi_1 + 8\pi^2 m/h^2\,(E_0 + \varepsilon^2/r_{a1})\,\varphi_1 = 0,\\ \Delta_2\varphi_2 + 8\pi^2 m/h^2\,(E_0 + \varepsilon^2/r_{a2})\,\varphi_2 = 0,\end{array}\right\} \quad (7)$$

where $E_0$ means – 13.5 volts, the *eigenvalue of the hydrogen ground state*. With the designation

$$E_1 = E - 2\,E_0 \quad (8)$$

for the eigenvalue perturbation we get from (1) the two differential equations for $\upsilon_\alpha$ and $\upsilon_\beta$:

$$\begin{array}{ll}\sqrt{(2 + 2S)}\,L\upsilon_\alpha + (E_1 - \varepsilon^2/r_{12} - \varepsilon^2/R)\,(\psi_1\varphi_2 + \psi_2\varphi_1) & \\ \qquad + (\varepsilon^2/r_{a1} + \varepsilon^2/r_{b2})\,\psi_2\varphi_1 + (\varepsilon^2/r_{b1} + \varepsilon^2/r_{a2})\,\psi_1\varphi_2 & (9\alpha)\\ \sqrt{(2 - 2S)}\,L\upsilon_\beta + (E_1 - \varepsilon^2/r_{12} - \varepsilon^2/R)\,(\psi_1\varphi_2 - \psi_2\varphi_1) & \\ \qquad + (\varepsilon^2/r_{a1} + \varepsilon^2/r_{b2})\,\psi_2\varphi_1 - (\varepsilon_2/r_{b1} + \varepsilon^2/r_{a2})\,\psi_1\varphi_2. & (9\beta)\end{array}$$

In order for these equations to be solved at all in $\upsilon$ inhomogeneous equations for an eigenvalue E of the homogeneous equation, it must be required that the inhomogeneities are orthogonal to all eigenfunctions of the homogeneous equations, i.e. to $\chi_\alpha$ and $\chi_\beta$, which belong to the same eigenvalue E.

First of all, we find that the inhomogeneity of (9α) is already orthogonal to β, and the inhomogeneity of (9β) is orthogonal to α (except for negligible quantities). This verifies the correctness of the approach (4a). The other two demands:

Inhomogeneity (9α) orthogonal to α,
Inhomogeneity (9β) orthogonal to β,

deliver the two eigenvalue perturbations

$$E_\alpha = E_{11} - (E_{11}S - E_{12})/(1 + S) \quad \} \qquad (10)$$
$$E_\beta = E_{11} + (E_{11}S - E_{12})/(1 - S) \quad \}$$

with the following designations:

$$E_{11} = \int [(\varepsilon^2/r_{12} + \varepsilon^2/R)(\psi_1{}^2\varphi_{22} + \psi_2{}^2\varphi_1{}^2)/2 - (\varepsilon^2/r_{a1} + \varepsilon^2/r_{b2})\,\psi_2{}^2\varphi_1{}^2/2 \qquad (11)$$
$$- (\varepsilon^2/r_{a2} + \varepsilon^2/r_{b1})\,\psi_1{}^2\varphi_2{}^2/2]\,d\tau_1\,d\tau_2,$$
$$E_{12} = \int (2\varepsilon^2/r_{12} + 2\varepsilon^2/R - \varepsilon^2/r_{a1} - \varepsilon^2/r_{a2} - \varepsilon^2/r_{b1} - \varepsilon^2/r_{b2})\,\psi_1\varphi_1\psi_2\varphi_2/2\,d\tau_1\,d\tau_2,$$
$$S = \int \psi_1\varphi_1\psi_2\varphi_2\,d\tau_1\,d\tau_2.$$

§ 2. ***Discussion of the result.***

**1.** So we get two perturbation energies corresponding to the inputs (5a):

$$E_\alpha = E_{11} - (E_{11}S - E_{12})/(1 + S)$$

belongs to $1/\sqrt{(2 + 2S)}\,(\psi_1\varphi_2 + \psi_2\varphi_1)$ (*symmetrical* in 1 and 2),

$$E_\beta = E_{11} + (E_{11}S - E_{12})/(1 - S)$$

belongs to $1/\sqrt{(2 - 2S)}\,(\psi_1\varphi_2 - \psi_2\varphi_1)$ (*antisymmetrical* in 1 and 2).

It is a result that can only be described very artificially in classical terms, *that two neutral atoms can interact with each other in two ways*. We are still a long way from a real understanding of this fact. But it is desirable at least to be clear about how this curious ambiguity comes about mathematically. *The essential point is evidently that the problem is originally twofold degenerate* (la and lb), *corresponding to the two ways of assigning the electrons to the neutral atoms*[1].

[1] In classical mechanics, this problem has not degenerated: the same energy value belongs to the two electron arrangements. But the criterion of a degeneration of classical mechanics is not $E_k - E_l = 0$ but $\nu \equiv \partial E/\partial J = 0$ (a "proximity relationship" in terms of the effects of variable J). In fact, the electrons in this case also have their full degree of periodicity. Only when the energy is sufficient to overcome the potential visual wave between the two atoms (or when the two atoms are sufficiently approached), does now $E_k - E_l = 0$ result in $\partial E/\partial J = 0$ and the problem is degenerate. *The electrons now describe orbits around both nuclei and are constantly exchanging information*. In *quantum mechanics* this distinction does not occur [cf. Hund, F. (1926). *Zeit. Phys*., 40, 442]. *No matter how high the (finite) potential threshold, there is always a certain probability of an exchange of electrons*.

While in classical mechanics there is a way to attach a label to electrons (place each electron in a sufficiently deep potential well and keep away from a large amount of energy), something similar is impossible in quantum mechanics. Even if you know an electron in a potential well at an instant, you are never sure whether it has not already exchanged with another in the next moment. *For this reason, a statistic which in principle discriminates against an individualization of electrons and takes into account only their numerical distribution among the states, such as the Bose or Fermi statistics, is so immensely adapted to the possibilities of quantum mechanical description.*

The abolition of this degeneracy is linked as can be seen from (10), the non-disappearance of the quantum function of the atom *a* at the place of the atom b and vice versa (otherwise $\psi_1\varphi_1 = 0$ and consequently $E_{12} = 0$ and $S = 0$); but *this means that there must be a finite probability for the electron of a to belong to the atom b*. In fact, the exponential behavior of $\psi$ and $\varphi$ always satisfies this condition. The magnitude $1/h\ (E_\beta - E_\alpha)$ will have to be interpreted as the frequency at which an exchange of the two electrons is effected on average. For long distances, this difference decreases as $\varepsilon^2/a_0\ e^{-2R/a0}$ so that distant atoms very rarely enter into an exchange [1].

[1] "Large" distances are already the mean gas-kinetic distances ($3.3\ .\ 10^{-7}$ cm); with them the period of exchange is of the order of magnitude $a_0/\varepsilon^2\ h.\ e^{-2/a0\ 3.3.10-7} \sim 10^{30}$ years; on the other hand, with the distances in the crystal lattice ($\sim 3\ .10^{-8}$ cm), an exchange takes $10^{-10}$ sec.

The whole phenomenon is almost intertwined with the quantum mechanical *resonance* phenomenon discussed by Heisenberg. But, while in *resonance*, the electrons of different motions of one and the same eigenfunction series exchange their energy, *here electrons of the same excitation levels (of the same energy), but at different eigenfunction systems ($\psi$ and $\varphi$), exchange their energies*. There, the general occurrence of the same jump frequency twice is characteristic (*resonanc*e phenomenon); here, on the other hand, there is no question of resonance.

**2.** *Let us now turn to the quantitative discussion of interaction energies* (10). Even without evaluating the integrals that occur, it can be said that the indication of $E_{11}S - E_{12}$ is always positive. In a Sturm-Liouville *eigenvalue* problem of any number of dimensions with homogeneous boundary conditions, the natural oscillation that has no *nodes* has the lowest *eigenvalue*. Every other orthogonal natural oscillation has *nodes* and a higher *eigenvalue*. Now, however, the solution $\alpha$ is apparently *nodeless*, while $\beta$ always has a *node* as an *antisymmetric eigenfunction*.

[In quantum mechanics, a *node* is a point or region where the probability of finding a particle is zero. *Nodes* are points of zero probability densities, and the regions in the neighborhood of *nodes* will have small probability densities. *Nodes* can be found in simple quantum mechanical systems, such as the behavior of a particle in a harmonic oscillator potential. In atomic orbitals, *nodes* are the regions or spaces around the nucleus where the probability of finding an electron is zero.]

So

$$E_\beta > E_\alpha. \tag{12}$$

The meaning of $E_{11}$ is immediately evident from (11): *it is the purely Coulomb interaction of the existing charge distribution*. The calculation results in

$$E_{11} = \varepsilon^2/a_0\, e^{-2R/a0}\, (a_0/R + 5/8 - 3/4\, R/a_0 - 1/6\, R^2/a_0^2) \tag{13}$$

...

The rest of the energies (11) are not so easy to interpret. Not all of the integrals found there could be evaluated. The calculation is:

$$S = (1 + R/a_0 + R^2/3a_0^2)^2\, e^{-2R/a0} \tag{14}$$

$$\int \psi_1\psi_2\varphi_1\varphi_2/2\, (\varepsilon^2/r_{a1} + \varepsilon^2/r_{a2} + \varepsilon^2/r_{b1} + \varepsilon^2/r_{b2} - 2\varepsilon^2/R)\, d\tau_1 d\tau_2$$
$$= 2\varepsilon^2/a_0\, (1 + 2R/a_0 + 4R^2/3a_0^2 + R^3/3a_0^3)\, e^{-2R/a0} - \varepsilon^2/R\ S,$$

while for the missing integral over $1/r_{12}$ we only have an upper limit. ...

...

The physical significance of the two solutions, we are concerned with is the following: *The antisymmetic solution with the interaction energy $E_\beta$ corresponds to the van der Waals repulsion (elastic reflection) of the two hydrogen atoms. But if α is excited, the two hydrogen atoms can join together to form a homopolar molecule, where the minimum of $E_\alpha$ indicates equilibrium.*

We will also show that in the interaction of two non-excited noble gas atoms (§ 3) the solution corresponding to the formation of molecules is prohibited by quantum theory. The substantiation of these allegations is set out in § 4.

The estimated curves give approximately the values of the respective atomic quantities: *atomic diameter*, *dissociation energy* and *moment of inertia* of the molecule. As far as the error caused by the estimation is concerned, it should be noted that in any case $E_\beta$ is actually higher, $E_\alpha$ lower, whereby the typical character of the two potentials would be even more pronounced. Since it is not our goal to calculate the most accurate numerical values possible, but to gain insight into the physical conditions of homopolar bonding, let us content ourselves with this estimation here.

## § 3. *Interaction of two He-atoms.*

*We use the same method to investigate the exchange of two neutral He atoms.* The two nuclei are called again *a* and b and are kept at a fixed distance R. The electrons are numbered 1 to 4. For each nucleus there exists an eigenfunction, $\psi$ for the nucleus *a*, $\varphi$ for the nucleus b, which are now functions of two electrons each, and Heisenberg's He-theory has shown that the eigenfunctions of the He-ground state are *symmetrical* in the spatial coordinates of the two electrons[1].

[1] Heisenberg, W. (1926). *Zeit. Phys.*, 38, 411.

By creating an $\psi$ and a $\varphi$-function, an eigenfunction is created (we write the arguments again as indices):

$$\psi_{ik} \cdot \varphi_{lj} \qquad (i \neq k \neq l \neq j = 1, 2, 3, 4) \qquad (16)$$

of the unperturbed problem (mostly separated neutral atoms). By permutation of the four electrons (taking into account that $\psi_{ik} = \psi_{ki}$, $\varphi_{ik} = \varphi_{ki}$) $\binom{4}{2} = 6$ new eigenfunctions, which all belong to the same eigenvalue. They are:

$$\begin{matrix} \psi_{12} \cdot \varphi_{34}, & \varphi_{34} \cdot \psi_{12}, & \} \\ \psi_{13} \cdot \varphi_{42}, & \varphi_{42} \cdot \psi_{13}, & \} \\ \psi_{14} \cdot \varphi_{23}, & \varphi_{23} \cdot \psi_{14}, & \} \end{matrix} \qquad (17)$$

*The electrons of the helium ground state must be divergent from each other with respect to their spin.* We choose the designation in such a way that the electrons 1 and 3 are of the same *spin*, also 2 and 4, but 1 and 2 have a different *spin*. Then, according to the *Pauli principle*, it is excluded that 1 and 3 or 2 and 4 are located at the same He nucleus. We can, therefore, exclude two of the six eigenfunctions (17) from the outset and limit ourselves to

$$\begin{matrix} \psi_{12} \cdot \varphi_{34}, & \varphi_{34} \cdot \psi_{12}, & \} \\ \psi_{14} \cdot \varphi_{23}, & \varphi_{23} \cdot \psi_{14}, & \} \end{matrix} \qquad (17a)$$

*The wave equation for the 2-He problem* is:

$$\Sigma_{i=1}^{4} [\Delta_i\chi + 8\pi^2 m/h^2 \{E_0 - \varepsilon^2(4/R + \Sigma_{k=1}^{4} 1/r_{ik} - 1/r_{ai} - 1/r_{bi}) \chi] = 0. \qquad (18)$$

It is *symmetrical* in all four electrons. Here, too, we do not have a perturbation problem of the usual form, since it is just as impossible to separate a perturbation potential as in § 1; but we infer from the symmetry of the differential equation that the eigenfunctions of the zero-approximation are the same as if there were a small *perturbation potential* in the many symmetrical equations. We call this H.

Of course, it is not the first-order eigenvalues that can be used to deduce your secular problem, but we can deduce the correct eigenfunctions of the first order, which are characterized solely by the *symmetry* character of the differential equation. ...

...

For the same reason as in the case of the *2-hydrogen problem*, $E_\alpha$ the lowest eigenvalue, must consequently be $H_{12} < 0$; then $E_\alpha$ belongs the *nodeless* symmetric eigenfunction α. In the following section, however, we shall show that this α, which would energetically necessitate the formation of molecules, is not permissible in quantum theory for neutral He, but that of the four solutions (20) only β can occur, which again means *elastic reflection*.

**§ 4. *The Pauli Principle and Molecular Formation.***

**1.** How to convince yourself easily; our solutions include (4a) or (20) systems which do not combine with each other[1].

[1] It is distinguished from a similar one, recently by Hund, F. (1927). *Zeit. Phys*., 40, 742, namely the two-center problem.

This circumstance opens up the possibility of applying the Pauli principle, which has proven itself so well in the discussion of the electron configurations of the individual atoms, in a broader sense to the system of two atoms interacting, in order to achieve a narrower selection of the modes of behavior of two atoms permitted by quantum theory. The following formulation of this regulation is sufficient for our purposes[2]:

[2] This formulation of the Pauli principle is not entirely correct. But a really general version of it is not yet known. We refer to the work of Heisenberg, W. (1927). *Zeit. Phys*., 41, 239, as well as especially on soon to be published reflections of F. Hund.

*the selected eigenfunctions of the system should change/maintain their sign, when two electrons are swapped, if the two electrons compare/have different spin.* (A so-called "eigenfunction of the spin" is not to be taken into account here.)

**2.** We apply this provision to the solutions of § 1 and 3. For hydrogen, the symmetrical α retains its sign when 1 and 2 are swapped, β changes. At α the two electrons have different spin, at β the same spin. However, since there is no constraint on the spin of separated atoms, both solutions can occur - depending on chance.

For the solutions (20) of the *2-He problem*, however, we know that 1 and 3 and 2 and 4 each had the same *spin*, 1 and 2, and 3 and 4 had different ones. As a result, there are limits to the interchangeability of the electrons from the outset. If the single He-atom is to remain viable (Pauli principle, applied to the single He-atom), then only the electrons 1 for 3 and 2 for 4 may be exchanged. With respect to each of these swaps, the selected eigenfunctions must be *antisymmetric*. If both pairs are swapped, it must reproduce itself again (without a change of sign). The eigenfunctions $\alpha$, $\beta$, $\gamma$, $\delta$ now merge into an electron in which solutions of the form $\varphi + \psi$ and $\varphi - \psi$ occur, which combine with each other.

We see, then, that in He's case the only solution to the demands suffices. To it, however, belongs the higher eigenvalue. ... Closer discussion shows that *it yields about the right size of the He gas kinetic-radius. The extraordinary difference that we find here in the case of He is evidently due to the fact that 2 He atoms (and the same applies to all noble gases) cannot differ in their spin* – in contrast to hydrogen (and all atoms with unfinished shells) -, so that 2 He atoms have only one possible mode of behaving.

**3.** We now want to show which of the solutions thus eliminated allow *molecule formation*.
...

### § 5. *Hydrogen and ion formation.*

Our considerations in the first paragraph cannot be complete insofar as the possibility of *ion formation* has not been taken into account.
...

We would like to believe that the categories of degeneracies and symmetry relations presented here are typical of a wide range of facts connected with the interactions of atomic systems with each other, especially with the discontinuities of their chemical modes of behavior[3].

[3] It will be considered whether the *exchange phenomenon*, which is so decisive for the interaction of atoms, is also noticeable in other branches of physics. We would like to point out two things here. In the case of collision processes, *the exchange of collision and atomic electrons also makes it possible to stimulate quantum leaps between optically non-combining term systems* (e.g. $1^1S$ — $2^3S$ for He). *According to the Born theory, such transitions are only possible due to the minimal magnetic interaction.* In the interpretation of the law of force of atomic nuclei, which is noticeable in the scattering experiments with $\alpha$ or H particles, one will have to take into account the exchange with nuclear building blocks as an essential influence.

**Heisenberg, W. (September, 1928). Zur Theory of Ferromagnetismus. (On the theory of ferromagnetism.)**

*Zeit. Phys*., 49, 619–36 (in German); https://doi.org/10.1007/BF01328601; translation by D. H. Delphenich; https://neo-classical-physics.info/uploads/3/0/6/5/3065888/ Heisenberg _-_on_the_theory_of_ferromagnetism.pdf.; also in Underwood, T. G. (2024). *Electricity & Magnetism.* Part II: pp. 370-9.

Received May 20, 1928.

Leipzig, Institut für theoretische Physik der Universität.

In another brilliant paper, Heisenberg noted that empirical results exhibit *ferromagnetism* as an entirely similar state of affairs to what was previously observed in the spectrum of the helium atom; and it seemed to follow from the levels in the helium atoms that a powerful interaction prevailed between the spin directions of two electrons that led to the splitting of the level structure into systems of singlets and triplets. He also noted that this was closely related to explaining ferromagnetic phenomena as being implied by the *exchange phenomenon* (resulting from *quantum entanglement*). He proceeded to show that the *Coulomb interaction*, together with the *Pauli principle*, succeeded in evoking the same effects as the molecular field that Weiss postulated, noting that it was only in recent times that mathematical methods were developed for the treatment of such a complicated problem in the important investigations of Wigner, Hund, and Heitler and London. As a first approximation, Heisenberg assumed that the lattice separations were very large, and that every electron belonged to its own atom, and applied Heitler-London's calculations to the case of 2n electrons in a state of interaction, finding 2n electrons in 2n positionally different quantum cells. Due to their smallness, he was able to leave the magnetic interactions outside of consideration, and showed that *the spin moments of all electrons become partly parallel and partly anti-parallel as a result of the exchange processes*. By adding the fundamental Pauli principle to this, viz., that the eigenfunctions of the total system should be *anti-symmetric* in all electrons, he showed that an entirely well-defined *total magnetic moment* belonged to each level value of the perturbed system, and there were (2n)! levels in the unperturbed system (ignoring the Pauli principle and spin). He then showed that a *statistical treatment of ferromagnetism was possible when all energy values had been calculated.* Heisenberg concluded that *an atom in a lattice can only be exchanged with its "neighbors"*; exchanges with atoms that lie further away that the "neighboring atoms" could then be neglected. The *number of "neighbors" of an atom* is, e.g., 1 in a molecular lattice of diatomic molecules, 2 in a linear chain, 4 in a quadratic surface lattice, 6 in a simple cubic lattice, 8 *in a cubic, space-centered lattice*, and 12 in a cubic, face-centered lattice. By assuming a distribution of energy values about the mean had the approximate form of a Gaussian error curve, Heisenberg showed that small or negative values of the constant

$\beta$ [= $zJ_0/kT$)] resulted in *paramagnetism*; and that *ferromagnetism was only possible for lattice types for which an atom had at least eight neighbors*, which was the case for Fe, Co, Ni, whose lattices were all cubic, some of which were space-centered ($z = 8$) and some of which were face-centered ($z = 12$). He concluded that two conditions were necessary for the appearance of *ferromagnetism*: the crystal lattice must be a type such that *any atom has at least 8 neighbors*; and the *principal quantum number* of the electrons that are responsible for magnetism must be $n \geq 3$.

---

***Abstract.***

Weiss's molecular forces will be attributed to a quantum-mechanical exchange phenomenon, and indeed, it will be treated as the *exchange process* that was successfully enlisted in recent times by Heitler and London in order to interpret homopolar valence forces.

***Introduction.***

*Ferromagnetic* phenomena have been interpreted in a formally satisfying way by the well-known Weiss theory*.

* Weiss, P. (1907). *Journ. de phys.*, 4, 6, 661 and (1908). *Phys. Zeit.*, 9, 358.

That theory is based upon the assumption that every atom in a crystal experiences a directed force from the remaining atoms of the lattice that should be proportional to the number of already-directed atoms. By contrast, the origin of these atomic fields was unknown, and several obstacles stood in the way of any interpretation of the Weiss forces on the basis of classical theory: magnetic interactions between atoms are always a few orders of magnitude smaller than the atomic fields that follow from ferromagnetic experiments. Indeed, electric interactions lead to the correct order of magnitude; however, one would rather expect that the electrical interactions of two atoms would be proportional to the square of the cosine of their mutual angle of inclination, rather than the cosine, which contradicts the assumptions of Weiss's theory. Other complications were discussed more thoroughly by Lenz**,

** Lenz, W. (1920). *Phys. Zeit.*, 21, 613.

and Ising***

*** Ising, E. (1925). *Zeit. Phys.*, 31, 253.

succeeded in showing that the assumption of directed, sufficiently large forces between any two neighboring atoms of a chain did not suffice to generate *ferromagnetism*.

The *ferromagnetic* complex of questions has entered a new arena with the Uhlenbeck-Goudsmit theory of *spin electrons*. In particular, it follows from the known factor g = 2 in the Einstein-de Haas effect (which was, in fact, measured for *ferromagnetic* substances) that in a *ferromagnetic* crystal *only the magnetic eigenmoment of the electrons is oriented*, but not, by any means, the atoms. Thus, the possibility of interpreting the Weiss forces as electrical interactions, independent of the relative *spin* directions of the *electrons*, goes away, since we know that such forces do not exist. Furthermore, by applying Pauli-Fermi-Dirac statistics, Pauli†

† Pauli, W. (1927). *Zeit. Phys*., 41, 81.

has been able to show that *paramagnetism or diamagnetism will always result from neglecting the interaction of the electrons in a metal*.

**§ 1. *A model for the foundations of the theory*.** The basic idea of the theory that we seek here is this: Empirical results exhibit *ferromagnetism* as an entirely similar state of affairs to what was previously observed in the spectrum of the helium atom. At the time, *it seemed to follow from the levels in the helium atoms that a powerful interaction prevailed between the spin directions of two electrons that led to the splitting of the level structure into systems of singlets and triplets*. At the time, this difficulty could be resolved by verifying that the apparently large interaction would emerge indirectly from a *resonance* or *exchange phenomenon* that would be characteristic of all quantum-mechanical systems of identical particles. *This is also closely related to explaining ferromagnetic phenomena as being implied by this exchange phenomenon*. We will attempt to show that the *Coulomb interaction*, together with the *Pauli principle*, succeeds in evoking the same effects as the molecular field that Weiss postulated. It was only in recent times that mathematical methods were developed for the treatment of such a complicated problem in the important investigations of Wigner*, Hund**, Heitler and London***.

* Wigner, E. (1927). *Zeit. Phys*. 40, 883; 43, 624; ** Hund, F. (1927). *ibid*., 43, 788; *** Heitler, W. & London, F. (1927). *ibid*., 44, 455, cited as I in what follows; Heitler, W. (1927). *ibid*., 46, 47 (cited as II); (1928). *ibid*., 47, 835 (cited as III); London, F. (1928). *ibid*., 46, 455.

Before I go into the actual calculations, I would like to give a brief overview of the methods of approximation that can be applied in the treatment of electronic motions in metals.

***Method I.*** From Pauli (*loc. cit.*) and Sommerfeld****, in the first approximation, *electrons can be assumed to be completely free*.

**** Sommerfeld, A. (1928). *Zeit. Phys.*, 47, 1; cf., also Houston, W. V. (1928). *ibid.*, 47, 33, and Eckart, C. (1928). *ibid.*, 47, 38.

In the second approximation, one might, perhaps, add in the *interactions with the lattice points* [Houston†].

† Houston, W. V. (1928). *Zeit. Phys.*, 48, 449.

*The interaction of electrons with each other is neglected completely.*

***Method II.*** As a first approximation, *one calculates the motion of an electron in a force field* (that, by no means, needs to be small) *that is periodic* (in three directions). In the next approximation, *one might perhaps consider the perturbations that arise from the deviations from periodicity in the lattice. The treatment of the interaction of electrons with each other encounters the same difficulties here as it does in method I.*

***Method III.*** In the first approximation, *one thinks of the lattice separations as being very large and assumes that every electron thus belongs to its own atom.* In the next approximation, *one considers the exchange of electrons that move in the unperturbed system with equal energies at different points*, which was first considered by Heitler and London (*loc. cit.* I). States in which more electrons are found in comparison to the number that is found in one atom in the unperturbed system will not be considered in this approximation.

The difference between these three methods becomes clearer when we explain it by another example, namely, the *hydrogen molecule*, which was treated rigorously by Heitler and London (*loc. cit.* In method I, *the electrons were, once more, first treated as free*, which would naturally not yield any suitable starting solution for the calculation. In method II, *one starts with the solutions of the two-center problem* [cf., Hund*].

* Hund, F. (1927). *Zeit. Phys.*, 40, 742.

A level that describes electron 1 as being in a 1s state around nucleus *a* and electron 2 as being a 1s state around nucleus b in the limiting case of infinite nuclear separations would split into four levels (1 to 4) that might be characterized by the table:

| | *Nucleus a* | *Nucleus b* |
|---|---|---|
| 1 | 1 | 2 |
| 2 | 2 | 1 |
| 3 | 1,2 | – |
| 4 | – | 1,2 |

The *interaction* of the two *electrons* will first be considered in higher-order approximations. Method II will be directly identical to the method that was employed by Heitler and London. Only levels 1 and 2 included in an unperturbed system. It will be assumed that levels 3 and 4 lead to substantially higher-lying energy values. The diversity of levels in unperturbed systems will then be more meager for method III than it is for methods I or II.

There is, indeed, no argument, a priori, for preferring any of the three approximation procedures over the other ones. Method I will be most closely applicable to metals of very large conductivity, while method III is most applicable to metals of very feeble conductivity. Method II is in the middle between these two limiting cases.

*I have based the following calculations upon method III*, since only it can permit a quantitative treatment of the electron interactions.

**§ 2. *The distribution of the level values*.** The following calculations define a simple generalization of the Heitler-London investigations (*loc. cit.* I) to the case of 2n electrons in a state of interaction (the number of electrons is now assumed to be even, upon purely formal grounds). One will then find 2n electrons in 2n different (indeed, they are not different energetically, but positionally) quantum cells.

We shall first assume only that the quantum numbers of the electrons in their atoms are the same for all atoms. Other stationary states of the unperturbed system will not be considered, since it will be assumed that they would lead to much high energy values.

*One is then dealing with the determination of the energy values of the stationary states of the total system*, which will belong to the state that was described above when the Coulomb interaction of the charges in an atom with the charges of any other atom is considered to be a perturbation. Due to the great computational complications that have appeared up to now, it will only be possible for us to attempt the perturbation calculations up to the first approximation. Whether this first approximation will actually be successful for the cases that nature presents must remain undecided. We take the eigenfunctions of the unperturbed system to be, say, products of the Schrodinger eigenfunctions of the *hydrogen atom*, or better yet, *the eigenfunctions that correspond to the rest of the atoms considered*, just like in the cited paper of Heitler and London; it is entirely superfluous to repeat those Ansätze here explicitly.

> [An *ansatz* is the establishment of the starting equation(s), the theorem(s), or the value(s) describing a mathematical or physical problem or solution. It typically provides an initial estimate or framework to the solution of a mathematical problem, and can also take into consideration the boundary conditions (in fact, an ansatz is

sometimes thought of as a "trial answer" and an important technique in solving differential equations).]

These eigenfunctions are certainly not orthogonal, but the deviation from the usual treatment that is required first differs in the terms of order two, so we can apply the usual method of treating things in the case of orthogonal eigenfunctions. *The electrons of an atom can be exchanged with those any other atom as a result of perturbations*. As long as one overlooks the perturbing terms of order two, only simple transpositions between two neighboring atoms will occur. If one chooses the simplest case to be the one in which any atom in the unperturbed system possesses one valence electron then the "exchange terms" for the perturbing energy will reduce to the expressions that were given by Heitler and London:

$$J_{(kl)} = 1/2 \int \psi_k^{\kappa}\psi_k^{\lambda}\psi_l^{\kappa}\psi_l^{\lambda}\,(2\varepsilon^2/r_{kl} + 2\varepsilon^2/r_{\kappa\lambda} - \varepsilon^2/r_{\kappa k} - \varepsilon^2/r_{\kappa l} - \varepsilon^2/r_{\lambda l})\,d\tau_k d\tau_l. \qquad (1)$$

Here, k and *l* mean the numbers of *two electrons*, while κ and λ mean the numbers of the *remaining atoms to which k and l belong in the unperturbed state*. A very important constant that enters into the perturbation calculations is the purely "static" interaction:

$$J_E = \int d\tau_1 d\tau_2 \ldots d\tau_{2n}\,(\psi_1^{1})^2(\psi_2^{2})^2 \ldots (\psi_{2n}^{2n})^2 \Big[\sum_{\substack{k,l\\ k>l}} \varepsilon^2/r_{kl} + \sum_{\substack{\kappa,\lambda\\ \kappa>\lambda}} \varepsilon^2/r_{\kappa\lambda} - \sum_{\substack{k,\lambda\\ k<\lambda}} \varepsilon^2/r_{k\lambda}\Big] \qquad (2)$$

Due to their smallness, we can leave the magnetic interactions completely outside of consideration. Nevertheless, *the spin moments of all electrons will become partly parallel and partly anti-parallel as a result of the exchange processes*. If one adds the fundamental Pauli principle to this, viz., that the eigenfunctions of the total system should be *anti-symmetric* in all electrons, then an entirely well-defined *total magnetic moment* will belong to each level value of the perturbed system that will be characterized by the rotational moment $s\hbar/2\pi$ of the system. In all, there will be (2n)! levels in the unperturbed system (if one ignores the Pauli principle and spin). *A statistical treatment of ferromagnetism will be possible when all energy values that belong to a given value of s have been calculated.* This problem is generally not soluble in this form, since 2n is a very large number. We can only hope to obtain a general insight into the distribution of the eigenvalues for a given s. In what follows, we will calculate the *number* of levels, the center of mass of the energy (thus, the mean value of energy for a given s), and the *mean-square variance* of the energy about that mean value. We shall then make the generally somewhat arbitrary assumption that, in the first approximation, the energy values are distributed around the mean in a Gaussian error curve, such that the breadth of the error curve is calculated from the mean-square variance.

From the investigations of Wigner, Hund, and Heitler (*loc. cit.*) and the assumption of the Pauli principle, every value s of the total *spin* moment belongs to one system of levels ("σ") that are characterized by the well-defined partitioning of 2n into summands:

$$2n = \underbrace{2 + 2 + \ldots \quad + 2}_{(n-s)\ times} + \underbrace{1 + 1 + \ldots + 1}_{2s\ times}. \tag{3}$$

The partition of the "reciprocal" system is then called simply:

$$2n = (n - s) + (n + s). \tag{4}$$

Heitler (*loc. cit.* II) has given the following formula for the mean value – i.e., the "center of mass of energy" – of the system σ:

$$E_\sigma = 1/f_\sigma \sum_P \chi_\sigma^P J_P. \tag{5}$$

In this, $\chi_\sigma^P$ means the group character that belongs to the permutation P, and $f_\sigma = \chi_\sigma^E$ is the *number* of levels in the system. The energy of the unperturbed levels is omitted, as an additive constant. We further calculate the mean-square variance $\Delta E^2$ of the energy about the value $E_\sigma$: The energy value is given by the square root of an equation of degree $f_\sigma$ that one obtains when one sets the following determinant equal to zero:

$$\ldots \tag{6}$$

... It next yields for the reciprocal system of levels:

$$\chi^E_{n-s,n+s} = (2n)!\,(2s + 1)/\{(n - s)!(n + s + 1)!\} \tag{13}$$
$$\chi^{(12)}_{n-s,n+s} = \ldots$$
$$\chi^{(123)}_{n-s,n+s} = \ldots$$
$$\chi^{(12)(34)}_{n-s,n+s} = \ldots$$

The characters of the system of levels that actually is present differ from the characters of their reciprocals that are employed here only by their signs. Indeed, the character of the reciprocal system is equal to (equal and opposite to, resp.) that of the system itself when the permutation P arises from an even (odd, resp.) number of transpositions.

Up to this point, everything is true in complete generality, with no relationship to any special assumptions that we might make about the crystal lattice or the atomic structure of the ferromagnetic substance.

In order to be able to calculate, we must now specialize our assumptions somewhat further. It follows from formula (1) and the calculations of Heitler and London that $J_{(12)}$ decreases exponentially with increasing distance. *For the most part, one can exchange an atom in a lattice only with its "neighbors"*; exchanges with atoms that lie further away that the "neighboring atoms" will then be neglected. The *number of "neighbors" of an atom* [z] is, e.g., 1 in a molecular lattice of diatomic molecules, 2 in a linear chain, 4 in a quadratic surface lattice, 6 in a simple cubic lattice, 8 in a cubic, space-centered lattice, and 12 in a cubic, face-centered lattice.

*We shall make only the assumption that all non-vanishing exchange terms $J_P$ should be equal (we call that value $J_0$).* That must be case when the remaining atoms are non-magnetic; i.e., centrally-symmetric. We then now calculate $E_\sigma$ and $\Delta E_\sigma^2$ for a lattice in which every atom has z neighbors. ...

...

§ **3.** ***Statistics: connection with Weiss's formulas***. The following arguments will be founded upon the aforementioned, generally somewhat arbitrary, assumption that the distribution of energy values about the mean has the approximate form of a Gaussian error curve. …

…

For small or negative values of the constant β [= $zJ_0/kT$)], one will get *paramagnetism*. *Ferromagnetism* enters in when the tangent of the curve II for y = 0 subtends a smaller angle with the x-axis than the tangent of I; the influence of the cubic terms is first ignored in this. The *condition for ferromagnetism* then reads:

$$\beta(1 - \beta/z) \geq 2. \tag{25}$$

This condition can be fulfilled only for high values of z. The maximal value on the left-hand side of (25) is ($\beta_{max} = z/2$) z/2(1 – 1/2), and it follows that:

$$z \geq 8. \tag{26}$$

*Ferromagnetism is then possible only for lattice types for which an atom has at least eight neighbors*. That is the case for Fe, Co, Ni, whose lattices are all cubic, some of which are space-centered (z = 8) and some of which are face-centered (z = 12). ...

...

One must improve the provisional theory that was attempted here by calculating the higher variances of the mean $\Delta E^3$, $\Delta E^4$, etc. and correspondingly construct improved distribution curves for the values of the terms. Corresponding higher powers of β would appear on the left-hand side of equation (25) in this improved theory; the left-hand side of (25) is thus actually a transcendental function of β. ... Such a more precise examination of the distribution curve would also most likely displace the limiting value (26) of z. However, nothing in our results would change very much qualitatively.

...

§ 4. ***Magnitudes and signs of the "molecular field"***. The constant $\beta$ must have order of magnitude 1, in order for *ferromagnetism* to be possible; one must then have $J_0 \sim kT$, where T will assume values on the order of $10^3$ degrees for Fe, Co, Ni. *It follows that $J_0 \sim 10^{-13}$ erg ~ 1/100 the energy of the hydrogen ground state. That is just the order of the energy contribution that one would expect for the exchange term* of the form (1) when the atoms lie close to each other. If the atomic separations become larger, then the exchange terms will decay exponentially. *That is the basis for the fact that iron or nickel salt solutions are never ferromagnetic.*

*The question of the sign of $J_0$ is much more difficult to answer.* In their theory of the homopolar bond, *Heitler and London make that assumption that $J_0$ is negative in complete generality, which would exclude ferromagnetism.* For the special case in which the electrons are found to be unperturbed in the 1s state, it follows, in fact, from general theorems that the energy values must lie in a way that would correspond to negative values of $J_0$. *Such an argument is, in turn, applicable only to electrons in the 1s state, and one can show that $J_0$ will generally be positive for high principal quantum numbers.* One must then deal with the expression:

$$J_0 = 1/2 \int \psi_k^{\kappa}\psi_k^{\lambda}\psi_l^{\kappa}\psi_l^{\lambda}\,(2\varepsilon^2/r_{kl} + 2\varepsilon^2/r_{\kappa\lambda} - \varepsilon^2/r_{\kappa k} - \varepsilon^2/r_{\kappa l} - \varepsilon^2/r_{\lambda l})\, d\tau_k d\tau_l. \qquad (1)$$

*in which $\kappa$ and $\lambda$ are the indices for the atomic nuclei, while k and l are the those of the electrons.* Initially, $\psi$ will be a hydrogen eigenfunction, but later on, it will be shown that the argument is just as valid for other central fields in the vicinity of the nucleus. *One can then say with certainty that $J_0$ will be positive for very small values of $r_{\kappa\lambda}$, since the term $1/r_{\kappa\lambda}$ will then outweigh all of the other ones.* However, that result does not need to have any physical meaning, since for very small values of $r_{\kappa\lambda}$, even the entire approximation becomes illusory (cf., the case of 1s terms!). One then comes to the values of $J_0$ for very large $r_{\kappa\lambda}$. When $J_0$ is positive there, one must assume that it remains positive for all values of $r_{\kappa\lambda}$, in general. We thus investigate further how a charge distribution of density $\psi_k^{\kappa}\psi_k^{\lambda}$ appears at large distances $r_{\kappa\lambda}$, first for perhaps the higher s terms.

The Schrödinger functions contain an e-function as the most important term, and $\psi_k^{\kappa}\psi_k^{\lambda}$ thus contains the factor $e^{-(r_{k\kappa} + r_{k\lambda})/a_0 n}$ ($a_0$ = Bohr hydrogen radius, n = principal quantum number; thus, no confusion with the electron number 2n should be created). *If one drops the remaining factors then the density will be constant on confocal ellipsoids of rotation around the two nuclei.* For increasing distance between the nuclei, the charge ellipsoid degenerates into a cylinder around the connecting line between the nuclei. (This happens for both values of the principal quantum number.) Furthermore, the e-function appears multiplied by a polynomial in $r_{\kappa\kappa}$ ($r_{\kappa\lambda}$, resp.) of degree n – 1. The zero locus of this polynomial lies entirely in the neighborhood of the nucleus; at greater distances from it, it will suffice to replace the polynomial with its highest power $r^{n-1}$. The behavior of the central force at distances of order $a_0$ from the nucleus is entirely inessential when only $r_{\kappa\lambda}$ is sufficiently large. The density distribution of the charge over the length of the aforementioned cylinder is therefore non-uniform, but otherwise approximately proportional to $r_{k\kappa}{}^{n-1} r_{k\lambda}{}^{n-1}$. For small values of n, this distribution is still quite uniform and one can easily see that the negative terms in $J_0$ can substantially predominate. For increasing n, by contrast, the density distribution assumes an ever steeper maximum at the midpoint between the two nuclei. In the limit of very large values of n, the mean value of the terms of type $1/r_{\kappa\kappa}$, when taken over the density distribution that was given above, tends to the value $2/r_{\kappa\lambda}$:

$$1/r_{k\kappa} = 1/r_{k\lambda} = 1/r_{l\kappa} = 1/r_{l\lambda} \rightarrow 2/r_{\kappa\lambda}.$$

By contrast, the term with $1/r_{kl}$ – viz., the "*self-potential*" of the density distribution – increases beyond all limits with increasing n. $J_0$ is then certainly positive *for sufficiently high principal quantum numbers*. One can easily show that nothing will change in this result when one carries out the calculations for p, d, ..., or any other higher state. The limiting value of n for which $J_0$ can become positive for the first time is difficult to determine exactly. A rough calculation yields n = 3. This limiting value will possibly depend upon values of the remaining quantum numbers. The fact that, e.g., the oxygen molecule empirically possesses a *magnetic moment* of $2 \cdot 1/2\ h/2\pi$ in the ground state seems to show that $J_0$ can already be positive for n = 2. On the other hand, it can follow from the many-times-observed critical temperatures (e.g., for γ-iron) that there are many times $J_0$ that can also be negative for higher principal quantum numbers.

***Concluding remarks***. The calculations that were described here lead to two conditions for the appearance of *ferromagnetism*:

1. The crystal lattice must be a type such that any atom has at least 8 neighbors.
2. The principal quantum number of the electrons that are responsible for magnetism must be $n \geq 3$.

Both conditions together do not reach far enough to single out Fe, Co, Ni from all other materials; however, Fe, Co, Ni do satisfy the conditions. It was certainly also to be expected that the theory that was contrived here can meanwhile serve as only a qualitative schema in which *ferromagnetic* phenomena will perhaps be classified later. The theory admits an extension for the case of several exchanges per atom; an incisive study of the $J_{(kl)}$ values, as well as the distribution curve of the term values, will be requisite. I hope to be able to go into these questions, as well as a thorough comparison of the theory with the experimental results later.

## 10. Elementary and composite particles can exist as different quantum states, referred to as isospin states, which create an attractive force – the strong interaction or strong force - through exchange interaction resulting from entanglement between two quantum isospin states.

**The origin of isospin.**

In 1932, Werner Heisenberg[1] introduced a new (unnamed) concept to explain binding of the *proton* and the then newly discovered *neutron*. His model resembled the bonding model for the molecular Hydrogen ion, $H_2^+$: a single *electron* was shared by two *protons*.

> [1] Heisenberg, W. (January, 1932). Über den Bau der Atomkerne. I. (About the construction of atomic nuclei. I.); (March, 1932). Über den Bau der Atomkerne. II. (About the construction of atomic nuclei. II.); (September, 1933). Über den Bau der Atomkerne. III. See below.) See Underwood, T. G. (2024). The Standard Model.

Heisenberg's theory had several problems, most notable it incorrectly predicted the exceptionally strong binding energy of $He^{+2}$, *alpha particles*. However, its equal treatment of the *proton* and *neutron* gained significance when several experimental studies showed these particles must bind almost equally. In response, Eugene Wigner used Heisenberg's concept in his 1937 paper where he introduced the term "*isotopic spin*" to indicate how the concept is similar to *spin* in behavior[2].

> [2] Wigner, E. (1937). On the Consequences of the Symmetry of the Nuclear Hamiltonian on the Spectroscopy of Nuclei. *Phys. Rev.*, 51, 2, 106–19. doi:10.1103/PhysRev.51.106. Also in Underwood, T. G. (2024). The Standard Model.

**Isospin.**

*Elementary* particles, such as *protons* and *neutrons*, can exist as different *quantum states*, referred to as *isospin states*. The name of the concept contains the term *spin* because its quantum mechanical description is mathematically similar to that of *angular momentum* (in particular, in the way it *couples*; for example, a *proton–neutron pair* can be *coupled* either in a *state* of *total isospin* 1 or in one of 0. But unlike angular momentum, it is a dimensionless quantity and is not actually any type of spin.

***Isospin invariance.***

To a good approximation the *proton* and *neutron* have the same *mass*: they can be interpreted as two *states* of the same particle. These *states* have different values for an *internal isospin coordinate*. The mathematical properties of this coordinate are completely analogous to *intrinsic spin angular momentum*. The component of the operator, $T_3$, for this coordinate has *eigenvalues* + ½ and − ½; it is related to the *charge operator*, Q:

$$Q = e(T_3 + \tfrac{1}{2})$$

which has *eigenvalues* e for the *proton* and zero for the *neutron*. For a system of n *nucleons*, the *charge operator* depends upon the *mass number* A:

$$Q = e(T_3 + \tfrac{1}{2} A)$$

*Isobars, nuclei with the same mass number* like $^{40}K$ and $^{40}Ar$, only differ in the value of the $T_3$ *eigenvalue*. For this reason, *isospin* is also called "*isobaric spin*".

The *internal* structure of these *nucleons* is governed by the *strong interaction*, but the Hamiltonian of the *strong interaction* is *isospin invariant*. As a consequence, the *nuclear forces* are *charge* independent. Properties like the stability of deuterium can be predicted based on *isospin* analysis. However, this invariance is not exact and the *quark model* gives more precise results.

**Heisenberg, W. (January, 1932). Über den Bau der Atomkerne. I. (About the construction of atomic nuclei. I.); (March, 1932). Über den Bau der Atomkerne. II. (About the construction of atomic nuclei. II.); (September, 1933). Über den Bau der Atomkerne. III.†**

*Zeit. Phys.*, 77, 1–11; https://doi.org/10.1007/BF01342433; *Ibid.*, 78, 156–64; https://doi.org/10.1007/BF01337585; *Ibid.*, 80, 587–96; https://doi.org/10.1007/BF01335696.

† Each part is hidden behind pay walls but can be purchased from the publisher.

Three-part paper by Heisenberg, which attempted to address recent observations that the *forces* between all pairs of constituents of the *nucleus* were approximately equal. Heisenberg introduced a new (unnamed) concept to explain binding of the *proton* and the then newly discovered *neutron.* He treated *protons* and *neutrons* on an equal footing by considering them *as different charge states* of the same particle. His model resembled the bonding model for the molecular Hydrogen ion, $H_2^+$: a single *electron* was shared by two *protons*. Heisenberg's theory had several problems, most notable it incorrectly predicted the exceptionally strong binding energy of $He^{+2}$, *alpha particles*. However, its equal treatment of the *proton* and *neutron* gained significance when several experimental studies showed these particles must bind almost equally. In response, Eugene Wigner used Heisenberg's concept in his 1937 paper where he introduced the term "*isotopic spin*" to indicate how the concept is similar to *spin* in behavior.

***Abstract of Part I.***

The consequences of the assumption that the *atomic nuclei* are made up of *protons* and *neutrons* without the participation of *electrons* are discussed. § 1. The Hamiltonian function of the *nucleus*. § 2. The relationship between *charge* and *mass* and the special stability of the *Helium* (He) *nucleus*. § 3 to 5; Stability of the *nuclei* and radioactive decay series. § 6. Discussion of the basic physical assumptions.

***Abstract of Part II.***

§ 1. Stability of even and odd *neutron* nuclei. § 2. Scattering of γ-rays at the atomic *nucleus*. § 3. The properties of the *neutron*.

***Abstract of Part III.***

The experiments of Curie, Joliot and Chadwick on the existence and stability of the *neutron* prompted the attempt made in Parts I and II of this thesis to define the role played by *neutrons* in the structure of *atomic nuclei* in very specific physical assumptions and to test the usefulness of these assumptions on the factual material of nuclear physics. The incompleteness of the empirical results available so far leads to a great uncertainty even of the foundations of any theory, and only in very few cases do the experiments force a certain interpretation. For this reason, *it seemed necessary to first put a certain hypothesis at the top and see how it is suitable for ordering experience*. In the following, however, it will also be discussed in detail which consequences are characteristic of the *chosen hypothesis* and at which points a different choice of the basic assumptions would lead to the same results. Before this discussion, the considerations of the first two parts will be supplemented and corrected in some places.

**Wigner, E. (January, 1937). On the Consequences of the Symmetry of the Nuclear Hamiltonian on the Spectroscopy of Nuclei.**

*Phys. Rev.* 51, 2, 106 (1937); https://journals.aps.org/pr/abstract/10.1103/PhysRev.51.106; also at https://harvest.aps.org/v2/journals/articles/10.1103/PhysRev.51.106/fulltext; also in Underwood, T. G. (2024). *The Standard Model*, Part I: pp. 174-80.

* A paper delivered at the Tercentenary Conference of Arts and Sciences at Harvard University, September, 1936.

Princeton University, Princeton, New Jersey.

Received October 23, 1936.

---

***Abstract.***

The structure of the *multiplets* of nuclear terms is investigated, using as *first approximation* a Hamiltonian which does not involve the ordinary *spin* and corresponds to equal forces between all nuclear constituents, *protons* and *neutrons*.

[A *multiplet* is the *state space* for 'internal' degrees of freedom of a particle, that is, degrees of freedom associated to a particle itself, as opposed to 'external' degrees of freedom such as the particle's position in space. Examples of such degrees of freedom are the *spin state* of a particle in *quantum mechanics*, or the *color*, *isospin* and *hypercharge state* of particles in the *Standard Model* of particle physics. Formally, this state space is described by a *vector space* which carries the *action* of a group of *continuous symmetries*.]

The *multiplets* turn out to have a rather complicated structure, instead of the S of atomic spectroscopy, one has three quantum numbers S, T, Y. The *second approximation* can either introduce *spin forces* (method 2), or else can discriminate between *protons* and *neutrons* (method 3). The *last approximation* discriminates between *protons* and *neutrons* as in method 2 and takes the *spin forces* into account as in method 3. The method 2 is worked out schematically and is shown to explain qualitatively the table of stable *nuclei* to about Mo.

---

**1.** Recent investigations[1] appear to show that the *forces* between all pairs of constituents of the *nucleus* are approximately equal.

[1] Tuve, M. A., Heydenburg, N. P. & Hafstad, L. R. (1936). *Phys. Rev*., 50, 806; Breit, G., Condon, E. U., & Present, R. D. (November, 1936). Theory of Scattering of Protons by Protons. *Phys. Rev*., 50, 825; https://journals.aps.org/pr/abstract/10.1103/ PhysRev.50.825.

This makes it desirable to treat the *protons* and *neutrons* on an equal footing. A scheme for this was devised in his original paper by W. Heisenberg[2] *who considered protons and neutrons as different states of the same particle.*

[2] Heisenberg, W. (January, 1932). Über den Bau der Atomkerne. I. (About the construction of atomic nuclei. I.) *Zeit. Phys*., 77, 1–11; see above.

Heisenberg introduced a variable $\tau$ which we shall call the *isotopic spin*, the value –1 of this variable can be assigned to the *proton state* of the particle, the value +1 to the *neutron state*. The assumption that the forces between all pairs of particles are equal is equivalent, then, to the assumption that they do not depend on $\tau$ or that the Hamiltonian does not involve the *isotopic spin.*

In addition to this *isotopic spin* $\tau$, we must keep, of course, the *ordinary spin* variable s also; s also can assume the two values +1 and –1. It has been pointed out lately[3] that the Pauli principle requires that the *wave function*

$$\Psi\ (r_1 s_1 \tau_1,\ r_2 s_2 \tau_2, \ldots\ r_n s_n \tau_n) \qquad (1)$$

be *antisymmetric* with respect to the simultaneous interchange of Cartesian, *spin* and *isotopic spin* variables of any pair of *heavy particles.*

[3] Bartlett, J. H. (1936). *Phys. Rev*., 49, 102; Elsasser, W. (1936). *J. de Phys. et Rad.* I, 312; and especially Cassen, B. & Condon, E. U. (1936). *Phys. Rev*., 50, 846.

This fact is quite analogous to the similar statement for *ordinary spin.*

Of course, if Eq. (1) is to represent the state of a given *nucleus*, say with $n_P$ *protons* and $n_N$ *neutrons*, it must vanish at every place where the sum of the $\tau$'s

$$\tau_1 + \tau_2 + \ldots + \tau_n \neq r_N - n_P \qquad (2)$$

is not equal to the "*isotopic number*" of this element. All *wave functions* which are finite for several sums of the $\tau$'s, refer to *states* which can be different elements with finite probabilities. No such *states* are known to be of any importance and the mathematical apparatus of the *isotopic spin* is, hence, somewhat redundant. It will turn out that it is very useful in spite of this.

In addition to the assumption of the approximate equality of *forces* between all pairs of particles, *it appears to be a useful approximation to neglect the forces involving the ordinary spin*. The Hamiltonian depends then *on the space coordinates alone*. By keeping both, one or none of these assumptions, one comes to four possible schemes;

(1) Take into account *forces* depending on *space coordinates alone*.
(2) Take into account *forces* depending on *space and ordinary spin coordinates*, assuming, however, *interactions* between all kinds of pairs to be equal.
(3) *Neglect ordinary spin forces*, take into account *forces* depending on space *coordinates and isotopic spin*, i.e., discriminate between proton-proton, proton-neutron and neutron-neutron interactions.
(4) Take *all kinds of interaction* into account.

The first is the roughest method, the last the most exact and it is probable that (2) is more accurate for light elements, (3) for heavy elements. On the other hand, of course, one can obtain most results from symmetry considerations for 1, fewest for 4. Approximation (1) is identical with the "all orbital forces equal" model[4].

[4] Feenberg, E. & Wigner, E. *Phys. Rev.* This issue.

The statement that an *operator* involves only one or another set of variables needs further amplification. As used in the ordinary theory of spectra, this expression means that the *operator* can be written in terms of these variables alone. It did not mean that it cannot be written in some other way as well. Thus, e.g., the *interchange* P of the space coordinates acts only on space coordinates, although it can be written by *Dirac's identity*,

$$P = -\tfrac{1}{2} - \tfrac{1}{2}(s_1 \cdot s_2)$$

entirely in terms of *spin operators* for *antisymmetric* functions. We shall keep this definition for the forces depending on Cartesian and ordinary *spin* coordinates for nuclei also.

The *operators* which involve τ are, however, somewhat specialized to begin with. Using Heisenberg's notation for *isotopic spin operators*

$$\tau = \tau_\zeta = \begin{Vmatrix} -1 & 0 \\ 0 & -1 \end{Vmatrix}, \qquad \tau_\xi = \begin{Vmatrix} 0 & i \\ -i & 0 \end{Vmatrix}, \qquad \tau_\eta = \begin{Vmatrix} 0 & 1 \\ 1 & 0 \end{Vmatrix}, \tag{4}$$

the *conservation law for electric charge* requires that all *operators* commute with

$$\tau_{\zeta 1} + \tau_{\zeta 2} + \ldots + \tau_{\zeta n} = n_N - n_P = 2T_\zeta. \tag{3}$$

In addition to this, one hardly would say that

$$\tau_{\xi 1}\tau_{\xi 2} + \tau_{\eta 1}\tau_{\eta 2} + \tau_{\zeta 1}\,\tau_{\zeta 2} = -\,1 - 2PQ. \qquad (5)$$

(P *interchange of space*, Q *interchange of spin coordinates*) does not involve the Cartesian or *spin* coordinates, since Eq. (5) is a rather artificial expression, $\tau_\xi$ and $\tau_\eta$ having no immediate physical significance. We shall assume hence for approximation (3) *only such operators which are equivalent to operators acting on the Cartesian coordinates alone, but in a different way for protons and neutrons*. This is equivalent to using only *operators* involving the space coordinates and the $\tau_\zeta$'s. If we do this, the results of method (3) must become equivalent to the usual theory (without $\tau$'s) which neglects the *spin*. As a matter of fact, for approximation (3), the introduction of $\tau$ is entirely useless and it is taken up here only in order to establish the transition from approximation (1) to (3).

**2.** The *interaction* in the *electronic shells* of atoms is a sum of terms containing two particles only and the *momenta* is no higher than the second power. The reason for the first is, that the *interaction* occurs through a *field* and this gives in first approximation only *interaction between two particles*. The reason that one can stop with the second power of the *momenta* is that these always enter in the combination p/mc which is a small quantity.

An advantage of introducing the variable $\tau$ is[2,3] *that one can take over these assumptions to nuclei*. If one does not use the variable $\tau$ the *interchange* of two particles if expressed as a power series of the *momenta* is an infinite series[5]

[5] Wheeler, J. A. (1936). The Dependence of Nuclear Forces on Velocity. *Phys. Rev.*, 50, 643.

$$\Sigma\; n_1 n_2 n_3 \;\{\; \dots \;\}\; x \;\{\; \dots \;\}$$

However, it can be expressed by means of *Dirac's identity* also entirely without the *momenta* by means of Eq. (5). It must be admitted, however, that the *spin* cannot be considered to be small as in the atomic theory. We shall determine here all *interaction* forms between two particles which do not contain higher than first power terms of *momenta*[6] as far as the dependence on s and $\tau$ goes.

[6] Some of these were given previously by Cassen and Condon, reference 3. The expressions given here are invariant only under Galilei transformations. G. Breit has shown that, in order to ensure *relativistic invariance*, correction terms must be added to the expressions derived here.

Nothing can be said, of course, on the dependence on the distance, and this factor will be omitted hence. It seems to be of lesser importance for the present.

The *interaction* must have *spherical symmetry*, depending on the differences of *coordinates* and *momenta* only, be *invariant under inversion*, substitution of – t for t and also be *symmetric in the particles*. The first requirements determine the dependence on s, x and p. From the two triples of *spin operators*, one can form two invariants

$$\text{(i)} \quad 1;\ \text{and (i')}\ \tfrac{1}{2} + \tfrac{1}{2}(s_{x1}s_{x2} + s_{y1}s_{y2} + s_{z1}s_{z2}) = Q_{12},$$

three axial vectors with Z components

$$\text{(v)} \quad s_{z1} + s_{z2}; \qquad s_{z1} - s_{z2}; \qquad s_{x1}s_{y2} - s_{y1}s_{x2}$$

respectively, and one axial tensor, with components

$$s_{x1}s_{y2} + s_{y1}s_{x2}; \qquad s_{y1}s_{z2} + s_{z1}s_{y2}; \qquad s_{z1}s_{x2} + s_{x1}s_{z2};$$
$$s_{x1}s_{x2} - s_{y1}s_{y2}; \qquad s_{x1}s_{x2} + s_{y1}s_{y2} - 2s_{z1}s_{z2}$$

The **first two of these**, (i) and (i'), can be used as they stand, cannot be combined with first power expressions of p, however, since these change sign under the t' = – t substitution. The last one (t) gives the familiar expression

$$\text{(i'')} \quad (s_1 \cdot r_{12})(s_2 \cdot r_{12}) - 3(s_1 \cdot s_2)r_{12}^2$$

if combined with the similar tensor of the coordinates[7].

[7] (ii) has the property that it is identical with $Q_{12}$(ii). It is an interaction which shows saturation.

It cannot be combined with the p either. The middle one must be combined with the vector $p_1 - p_2$ which gives a useless axial *invariant* and tensor and an ordinary vector. This combined with the distance vector gives the familiar

$$\text{(ia) (ib) (ic)} \quad \begin{vmatrix} s_x & s_y & s_z \\ x_1 - x_2 & y_1 - y_2 & z_1 - z_2 \\ p_{x1} - p_{x2} & p_{y1} - p_{y2} & p_{z1} - p_{z2} \end{vmatrix}$$

Here $s_x$, $s_y$, $s_z$, can be the components of one of the three vectors (v). On the whole, we have 6 invariants. These *invariants* can be multiplied with one of the six expressions in τ which commute with $\tau_{\xi 1} + \tau_{\xi 2}$. These are, first of all

$$(\tau_0)\ 1 \quad \text{and} \quad (\tau_0')\ -\tfrac{1}{2} - \tfrac{1}{2}(\tau_1 \cdot \tau_2) = P_{12}Q_{12},$$

which give the same *interaction* between all pairs of particles. In addition to these, we have

$(\tau_1)$ ½ + ½ $\tau_{\xi 1}\tau_{\xi 2}$ and $(\tau_2')$ ½ $(\tau_{\xi 1} = \tau_{\xi 2})$.

The first of these gives ordinary *interaction* but only between like particles; the second gives a negative *interaction* for *proton* pairs, a positive for *neutron* pairs, none for unlike particles. These *interactions* are *symmetric* in the particles and can be combined with (i), (i'), (i") and (i*a*), giving in the whole 16 different forms.

Finally, we have

$(\tau_{\xi 2})$ $\tau_{\xi 1} - \tau_{\xi 2}$ and $(\tau_{\xi 2}')$ ½ $(\tau_{\xi 1}\tau_{\eta 2} - \tau_{\eta 1}\tau_{\xi 2})$,

which can be combined with (ib) and (ic) giving 4 more types of interaction.

In *approximation* (1) we can have only (i)$(\tau_0)$ and (i')$(\tau_0')$, i.e., *ordinary* and *Majorana exchange forces*.

[There are three types of exchange forces; the *Majorana exchange force*, which arises from the *spatial exchange*; the *Bartlett force*, arising from the *spin exchange*; and the *Heisenberg force*, arising from the *space-spin exchange*. Ettore Majorana (born 5 August 5, 1906 – likely dying in or after 1959) was an Italian theoretical physicist who worked on neutrino masses. On 25 March 1938, he disappeared under mysterious circumstances after purchasing a ticket to travel by ship from Naples to Palermo.]

In *approximation* (2), all 8 forms derived from $(\tau_0)$, $(\tau_0')$ and (i) (i') (i") and (i*a*). These are, in addition to the previous ones, *spin-spin* (i")$(\tau_0)$, *spin-orbit* (242) $(\tau_0)$ *ordinary forces*, *Heisenberg forces* (i)$(\tau_0')$. Furthermore *spin-spin exchange forces* (i")$(\tau_0')$ and *spin-orbit exchange forces* (i*a*)$(\tau_0')$ of the *Heisenberg type*. The *Majorana exchange forces* of these types are identical with the *ordinary forces*. Finally, we have the *spin-exchange forces* (2') $(\tau_0)$ of *Bartlett*[8].

[8] The content of this section is based on the fundamental mathematical works of Cartan, E. (1913). *Bull. Soc. Math. de France*, 41, 43; (1914). *J. de Math.*, 10, 149; Schur, I. (1924). *Berl. Ber.*, pp. 189, 297, 346; and particularly, Weyl, H. (1925). *Math. Zs.*, 23, 271. I attempted to compile in this section—often without giving rigorous proofs— those results which suffice for the discussion of the physical problems in question.

In *approximation* (3) we must permit according to the preceding section, in addition to those of 1, only (i)$(\tau_l)$ and (i)$(\tau_l')$, allowing for different interactions between different kinds of pairs. The coefficient of (i)$(\tau_l')$ is certainly very small, the *proton-proton interaction* being very nearly equal to the *neutron-neutron interaction*.

In *approximation* (4), all 20 types become possible.

**3.** We next go over to *approximation* (1), and try to define the analog of the *multiplet* system.

This can be defined in two ways: either by considering the functional dependence of the *wave functions* on the *spins* or else by considering their dependence on the space coordinates. We shall first consider the *spin function*.

The great difference between the ordinary *spin* and the spin considered here is that we have, for every particle, two *spin* coordinates s and $\tau$ giving in the whole four different sets of values –1, –1; –1, 1; 1, –1; 1, 1. Instead of two two-valued *spins*, one can introduce one four-valued *spin* $\eta$, which has the values 1, 2, 3, 4 for the four different doublets of values of s and $\tau$, respectively. This $\eta$ plays the same role which the two-valued *spin* plays in the ordinary *spin* theory. However, because of the four-valuedness of $\eta$, instead of the *representations* of the two-dimensional *unitary group* (or the equivalent three-dimensional *rotation group*), the *representations* of the four-dimensional *unitary group* will characterize the *multiplet* systems.

Since the Hamiltonian does not contain the *spin* coordinates, any transformation which affects only these, will bring an *eigenfunction* into an *eigenfunction*. We can consider first, the *permutations* of the $\eta_i$ and second, simultaneous *unitary transformations* of all the $\eta$ …
…

# PART III Comparison between the Standard Model and New Physics.

## (1) The universe is composed of elementary particles.

Although the *Standard Model* is believed to be theoretically self-consistent and has demonstrated some success in providing experimental predictions, *it leaves some physical phenomena unexplained and so falls short of being a complete theory of fundamental interactions*. Although the physics of *special relativity* is included, *general relativity is not*, and it will fail at energies or distances where the *graviton* is expected to emerge. It does not fully explain *baryon asymmetry*, or account for the *universe's accelerating expansion as possibly described by dark energy.* The model does *not contain any viable dark matter particle* that possesses all of the required properties deduced from observational cosmology. It also does not incorporate *neutrino oscillations* and their non-zero masses.

Ideally, *New Physics* should confine itself to *elementary particles* that can be *observed.* On this basis *quarks*, *gluons*, and the $W^+$, $W^-$, $Z^0$ and *Higgs bosons*, should be removed from this list.

This restores the *proton* and the *neutron* to the list of *elementary particles.* Although the *proton* and *neutron* can be considered as different *isospin quantum states* of the same particle, they will be treated as separate *elementary particles.*

(1) The *Standard Model.* **Includes *quarks*, *gluons*, and the $W^+$, $W^-$, $Z^0$, and *Higgs boson*.**

**The introduction of quarks**.

Gell-Mann. M. (February, 1964). *A Schematic Model of Baryons and Mesons*, proposed that *baryons*, which include *protons* and *neutrons*, and *mesons* were composed of *elementary particles* called "quarks". A mathematical model based on *field theory* was described.

Makoto Kobayashi – Nobel Lecture, December 8, 2008. *CP Violation and Flavor Mixing*. In his Nobel Prize lecture Makoto Kobayashi provided a brief history of the development of the *six-quark model.*

(1) *New Physics*. **Confined to *elementary particles* that can be *observed*. Proton and *neutron* restored to list of *elementary particles*. No *quarks*.**

***Antiparticles and antimatter*.**

**Symmetry in elementary particles between positive and negative charge.**

In his 1933 Nobel Lecture, on December 12, 1933 [*Theory of electrons and positrons*], Dirac speculated that there was a *complete symmetry in elementary particles between positive and negative charge*. In the Standard Model *elementary particles* with the same *mass* but opposite *electric charge* are described as *antiparticles*, a different form of matter.

> (1) The *Standard Model*. **Existence of antiparticles with *differences in quantum numbers* in additional to electric charge, which form *antimatter*, a different form of matter. Unexplained asymmetry of *matter* and *antimatter* in the visible universe.**

*Antimatter* is defined as *matter* composed of the *antiparticles* of the corresponding particles in "ordinary" matter, and in the *Standard Model* can be thought of as *matter with reversed electric charge, parity, and time, known as CPT reversal*. *Antimatter* particles carry the same *electric charge* as *matter* particles, but of opposite sign. That is, an *antiproton* is negatively charged and an *antielectron* (*positron*) is positively charged. *Neutrons* do not carry a net charge, but *according to quark formulation of the Standard Model* their constituent *quarks* do. In this theory, a *particle* and its *antiparticle* (for example, a *proton* and an *antiproton*) have the same *mass*, but opposite *electric charge*, and, *according to the Standard Model, other differences in quantum numbers*. *Protons* and *neutrons* have a *baryon number* of +1, while *antiprotons* and *antineutrons* have a *baryon number* of –1. Similarly, *electrons* have a *lepton number* of +1, while that of *positrons* is –1.

When a *particle* and its corresponding *antiparticle* collide, they are both converted into energy. In this view, a collision between any *particle* and its *anti-particle* partner is seen to lead to their mutual *annihilation* giving rise to various proportions of intense *photons* (*gamma rays*), *neutrinos*, and sometimes less-massive *particle–antiparticle* pairs. The majority of the total energy of annihilation emerges in the form of ionizing radiation. If surrounding *matter* is present, the energy content of this radiation will be absorbed and converted into other forms of energy, such as heat or light. The amount of energy released is usually proportional to the total mass of the collided *matter* and *antimatter*, in accordance with the notable mass–energy equivalence equation, $E=mc^2$.

*Antiparticles* bind with each other to form *antimatter*; just as ordinary particles bind to form normal *matter*. …

There is no difference in the gravitational behavior of *matter* and *antimatter*. In other words, *antimatter* falls down when dropped, not up. There are compelling theoretical reasons to believe that, aside from the fact that *antiparticles* have different signs on all charges (such as *electric* and *baryon charges*), *matter* and *antimatter* have exactly the same properties. This means a *particle* and its corresponding *antiparticle* must have identical *masses* and *decay lifetimes*.

There is strong evidence that the observable universe is composed almost entirely of ordinary *matter*, as opposed to an equal mixture of *matter* and *antimatter*. According to this theory, this asymmetry of *matter* and *antimatter* in the visible universe is one of the great unsolved problems in physics. The process by which this inequality between *matter* and *antimatter* particles developed is called *baryogenesis*.

> (1) *New Physics*. **Elimination of notion of *antimatter* and problem of asymmetry of *matter* and *antimatter* in the visible universe. *Antiparticles* are simply the less stable *particles* of similar *mass* but opposite *electric charge*.**

Does not accept that a *particle* and its corresponding *antiparticle* must have identical *decay lifetimes*.

*Antiparticles* are simply the less stable *particles* of similar *mass* but opposite *electric charge*, just as the *positron* is the less stable *antiparticle* of the *electron* and the *antiproton* the less stable *antiparticle* of the *proton*. There is no longer any need to rely on a *lepton number* or *baryon number*. Then *antimatter* is just a less stable form of *matter*, which would account for the asymmetry of *matter* and *antimatter* in the visible universe.

***Elementary particles*.**

> (1) *Standard Model*. **The universe is composed of 52 *elementary particles* and *antiparticles* of which only the *electron* and the *photon* are stable.**

The *Standard Model* of particle physics evolved from the Bohr model of the atom in 1913, based on what were believed to be 3 stable particles, *electrons* in orbit around a *nucleus* comprised of *protons* and *neutrons*, to its emergence in 1973 as the *six-quark model*, comprising 26 *elementary particles* or a total of 52 *elementary particles* and their *anti-particles* of which only the *electron* and *photon* are stable.

> (1) *New Physics*. **The universe is composed of 14 *elementary particles* including the *proton*, *neutron* and *graviton*. *Elementary particles* are confined to those that have the possibility of being *observed*.**

**(2) The speed of light in vacuum.**

(2) *Standard Model.* **The speed of light in a vacuum is constant for all observers, regardless of the motion of light source or observer. Results in *length contraction* and *time dilation* for a moving observer.**

**Einstein's theory of Special Relativity**. Einstein, A. (September, 1905). *Zur Elektrodynamik bewegter Körper*. (On the electrodynamics of moving bodies.) In 1905 Albert Einstein published his *theory of Special Relativity*, based on two postulates:

(i) The laws of physics are invariant (that is, identical) in all *inertial frames of reference* (that is, frames of reference with no acceleration), known as the *principle of relativity*.

(ii) The *speed of light in vacuum is the same for all observers*, regardless of the motion of the light source or observer.

Underwood, T. G. (2023). *General Relativity*, Conclusion, p. 474: "Einstein's *theory of general relativity* attempted to extend his *theory of special relativity* beyond space and time, to include *matter* and *gravitational fields*. *Gravitation* was introduced through the "*equivalence principle*", the equivalence of the outcome of the force of gravity and the acceleration of matter, first recognized in *Newton's Principia*. This allowed Einstein to construct a *relativistic theory* of the effect of a *gravitational field* on *matter*, but it also resulted in him *rejecting his postulate on the constancy of light* in the presence of a gravitational field, …"

Einstein, A. (May, 1912). *Lichtgeschwindigkeit und Statik des Gravitionsfeldes.* (The Speed of Light and the Statics of the Gravitational Field.). Einstein (1911) showed that the validity of one of the fundamental laws of his theory of special relativity, namely, the law of the *constancy of the speed of light*, could claim to be valid only for space-time domains of constant gravitational potential. Einstein noted that despite the fact that this result *excluded the general applicability of the Lorentz transformation*, it should not deter us from pursuing the consequences of that path. Here he took that further by demonstrating that the Lorentz transformation could not be established for infinitely-small space-time regions either *as soon as one abandons the universal constancy of c*.

Underwood, T. G. (2023). *Special Relativity*, Conclusion, p. 381: "There is no evidence, based directly or indirectly, on the observation of the speed of electromagnetic radiation in a vacuum emitted by an inertial body, or as observed by an inertial observer, moving in a straight line and not involving mirrors.

However, by now it may be possible to achieve this in a laboratory experiment in a vacuum without mirrors, using electromagnetic radiation emitted by two sources of the same frequency, one stationary and the other moving at a constant velocity in a straight line; either directly, or by measuring the observed frequency of the radiation. …

Recognizing that evidence based on *celestial observations*, experiments with *light passing through a medium*, and observations on *rotating platforms* and other *accelerated* systems, is suspect, the evidence in support of the *theory of special relativity* reduces to extremely slim pickings. The Ehrenfest paradox, the *non-relativistic* Doppler red shift and blue shift for light, the known physics of the emission of electromagnetic radiation and of the electron, and the success of *non-relativistic* quantum electrodynamics in explaining the interaction of the electromagnetic field with electrically charged particles, comprise the strongest evidence against Einstein's *second postulate*, the *constancy of the speed of light*.

Quite apart from enormity of the consequences of Einstein's two postulates taken together, including *length contraction*, *time dilation*, and the requirement to assume a *point electron* in the unsuccessful attempt to introduce special relativity into quantum electrodynamics, the evidence in support of Einstein's *second postulate* on the constancy of the speed of light is far outweighed by the evidence against it.

For this reason, until more satisfactory evidence in support of Einstein's *second postulate*, a refutation of the Ehrenfest paradox, and an explanation for the observed Doppler red shift and blue shift consistent with Einstein's two postulates, is provided, under any normal measure of a theory in physics, *Einstein's second postulate, and consequently his theory of special relativity, must be rejected. …*"

(2) *New Physics*. **The speed of light in a vacuum is constant relative to the emitter. Replaces *Einstein's theory of Special Relativity* with *Walter Ritz's emission theory*. Avoids *length contraction* and *time dilation* for a moving observer.**

*Walter Ritz's emission theory*, also called emitter theory or ballistic theory of light, was a competing theory for the *special theory of relativity*, explaining the results of the Michelson–Morley experiment of 1887. Emission theories obey the principle of relativity by having no preferred frame for light transmission, but say that *light is emitted at speed "c" relative to its source* instead of applying the invariance postulate. Thus, emitter theory combines electrodynamics and mechanics with a simple Newtonian theory.

Ritz, W. (December, 1908). *Über die Grundlagen der Elektrodynamik und die Theorie der schwarzen Strahlung*. (On the basics of electrodynamics and the theory of black body radiation.): Ritz notes that the differential equations in the Maxwell-Lorentz formulation of electrodynamics permit infinite solutions, including those with both *retarded* and *advanced* potentials, on which Einstein (1905) relied in deriving the consequences of his two postulates. Ritz shows that *advanced potentials* are inadmissible and how this could be addressed based on *retarded potentials* alone, i.e. by an *emission theory*. He also demonstrates why the role of an ether must be removed from the theory of electrodynamics.

**(3) All elementary particles, including electromagnetic waves (photons), are quantized.**

(3) *Standard Model.* **Lack of convergence in current formulations of quantum electrodynamics due to the interaction of the electromagnetic and matter fields with their own vacuum fluctuations. The question is whether all *divergencies* can be isolated in unobservable *renormalization* factors.**

Einstein, A., Podolsky, B. & Rosen, N. (May, 1935). Can Quantum-Mechanical Description of Physical Reality Be Considered Complete?: Suggests that the description of reality as given by a wave function in quantum mechanics is not complete.

Schwinger, J. (November, 1948). Quantum Electrodynamics. I. A Covariant Formulation: Notes that the *lack of convergence in current formulations of quantum electrodynamics indicates that revision of electrodynamic concepts at ultra-relativistic energies is necessary*. Elementary phenomenon in which *divergences* occur as a result of virtual transitions involving particles with unlimited energy are *polarization of the vacuum* and *self-energy of the electron* which express *the interaction of the electromagnetic and matter fields with their own vacuum fluctuations*. This alters the constants characterizing the properties of the individual fields and their mutual coupling by infinite factors, the question is whether all *divergencies* can be isolated in such unobservable *renormalization* factors.

(3) *New Physics.* **Only relationships among observable quantities occur. Avoids requirement to assume a *point electron*, and address through a process of *renormalization* the still unresolved *divergencies*.**

**Non-relativistic Quantum Mechanics and Quantum Electrodynamics.**

Heisenberg, W. (July, 1925). Über quantentheoretische Umdeutung kinematischer und mechanischer Beziehungen. (On the quantum-theoretical re-interpretation of kinematic and mechanical relations.): Heisenberg proposes a *quantum mechanics* in which *only relationships among observable quantities occur*. Not possible to assign to the electron a point in space as a function of time. Builds on Kramer's dispersion theory and instead assigns to the electron an *emitted radiation*. Substitutes *frequencies* and *amplitudes* of Fourier components of emitted radiation of electron. Assigns *transition frequencies* and *transition amplitudes* as observables. Replaces classical component by *transition* component corresponding to the quantum jump from state *n* to state *n* – α. Translates the old *quantum condition* that fixes the properties of the *states* to a new condition to calculate the amplitude of a *transition* between two states by replacing the differential by a difference.

Dirac, P. A. M. (March, 1927). *The quantum theory of the emission and absorption of radiation*: addresses *non-relativistic quantum electrodynamics* by treating problem of an assembly of similar systems satisfying the Einstein-Bose statistical mechanics which interact with another different system by obtaining a Hamiltonian function to describe the motion. Theory of system in which *forces are propagated with velocity of light* instead of instantaneously. Time counted as a c-number instead of being treated symmetrically with the space co-ordinates. Addition of *interaction term*, production of electromagnetic field (emission of radiation) by moving electron, reaction of radiation field on emitting system. Applies to the interaction of an assembly of *light-quanta* with an atom. Shows that it leads to *Einstein's laws for the emission and absorption of radiation*. The interaction of an atom with *electromagnetic waves* is then considered. Treats *field* of radiation as a dynamical system whose interaction with an ordinary atomic system may be described by a Hamilton function. Dynamical variables specifying the *field* are the *energies* and *phases* of the harmonic components of the waves. Shows that if one takes the *energies* and *phases* of the waves to be *q-numbers* satisfying the proper quantum conditions instead of *c-numbers*, the Hamiltonian function for the interaction of the *field* with an atom takes the same form as that for the interaction of an assembly of *light-quanta* with the atom. Provides a complete formal reconciliation between the wave and light-quantum point of view.

**(4) All elementary particles have mass, apart from the photon and gluons.**

(4) *Standard Model*. **The large *masses* of the $W^+$, $W^-$, $Z^0$ and *Higgs bosons* and the *top quark* (respectively 85.7, 85.7, 97.2, 133.3 and 184.9 times the *mass* of the *proton*) raise questions regarding whether they are really *elementary particles*, in particular in view of how they were created by collisions of high energy *protons* in *proton-antiproton* and *hadron* colliders.**

Higgs, P. W. (October, 1964). *Broken Symmetries and the Masses of Gauge Bosons*: In a previous paper, Higgs had shown that the *Goldstone theorem*, that Lorentz-covariant field theories in which spontaneous breakdown of symmetry under an internal Lie group occurs contain zero-mass particles, *failed if and only if the conserved currents associated with the internal group were coupled to gauge fields*. The purpose of the present note was to report that, *as a consequence of this coupling, the spin-one quanta of some of the gauge fields acquired mass*; the longitudinal degrees of freedom of these particles (which would be absent if their *mass* were zero) go over into the *Goldstone bosons when the coupling tends to zero. The model was discussed mainly in classical terms*; nothing was proved about the quantized theory. Higgs noted that it should be understood, therefore, that *the conclusions which were presented concerning the masses of particles were conjectures based on the quantization of linearized classical field equations.*

(4) *New Physics.* **Removal of the $W^+$, $W^-$, $Z^0$ and *Higgs bosons* and the *quarks* avoids the problem of overweight *elementary particles.***

**(5) Elementary and composite particles can have electric charge or be neutral.**

(5) *Standard Model.* ***Up*, *charm*, and *top quarks* are assumed to have *electric charges* equal to 2/3 of the charge of the *electron* and *proton*; and *down*, *strange*, and *bottom quarks* to have *electric charges* equal to – 1/3 of the charge of the *electron* and *proton*, but this cannot be observed.**

(5) *New Physics.* **Removal of *quarks* avoids the problem of elementary particles with unobserved fractional *electric charges.***

**(6) Elementary particles have a quantum state called spin.**

**Spin.**

The *spin quantum state* is loosely related to the *angular momentum* of the particle. *Fermions* have ½-integer values, and *bosons* have integer values. These values have two directions, + or –.

Compton, A. H. (August, 1921). *The Magnetic Electron*: Compton's paper on investigations of ferromagnetic substances with X-rays was the first to introduce the idea of *electron spin.* Compton hypothesized that the electron's *magnetic moment* was intrinsically connected to the electron's *spin* and pointed out the possible bearing of this idea on the origin of the natural unit of magnetism.

Uhlenbeck, G. E. & Goudsmit, S. (November, 1925). *Ersetzung der Hypothese vom unmechanischen Zwang durch eine Forderung bezuglich des inneren Verhaltens jedes einzelnen Elektrons.* (Replacement of the hypothesis of unmechanical coercion by a requirement regarding the internal behavior of each individual electron.): The idea of a *quantized spinning of the electron* was put forward for the first time by Compton in August 1921. Without being aware of Compton's suggestion Uhlenbeck and Goudsmit noted doublets in the alkali spectra that did not conform to current models of the atom. They proposed applying the model of the *spinning electron* to interpret a number of features of the quantum theory of the *anomalous Zeeman effect*, and applied the classical formula for spherical rotating electron with finite radius and surface charge.

The first direct experimental evidence of the *electron spin* was the Stern–Gerlach experiment of 1922. However, the correct explanation of this experiment was only given in 1927. The original interpretation assumed the two spots observed in the experiment were due to *quantized orbital angular momentum*. However, in 1927 Ronald Fraser showed that Sodium atoms are isotropic with no *orbital angular momentum* and suggested that the observed magnetic properties were due to *electron spin*. In same year, Phipps and Taylor applied the *Stern-Gerlach technique* to hydrogen atoms; the ground state of hydrogen has zero *angular momentum* but the measurements again showed two peaks. Once the quantum theory became established, it became clear that *the original interpretation could not have been correct*: the possible values of *orbital angular momentum* along one axis is always an odd number, unlike the observations. Hydrogen atoms have a single electron with *two spin states* giving the two spots observed; silver atoms have closed shells which do not contribute to the *magnetic moment* and only the unmatched outer electron's *spin* responds to the field.

(6) *Standard Model*. **The *spin* of an elementary particles is a quantum state and consequently a *non-relativistic* concept.**

We could try to determine the behavior of spin under general Lorentz transformations, but we would immediately discover a major obstacle. Unlike SO(3), the group of Lorentz transformations SO(3,1) is *non-compact* and therefore does not have any faithful, unitary, finite-dimensional representations.

(6) *New Physics*. ***Spin* obeys the mathematical laws of angular momentum quantization.**

The specific properties of *spin angular momenta* include:

- *Spin quantum numbers may take either half-integer or integer values.*

- Although the *direction* of its *spin* can be changed, *the magnitude of the spin of an elementary particle cannot be changed.*
- *The spin of a charged particle is associated with a magnetic dipole moment with a g-factor that differs from 1*. (In the classical context, this would imply the internal charge and mass distributions differing for a rotating object.).

The conventional definition of the *spin quantum number* is s = n/2, where n can be any non-negative integer. Hence the allowed values of s are 0, 1/2, 1, 3/2, 2, etc. The value of s for an *elementary particle* depends only on the type of particle and cannot be altered in any known way (in contrast to the spin direction described below). The *spin angular momentum* S of any physical system is quantized. The allowed values of S are

$$S = \hbar\sqrt{s(s + 1)} = h/2\pi \sqrt{n/2\ (n + 2)/2} = h/4\pi \sqrt{n(n + 2)},$$

where h is the Planck constant, and $\hbar = h/2\pi$ is the reduced Planck constant. In contrast, *orbital angular momentum* can only take on integer values of s; i.e., even-numbered values of n.

*Those particles with half-integer spins, such as 1/2, 3/2, 5/2, are known as fermions, while those particles with integer spins, such as 0, 1, 2, are known as bosons*. The two families of particles obey different rules and broadly have different roles in the world around us. A key distinction between the two families is that *fermions obey the Pauli exclusion principle*: that is, *there cannot be two identical fermions simultaneously having the same quantum numbers* (meaning, roughly, having the same position, velocity and spin direction). *Fermions obey the rules of Fermi–Dirac statistics*. In contrast, *bosons obey the rules of Bose–Einstein statistics* and have *no such restriction*, so they may "bunch together" in identical states.

(6) *New Physics*. ***Non-relativistic theory.***

Pauli, W. (February, 1925). *Über den Zusammenhang des Abschlusses der Elektronengruppen im Atom mit der Komplexstruktur der Spektren.* (On the connection between the completion of electron groups in an atom and the complex structure of spectra.): Pauli noted a serious difficulty with the former is the connection of these ideas with the *correspondence principle*, which was well known to be a necessary means to explain the selection rules for the *quantum numbers* $k_1$, j, and m and the polarization of the Zeeman components, in particular, that *it was necessary that the totality of the stationary states of an atom corresponded to a collection (class) of orbits with a definite type of periodicity properties*. The dynamic explanation of this kind of motion of the *optically active electron*, which was based upon the assumption of deviations of the forces between the *atom* core and the *electron* from central symmetry, *seemed to be incompatible with the possibility to represent the alkali doublet (and thus also the magnitude of the corresponding precession frequency) by relativistic formulae.* Consequently, Pauli, decided to pursue instead the alternative *non-relativistic* theory to the problem of *completion of electron groups in an atom*, in order to draw conclusions only about the *number of possible stationary states* of an *atom* when several equivalent *electrons* are present. But this did not address the position and relative order of the term values. On the basis of these results, Pauli obtained a general classification of every *electron* in the *atom* by the principal quantum number n and two auxiliary quantum numbers $k_1$ and $k_2$ to which he added a further quantum number $m_1$ in the presence of an external field, in agreement with experiments.

Pauli, W. (September, 1927). *Zur Quantenmechanik des magnetischen Elektrons.* (On the quantum mechanics of magnetic electrons.); it will be shown how one can arrive at a formulation of the quantum mechanics of the *magnetic electron* by the Schrödinger method of eigenfunctions, with no use of double-valued functions, when one, on the basis of the Dirac-Jordan general theory of transformations, introduces the components of its *proper impulse moment* in a fixed direction as further independent variables in order to carry out the computations of its rotational degrees freedom, along with the position coordinates of any *electron.* In contradiction to classical mechanics, these variables can assume only the variables $+ \frac{1}{2}\, h/2\pi$ and $- \frac{1}{2}\, h/2\pi$, which is completely independent of any sort of external field.

**(7) Elementary and composite particles with mass attract each other through the gravitational interaction or gravitational force.**

**Gravity.**

*Newton's law of gravitation* states that every point mass in the universe attracts every other point mass with a force that is directly proportional to the product of their masses, and inversely proportional to the square of the distance between them.

Underwood, T. G. (2023). *Gravity*: Newton's universal law of gravitation, pp. 74-7: "While Newton was able to articulate his *Law of Universal Gravitation* and verify it experimentally, he could only calculate the relative *gravitational force* in comparison to another force. It was not until Henry Cavendish's verification of the *Gravitational Constant* that the *Law of Universal Gravitation* received its final form:

$$F = GMm/r^2 = 6.674 \times 10^{-11}\ Mm/r^2\ N\ (\textit{SI units})$$

where F represents the force in Newtons, M and m represent the two masses in kilograms, and r represents the separation in meters. G represents the Gravitational Constant, which has a value of $6.674 \times 10^{-11}$ N $(m/kg)^2$. Because of the magnitude of G, gravitational force is very small *unless large masses or short distances are involved.*"

Newton, I. (July, 1687). *Philosophiæ Naturalis Principia Mathematica*. (The Mathematical Principles of Natural Philosophy.) In Book I, *The Motion of Bodies*, Newton addresses the motion of bodies attracted to each other by centripetal forces. In Book III, *Of the System of the World*, Newton notes that the *centripetal force* which arises between planets is the same as the *gravitational force* attracting matter to the Earth and focusses on gravitational attraction. He then proposes that "all bodies gravitate towards; every Planet and that the Weights of bodies towards any the same Planet, at equal distances from the center of the Planet, are proportional to the quantities of matter which they severally contain; and that there is a power of gravity tending to all bodies, proportional to the several quantities of matter which they contain; and that the force of gravity towards the several equal particles of any body, is reciprocally as the square of the distance of places from the particles".

(7) *Standard Model*. **Einstein's theory of General Relativity.**

*General relativity* is claimed to generalize *special relativity* and refine Newton's *law of universal gravitation*, providing a unified description of *gravity* as a geometric property of *space and time* or four-dimensional *spacetime*. In particular, the *curvature of spacetime* is directly related to the *energy* and *momentum* of whatever *matter* and *radiation* are present. The relation is specified by the *Einstein field equations*, a system of second-order partial differential equations.

However, reconciliation of *general relativity* with the laws of *quantum physics* remains a problem *as there is a lack of a self-consistent theory of quantum gravity*. It is not yet known how gravity can be unified with the three non-gravitational forces: strong, weak and electromagnetic.

A detailed examination of Einstein's *theory of general relativity* reveals that it is not a *theory of gravity*; it is a *relativistic* theory about the *effects* of gravitation, or more strictly, of a uniformly accelerated reference frame. There is nothing in any version of this theory that represents or explains or provides any connection to the weak attractive gravitational force between matter. We are no further forward in understanding the origin of this fundamental force.

Einstein, A. (November 25, 1915). Die Feldgleichungen der Gravitation. (The Field Equations of Gravitation.) Einstein's *theory of general relativity* attempted to extend his *theory of special relativity* beyond space and time, to include *matter* and *gravitational fields. Gravitation* was introduced through the "*equivalence principle*", the equivalence of the *outcome* of the force of *gravity* and the acceleration of *matter*, first recognized in Newton's *Principia.* In order to make calculations with his theory, Einstein had to import *Newton's law of gravitation*, which itself is an empirical law with no fundamental foundation. Consequently, the only evidence that Einstein could provide for his *theory of general relativity* was effectively Newtonian.

Einstein's unsuccessful attempts at producing a *classical unified field theory* between 1923 until he died in 1955, during which time Einstein published 31 papers on a *unified theory of electromagnetism and gravity.*

In the light of the continued failure of Einstein's efforts to overcome the main objections to his *theory of special relativity* - the Ehrenfest paradox, and its failure to explain the observed Doppler redshift and blueshift of light – or to provide any evidence for it, and in the absence of any supportive evidence for his *theory of general relativity*, both theories must be rejected until such objections are overcome and such evidence is provided

Einstein, A. (February, 1917). *Kosmologische Betrachtungen zur allgemeinen Relativitätstheorie.* (Cosmological Considerations in the General Theory of Relativity.): describes Einstein's struggles with supplementing the *relativistic differential equations* by *limiting conditions* at *spatial infinity* in order to regard the universe as being of infinite spatial extent. As he noted, "*we admittedly had to introduce an extension of the field equations of gravitation which is not justified by our actual knowledge of gravitation*".

Weyl, H. (May, 1929). *Elektron und Gravitation.* (Electron and gravity.): Heinrich Weyl's attempt in 1929 to incorporate Dirac theory into the scheme of *general relativity* by introducing *gauge invariance* of *theory of coupled electromagnetic potentials* and Dirac *matter waves.* Weyl claimed that the barrier which hems progress of quantum theory is *quantization of the field equations.*

(7) *New Physics.* ***Quantum entanglement*** **between matter.**

In addition to *quantum entanglement* between the *spin states* of elementary and composite particles which creates the *weak interaction* or attractive *weak force*, it is possible that there may also be *quantum entanglement* between *matter* (*gravitons*) creating the *gravitational interaction* or attractive *gravitational force*.

> (7) *New Physics*. ***Gravity* is explained by a *quantum theory of gravity* based on *quantum entanglement* between *quantum states* of *matter* (*gravitons*).**

Unburdened by Einstein's theories of relativity, a *quantum theory of gravity* can easily be developed by following Dirac's procedure for *non-relativistic quantum electrodynamics* in Dirac, P. A. M. (March, 1927).

**(8) Elementary and composite particles with the same electric charge attract each other, and elementary and composite particles with opposite electric charge are repulsed, through the electromagnetic interaction or electromagnetic force, according to Coulomb's law.**

**Electromagnetism.**

*Electromagnetism* is an *interaction* that occurs between particles with *electric charge* via *electromagnetic fields*. The *electromagnetic force* is one of the four fundamental forces of nature. It is the dominant force in the interactions of atoms and molecules. *Electromagnetism* can be thought of as a combination of *electrostatics* and *magnetism*, which are distinct but closely intertwined phenomena. *Electromagnetic forces* occur between any two *charged particles*. *Electric forces* cause an attraction between particles with opposite charges and repulsion between particles with the same charge, while *magnetism* is an *interaction* that occurs between *charged particles in relative motion*. These two forces are described in terms of *electromagnetic fields*. Macroscopic charged objects are described in terms of *Coulomb's law for electricity* and *Ampère's force law for magnetism*.

> *Coulomb, C-A. (1785). Premier mémoire sur l'électricité et le magnétisme.*(First memoir on electricity and magnetism.)*:* Coulomb described the construction of a torsion balance and used this to demonstrate what he described as the fundamental law of electricity, now known as *Coulomb's Law*. *Coulomb's Law for Electricity* states that the magnitude, or absolute value, of the attractive or repulsive electrostatic force between two point-charges is directly proportional to the product of the magnitudes of their charges and inversely proportional to the squared distance between them.

> Ampère, A-M. (1822). *Memoire sur la Determination de la formule qui represente l'action mutuelle de deux portions infiniment petites de conducteurs voltaiques.* (Memoir on the Determination of the Formula which Represents the Mutual Action of Two Infinitely Small Portions of Voltaic Conductors.): In this paper Ampère

derived his force law. *Ampere's Force Law* is a relationship between the magnetic field of a closed path and the current around this path. It states that there is an attractive or repulsive force between two parallel wires carrying an electric current which is proportional to their lengths and to the intensities of their currents.

**(9) The spin of elementary and composite particles creates an attractive force – the weak interaction or weak force - through exchange interaction or quantum entanglement between two spin states.**

(9) *Standard Model*. ***Exchange interaction.***

A *weak interaction* occurs when *quarks* swap their flavor for another, mediated by exchanging *integer-spin*, force-carrying *bosons*.

According to the *quark formulation* in the *Standard Model*, a *weak interaction occurs when two particles (typically, but not necessarily, half-integer spin fermions) exchange integer-spin, force-carrying bosons*. In the *weak interaction*, *fermions* can *exchange* three types of force carriers, namely W+, W−, and Z *bosons*. The *masses* of these *bosons* are far greater than the *mass* of a *proton* or *neutron*. The *weak interaction* is the only fundamental *interaction* that breaks *parity symmetry*, and similarly, but far more rarely, the only *interaction* to break *charge–parity symmetry*. The *weak interaction* is considered unique in that it allows *quarks* to *swap* their flavor for another. *Quarks*, which make up composite particles like *neutrons* and *protons*, come in six "*flavors*" – *up*, *down*, *charm*, *strange*, *top* and *bottom* – which give those composite particles their properties. The *swapping* of those properties is mediated by the force carrier *bosons*. For example, during *beta-minus decay*, a *down quark* within a *neutron* is changed into an *up quark*, thus converting the *neutron* to a *proton* and resulting in the emission of an *electron* and an *electron antineutrino*.

(9) *Standard Model*. **Weak isospin and the weak hypercharge.**

The *hypercharge* of a particle is a quantum number conserved under the *strong interaction*. The concept of *hypercharge* provides a single *charge operator* that accounts for properties of *isospin*, *electric charge*, and *flavor*. The *hypercharge* is useful to classify *hadrons*; the similarly named *weak hypercharge* has an analogous role in the *electroweak interaction*. *Hypercharge* is one of two *quantum numbers* of the SU(3) model of *hadrons*, alongside *isospin* $I_3$. The *isospin* alone was sufficient for two *quark flavors* — namely u and d — whereas presently 6 *flavors* of *quarks* are known.

Glashow, S. L. (February, 1961). *Partial-symmetries of weak interactions*: In the 1960s, Sheldon Glashow, Abdus Salam and Steven Weinberg unified the *electromagnetic force* and the *weak interaction* by showing them to be two aspects of a single force, now termed the *electroweak force*. In Glashow, S. L. (February, 1961). Partial-symmetries of weak interactions. *Nuclear Physics*, 22, 4, 579–88, Glashow proposed that a relation similar to the Gell-Mann–Nishijima formula for charge to isospin would also apply to the weak interaction. Here the charge is related to the projection of *weak isospin* and the *weak hypercharge*. Glashow combined the *electromagnetic* and *weak interactions* and extended *electroweak unification models* due to Schwinger by including a short-range *neutral current*, the $Z_0$. The resulting *symmetry structure* that Glashow proposed, SU(2) × U(1), forms the basis of the accepted *theory of the electroweak interactions*.

*Isospin* and *weak isospin* are related to the same symmetry but for different forces. *Weak isospin* is the *gauge symmetry* of the *weak interaction* which connects *quark* and *lepton doublets* of left-handed particles in all generations; for example, *up* and *down quarks*, *top* and *bottom quarks*, *electrons* and *electron neutrinos*. By contrast (strong) *isospin* connects only *up* and *down quarks*, acts on both *chiralities* (left and right) and is a *global* (not a *gauge*) *symmetry*.

(9) *New Physics.* ***Quantum entanglement*** **between** ***spin states*****.**

*Quantum entanglement* is the phenomenon of a group of particles being generated, interacting, or sharing spatial proximity in such a way that the *quantum state* of each particle of the group cannot be described independently of the state of the others, including when the particles are separated by a large distance. The topic of *quantum entanglement* is at the heart of the disparity between classical and quantum physics: *entanglement* is a primary feature of *quantum mechanics* not present in *classical mechanics*.

One of the most common forms of *quantum entanglement* is between the two *quantum spin states* of a particle. For example, if a pair of *entangled* particles is generated such that their *total spin* is known to be zero, and *one particle is found to have clockwise spin on a first axis, then the spin of the other particle, measured on the same axis, is found to be anticlockwise*. However, this behavior gives rise to seemingly paradoxical effects: *any measurement of a particle's properties results in an apparent and irreversible wave function collapse of that particle and changes the original quantum state*.

(9) *New Physics.* **Exchange phenomenon (*****quantum entanglement*****).**

Heitler, W. & London, F. (June, 1927). Wechselwirkung neutraler Atome und homöopolare Bindung nach der Quantenmechanik. (Interaction of neutral atoms and homeopolar bonding according to quantum mechanics.): Heitler and London examined the interaction between *neutral atoms* though non-polar bonds, in what is known as valance bonds; and applied quantum mechanics to calculate the *interaction energy* of the atoms when they move closer together. They found that two neutral atoms could interact with each other in two ways; *the problem was twofold degenerate, corresponding to the two ways of assigning the electrons to the neutral atoms* (known as *quantum entanglement*). Examination of the different cases of two H atoms and two He atoms showed that by applying the *Pauli principle*, the selected eigenfunctions of the system should change or maintain their sign respectively, when two electrons were swapped, if the two electrons compared had the same or different *spin*.

Heisenberg, W. (September, 1928). Zur Theory of Ferromagnetismus. (On the theory of ferromagnetism.): Heisenberg noted that empirical results exhibit *ferromagnetism* as an entirely similar state of affairs to what was previously observed in the spectrum of the helium atom; and it seemed to follow from the levels in the helium atoms that a *powerful interaction prevailed between the spin directions of two electrons* that led to the splitting of the level structure into systems of singlets and triplets. He also noted that this was closely related to explaining ferromagnetic phenomena as being implied by the *exchange phenomenon* (resulting from *quantum entanglement*). Heisenberg concluded that *an atom in a lattice can only be exchanged with its "neighbors"*; exchanges with atoms that lie further away that the "neighboring atoms" could then be neglected. Then two conditions were necessary for the appearance of *ferromagnetism*: the crystal lattice must be a type such that *any atom has at least 8 neighbors*; and the *principal quantum number* of the electrons that are responsible for magnetism must be $n \geq 3$.

**(10) Elementary particles, such as protons and neutrons, can exist as different quantum states, referred to as isospin states, which create an attractive force – the strong interaction or strong force - through exchange interaction or quantum entanglement between two isospin states.**

**Isospin.**

*Elementary* and composite particles can exist as different *quantum states*, referred to as *isospin states*, which can be *exchanged* or *entangled*. The name of the concept contains the term *spin* because its quantum mechanical description is mathematically similar to that of *angular momentum* (in particular, in the way it *couples*). But unlike angular momentum, it is a dimensionless quantity and is not actually any type of spin.

(10) *Standard Model*. ***Exchange interaction.***

A *strong interaction* between *nucleons* results from *quarks* with unlike *color charge* attracting one another, mediated by a *gluon*.

Under the *six-quark* formulation of the *Standard Model*, *isospin* (*I*) is a *quantum number*, referred to as the *baryon number*, related to the *up- and down quark* content of the particle. *Isospin* is also known as *isobaric spin* or *isotopic spin*. *Isospin* symmetry is a subset of the *flavor* symmetry seen more broadly in the *interactions* of *baryons* and *mesons*.

*Protons* and neutrons are *baryons*, a type of composite subatomic particle *that contains an odd number of valence quarks and antiquarks*, conventionally three. *Baryons* participate in the residual *strong force*, which is mediated by particles known as *mesons*. A *meson* is a type of hadronic subatomic particle composed of *an equal number of quarks and antiquarks, usually one of each, bound together by the strong interaction*. Because *mesons* are composed of *quarks*, they participate in both the *weak interaction* and *strong interaction*.

Before the concept of *quarks* was introduced, particles that are affected equally by the *strong force* but had different *electric charges* (e.g. *protons* and *neutrons*) were considered different states of the same particle, but having *isospin* values related to the number of *charge states*.

(10) *Standard Model*. ***Gauged isospin symmetry.***

Attempts have been made to promote *isospin* from a *global* to a *local symmetry*. In 1954, Chen Ning Yang and Robert Mills suggested that the notion of *protons* and *neutrons*, which are continuously rotated into each other by *isospin*, should be allowed to vary from point to point. To describe this, the *proton* and *neutron* direction in *isospin space* must be defined at every point, giving *local* basis for *isospin*. A *gauge* connection would then describe how to transform *isospin* along a path between two points.

> Yang, C. N. & Mills, R. (October, 1954). Conservation of Isotopic Spin and Isotopic Gauge Invariance: Chen-Ning Yang and Robert Mills extended the concept of *gauge theory* for *abelian* groups, e.g. *quantum electrodynamics*, to *nonabelian* groups to provide an explanation for *strong interactions*. The *Yang–Mills theory* is a *quantum field theory* for *nuclear binding*. It is a *gauge theory* based on a *special unitary group* SU(n), or more generally any compact Lie group. It seeks to describe the behavior of *elementary particles* using these non-abelian Lie groups and *is at the core of the unification of the electromagnetic force and weak forces* (i.e. U(1) × SU(2)) as well as *quantum chromodynamics*, the theory of the *strong force* (based on SU(3)). Thus, it forms the basis of the understanding of the *Standard Model* of particle physics.

(10) *New Physics*. ***Quantum entanglement* between *isospin states*.**

Heisenberg, W. (January, 1932). Über den Bau der Atomkerne. I; (March, 1932). Über den Bau der Atomkerne. II.; (September, 1933). Über den Bau der Atomkerne. III. (About the construction of atomic nuclei. I; II; III.): three-part paper by Heisenberg, which attempted to address recent observations that the *forces* between all pairs of constituents of the *nucleus* were approximately equal. Heisenberg introduced a new (unnamed) concept to explain binding of the *proton* and the then newly discovered *neutron*. He treated *protons* and *neutrons* on an equal footing by considering them *as different charge states* of the same particle. His model resembled the bonding model for the molecular Hydrogen ion, $H_2^+$: a single *electron* was shared by two *protons*. Heisenberg's theory had several problems, most notable it incorrectly predicted the exceptionally strong binding energy of $He^{+2}$, *alpha particles*. However, its equal treatment of the *proton* and *neutron* gained significance when several experimental studies showed these particles must bind almost equally. In response, Eugene Wigner used Heisenberg's concept in his 1937 paper where he introduced the term "*isotopic spin*" to indicate how the concept is similar to *spin* in behavior.

Wigner, E. (1937). *On the Consequences of the Symmetry of the Nuclear Hamiltonian on the Spectroscopy of Nuclei*: "recent investigations appear to show that the *forces* between all pairs of constituents of the *nucleus* are approximately equal. This makes it desirable to treat the *protons* and *neutrons* on an equal footing. A scheme for this was devised by Heisenberg *who considered protons and neutrons as different states of the same particle*. In this paper, the structure of the *multiplets* of nuclear terms is investigated, using as *first approximation* a Hamiltonian which does not involve the ordinary *spin* and corresponds to equal forces between all nuclear constituents, *protons* and *neutrons*.

The *multiplets* turn out to have a rather complicated structure, instead of the S of atomic spectroscopy, one has three *quantum numbers* S, T, Y. The *second approximation* can either introduce *spin* forces, or else can discriminate between *protons* and *neutrons*. The *last approximation* discriminates between *protons* and *neutrons* and takes the *spin* forces into account."

www.ingramcontent.com/pod-product-compliance
Lightning Source LLC
Chambersburg PA
CBHW061322190326
41458CB00011B/3867